Francesco Bottacin

Exercises in Linear Algebra and Geometry

Società Editrice
ESCULAPIO

ISBN 978-88-9385-411-5

Società Editrice Esculapio s.r.l.
Via Terracini, 30 - 40131 Bologna
www.editrice-esculapio.com - info@editrice-esculapio.it

Cover layout: Carlotta Lenzi

Società Editrice
ESCULAPIO

Foreword

This book contains a collection of exercises designed to support and complement the theoretical exposition of the standard topics for a first course in linear algebra and geometry, at university level.

In this sense, this exercise book is a natural completion of the text

F. BOTTACIN, *Linear Algebra and Geometry*, Esculapio, Bologna

which will be referred to for the necessary theoretical notions.

In presenting the solutions to the proposed exercises, my main goal is to try to *teach* how to approach and solve a problem. For this reason, in many cases, theoretical considerations have been included and different solution methods have been proposed for the same type of exercise. The reader will thus be able to evaluate, case by case, the advantages and disadvantages of each method.

Francesco Bottacin

Contents

1

Vector Spaces

1.1 Vector spaces and subspaces

► **Exercise 1.1.1.** Check if the following sets are vector spaces:

(a) The set of non-negative, continuous real functions defined in the interval $[0, 1]$, with the operations of addition and multiplication by a real number.

(b) The set of real functions f defined in $\mathbb{R}$, such that

$$\lim_{x \to +\infty} f(x) = 0,$$

with the operations of addition and multiplication by a real number.

(c) The set $A = \{ x \in \mathbb{R} \,|\, x > 0 \}$, with the operations of addition and multiplication by a real number defined as follows:

$$x \oplus y = xy, \quad \forall\, x, y \in A$$
$$\lambda \cdot x = x^\lambda, \quad \forall\, x \in A, \lambda \in \mathbb{R}.$$

(d) The set of functions from $\mathbb{R}$ to $\mathbb{R}$ that vanish in 1 or in 4.

(e) The set of polynomials of degree equal to n (n a positive integer).

(f) The set of C^2-functions from $\mathbb{R}$ to $\mathbb{R}$ such that

$$f'' + \omega^2 f = 0,$$

with $\omega \in \mathbb{R}$.

(g) The set of continuous real functions $f(x)$ defined in the interval $[0, 1]$ such that

$$\int_0^1 f(x) \sin x \, dx = 0.$$

Solution. It is only a matter of verifying, case by case, whether all the conditions necessary for the definition of a vector space are satisfied.

(a) In this case the set is not a vector space. In fact it is not even an abelian group with respect to the operation of addition, as it does not contain the opposites of its elements: if f is a positive continuous function the opposite function $-f$ will then be negative, and therefore will not belong to this set.

(b) It is easy to verify that this set is a vector space, remembering that if $\lim_{x\to+\infty} f(x)$ and $\lim_{x\to+\infty} g(x)$ exist, then

$$\lim_{x\to+\infty}(f+g)(x) = \lim_{x\to+\infty} f(x) + \lim_{x\to+\infty} g(x)$$

and

$$\lim_{x\to+\infty}(fg)(x) = \big(\lim_{x\to+\infty} f(x)\big)\big(\lim_{x\to+\infty} g(x)\big).$$

(c) Also in this case the set is a vector space. It is advisable to carefully check all the required properties. As an example, let us verify that, given two elements $x, y \in A$ and a scalar $\lambda \in \mathbb{R}$, we have

$$\lambda \cdot (x \oplus y) = (\lambda \cdot x) \oplus (\lambda \cdot y).$$

Based on the given definitions, we have: $\lambda\cdot(x\oplus y) = (xy)^\lambda$, and $(\lambda\cdot x)\oplus(\lambda\cdot y) = x^\lambda y^\lambda$. Equality then follows from well-known properties.

(d) This set is not a vector space, as it is not closed under the addition operation: in fact, let for example f be a function such that $f(1) = 0$ and $f(4) = 1$, and g be a function such that $g(1) = 2$ and $g(4) = 0$. These two functions belong to this set, but their sum does not belong to the given set, since it vanishes neither in 1 nor in 4.

(e) This set is also not a vector space, since it is not closed under the addition operation: in fact, let, for example, $p(x) = 2x^n + 1$ and $q(x) = -2x^n + 3$ be two polynomials of degree n. Their sum is $p(x) + q(x) = 4$ whose degree is not n (at least if $n \neq 0$).

Remark: on the other hand, the set of polynomials of degree less than or equal to n is a vector space.

(f) This set is a vector space (this is due to the fact that the differential equation is linear). Let's just verify that it is closed with respect to the addition operation: let f and g be two functions that satisfy the differential equation. We have:

$$\begin{aligned}(f+g)'' + \omega^2(f+g) &= f'' + g'' + \omega^2 f + \omega^2 g \\ &= (f'' + \omega^2 f) + (g'' + \omega^2 g) = 0 + 0 = 0.\end{aligned}$$

(g) This set is also a vector space (this is due to the integration operation being linear). Let's just verify that it is closed with respect to the addition operation: let f and g be two functions that satisfy the integral equation. We have:

$$\int_0^1 (f+g)(x)\sin x\,dx = \int_0^1 f(x)\sin x\,dx + \int_0^1 g(x)\sin x\,dx = 0+0=0.$$

► **Exercise 1.1.2.** Check whether the following subsets of $\mathbb{R}^2$ are vector subspaces:

(a) $S_1 = \{(x_1,x_2)\in\mathbb{R}^2 \mid 2x_1 - 3x_2 = 0\}$

(b) $S_2 = \{(x_1,x_2)\in\mathbb{R}^2 \mid x_1 + 4x_2 = 3\}$

(c) $S_3 = \{(x_1,x_2)\in\mathbb{R}^2 \mid 2x_1 + x_2^2 = 0\}$

Solution. (a) Let us consider two elements (x_1,x_2) and (y_1,y_2) of S_1. Since $(x_1,x_2)\in S_1$, we have $2x_1-3x_2=0$ and similarly we have $2y_1-3y_2=0$, because $(y_1,y_2)\in S_1$. For every $\lambda_1,\lambda_2\in\mathbb{R}$ we consider the linear combination

$$(z_1,z_2)=\lambda_1(x_1,x_2)+\lambda_2(y_1,y_2).$$

We have $z_1=\lambda_1x_1+\lambda_2y_1$ and $z_2=\lambda_1x_2+\lambda_2y_2$.

Verifying that S_1 is a vector subspace is equivalent to verifying that $(z_1,z_2)\in S_1$ for every $(x_1,x_2),(y_1,y_2)\in S_1$ and every $\lambda_1,\lambda_2\in\mathbb{R}$.

From the definition of S_1 we have that $(z_1,z_2)\in S_1$ if and only if $2z_1-3z_2=0$. Substituting, we get:

$$\begin{aligned}2z_1-3z_2&=2(\lambda_1x_1+\lambda_2y_1)-3(\lambda_1x_2+\lambda_2y_2)\\&=\lambda_1(2x_1-3x_2)+\lambda_2(2y_1-3y_2)\\&=\lambda_1\cdot 0+\lambda_2\cdot 0=0.\end{aligned}$$

Since this is true for every $(x_1,x_2),(y_1,y_2)\in S_1$ and every $\lambda_1,\lambda_2\in\mathbb{R}$, we conclude that S_1 is a vector subspace of $\mathbb{R}^2$.

(b) In this case the same reasoning used for the previous point leads to the conclusion that S_2 is not a vector subspace. In fact, let us consider two elements (x_1,x_2) and (y_1,y_2) of S_2. Since $(x_1,x_2)\in S_2$, we have $x_1+4x_2=3$ and similarly $y_1+4y_2=3$, because $(y_1,y_2)\in S_2$. For every $\lambda_1,\lambda_2\in\mathbb{R}$ we consider the linear combination

$$(z_1,z_2)=\lambda_1(x_1,x_2)+\lambda_2(y_1,y_2).$$

We have $z_1 = \lambda_1 x_1 + \lambda_2 y_1$ and $z_2 = \lambda_1 x_2 + \lambda_2 y_2$.

Verifying that S_2 is a vector subspace is equivalent to verifying that $(z_1, z_2) \in S_2$ for every $(x_1, x_2), (y_1, y_2) \in S_2$ and every $\lambda_1, \lambda_2 \in \mathbb{R}$.

From the definition of S_2 we have that $(z_1, z_2) \in S_2$ if and only if $z_1+4z_2 = 3$. Substituting, we get:

$$\begin{aligned} z_1 + 4z_2 &= (\lambda_1 x_1 + \lambda_2 y_1) + 4(\lambda_1 x_2 + \lambda_2 y_2) \\ &= \lambda_1(x_1 + 4x_2) + \lambda_2(y_1 + 4y_2) \\ &= \lambda_1 \cdot 3 + \lambda_2 \cdot 3 \neq 3. \end{aligned}$$

So, in general, $z_1 + 4z_2 \neq 3$ and therefore $(z_1, z_2) \notin S_2$. This means that S_2 is not a vector subspace of $\mathbb{R}^2$.

Remark: much more simply it was enough to observe that $(0,0) \notin S_2$. Every vector subspace must necessarily contain the zero vector, since S_2 does not contain the vector $(0,0)$ it cannot be a vector subspace.

(c) Also in this case it can be easily verified that S_3 is not a vector subspace (in this case the problems are due to the presence of the quadratic term x_2^2 in the equation that defines the set S_3).

In fact, we have $(-2,2) \in S_3$, $(-8,4) \in S_3$, but the sum $(-2,2)+(-8,4) = (-10,6)$ does not satisfy the equation that defines S_3, hence $(-10,6) \notin S_3$. The set S_3 is therefore not closed for the addition operation, therefore it is not a vector subspace.

▶ **Exercise 1.1.3.** Check whether the following subsets of $\mathbb{R}^3$ are vector subspaces:

(a) $S_1 = \{(x_1, x_2, x_3) \in \mathbb{R}^3 \mid x_1 - 2x_2 + x_3 = 0 \text{ and } 2x_1 - x_3 = 0\}$

(b) $S_2 = \{(x_1, x_2, x_3) \in \mathbb{R}^3 \mid 2x_1 - x_2 = 0 \text{ and } x_1 + 4x_3 = 1\}$

(c) $S_3 = \{(x_1, x_2, x_3) \in \mathbb{R}^3 \mid 3x_1 - x_2 x_3 = 0\}$

Solution. (a) Let us consider two elements (x_1, x_2, x_3) and (y_1, y_2, y_3) of S_1. Since $(x_1, x_2, x_3) \in S_1$, we have $x_1 - 2x_2 + x_3 = 0$ and $2x_1 - x_3 = 0$. Similarly we have $y_1 - 2y_2 + y_3 = 0$ and $2y_1 - y_3 = 0$, because $(y_1, y_2, y_3) \in S_1$. For every $\lambda_1, \lambda_2 \in \mathbb{R}$ we consider the linear combination

$$(z_1, z_2, z_3) = \lambda_1(x_1, x_2, x_3) + \lambda_2(y_1, y_2, y_3).$$

We have $z_1 = \lambda_1 x_1 + \lambda_2 y_1$, $z_2 = \lambda_1 x_2 + \lambda_2 y_2$ and $z_3 = \lambda_1 x_3 + \lambda_2 y_3$. Now we need to verify that $(z_1, z_2, z_3) \in S_1$ for every $(x_1, x_2, x_3), (y_1, y_2, y_3) \in S_1$ and every $\lambda_1, \lambda_2 \in \mathbb{R}$.

From the definition of S_1 we have $(z_1, z_2, z_3) \in S_1$ if and only if $z_1 - 2z_2 + z_3 = 0$ and $2z_1 - z_3 = 0$. Substituting we get:

$$\begin{aligned} z_1 - 2z_2 + z_3 &= (\lambda_1 x_1 + \lambda_2 y_1) - 2(\lambda_1 x_2 + \lambda_2 y_2) + (\lambda_1 x_3 + \lambda_2 y_3) \\ &= \lambda_1(x_1 - 2x_2 + x_3) + \lambda_2(y_1 - 2y_2 + y_3) \\ &= \lambda_1 \cdot 0 + \lambda_2 \cdot 0 = 0 \end{aligned}$$

and

$$\begin{aligned} 2z_1 - z_3 &= 2(\lambda_1 x_1 + \lambda_2 y_1) - (\lambda_1 x_3 + \lambda_2 y_3) \\ &= \lambda_1(2x_1 - x_3) + \lambda_2(2y_1 - y_3) \\ &= \lambda_1 \cdot 0 + \lambda_2 \cdot 0 = 0. \end{aligned}$$

Since this holds for every $(x_1, x_2, x_3), (y_1, y_2, y_3) \in S_1$ and every $\lambda_1, \lambda_2 \in \mathbb{R}$, we conclude that S_1 is a vector subspace of $\mathbb{R}^3$.

(b) In this case we could use the same reasoning used in the previous point, to conclude that S_2 is not a vector subspace. However, it is much easier to observe that $(0, 0, 0) \notin S_2$ (the second equation $x_1 + 4x_3 = 1$ is not verified). Every vector subspace must contain the zero vector but S_2 does not contain the vector $(0, 0, 0)$, hence it cannot be a vector subspace.

(c) Also in this case it can be easily verified that S_3 is not a vector subspace (in this case the problems are due to the presence of the quadratic term x_2x_3 in the equation that defines the set S_3). In fact, we have $(1, 1, 3) \in S_3$, $(2, 6, 1) \in S_3$, but the sum $(1, 1, 3) + (2, 6, 1) = (3, 7, 4)$ does not verify the equation that defines S_3, hence $(3, 7, 4) \notin S_3$. The set S_3 is not closed under the addition operation, so it is not a vector subspace.

► **Exercise 1.1.4.** Let U and W be two vector subspaces of a vector space V. Prove that $U \cup W$ is a vector subspace of V if and only if $U \subseteq W$ or $W \subseteq U$.

Solution. If $U \subseteq W$ (resp. $W \subseteq U$) then $U \cup W = W$ (resp. $U \cup W = U$), hence it is a vector subspace.

Conversely, suppose $U \cup W$ is a vector subspace, but $U \not\subseteq W$ and $W \not\subseteq U$. This means that there exists a vector $u \in U$ such that $u \notin W$ and a vector $w \in W$ such that $w \notin U$. Since $u, w \in U \cup W$ and this is a vector space, we must have $u + w \in U \cup W$, hence $u + w \in U$ or $u + w \in W$. If $u + w \in U$, we have $u + w = u' \in U$, hence $w = u' - u \in U$, which contradicts the hypothesis that $w \notin U$. Similarly if $u + w \in W$, we would conclude that $u \in W$, which contradicts the hypothesis. It follows that the hypothesis is false, hence $U \cup W$ cannot be a vector subspace if $U \not\subseteq W$ and $W \not\subseteq U$.

▶ **Exercise 1.1.5.** Let U, V and W be three vector subspaces of the same vector space. Is it true that

$$U \cap (V + W) = (U \cap V) + (U \cap W)\,?$$

Solution. This is false. To prove it, it suffices to provide a counterexample: in $\mathbb{R}^2$ let us consider the subspaces $U = L\{(1,1)\}$, $V = L\{(1,0)\}$ and $W = L\{(0,1)\}$. We have $U \cap (V + W) = U \cap \mathbb{R}^2 = U$, and $(U \cap V) + (U \cap W) = \{\mathbf{0}\} + \{\mathbf{0}\} = \{\mathbf{0}\}$.

▶ **Exercise 1.1.6.** Check whether the sum of the following vector subspaces of $\mathbb{R}^4$ is direct: $U = L\{(1,0,1,0),(1,2,3,4)\}$ and $V = L\{(0,1,1,1),(0,0,0,1)\}$.

Solution. The sum of two vector subspaces is direct if and only if their intersection is the set that contains only the zero vector. A vector belongs to the intersection of U and V if and only if it can be written as follows:

$$w = \lambda_1(1,0,1,0) + \lambda_2(1,2,3,4) = \mu_1(0,1,1,1) + \mu_2(0,0,0,1).$$

By solving the system thus obtained, we find:

$$\begin{cases} \lambda_1 = -\lambda_2 \\ \mu_1 = 2\lambda_2 \\ \mu_2 = 2\lambda_2 \\ \lambda_2 \text{ any value.} \end{cases}$$

By setting, for example, $\lambda_2 = 1$, we get $\lambda_1 = -1$, $\mu_1 = 2$, $\mu_2 = 2$, hence

$$w = \lambda_1(1,0,1,0) + \lambda_2(1,2,3,4) = -(1,0,1,0) + (1,2,3,4) = (0,2,2,4).$$

The vector $w = (0,2,2,4)$ belongs to $U \cap V$, hence the sum of U and V is not direct.

▶ **Exercise 1.1.7.** Let us consider the following vector subspaces of $\mathbb{R}^4$:

$$U = L\{(1,0,1,0),(0,1,1,1),(0,0,0,1)\}$$
$$V = L\{(1,0,1,0),(0,1,1,0)\}.$$

Find a subspace $W \subset \mathbb{R}^4$ such that $U = V \oplus W$. Is such a subspace W unique?

Solution. We set $u_1 = (1,0,1,0)$, $u_2 = (0,1,1,1)$, $u_3 = (0,0,0,1)$, $v = (0,1,1,0)$, so that $U = L\{u_1, u_2, u_3\}$ and $V = L\{u_1, v\}$. The vectors u_1, u_2, u_3 are linearly independent, hence $\dim U = 3$. The vectors u_1 and v are linearly independent too, hence $\dim V = 2$. Note that $v = u_2 - u_3$, hence $V \subset U$.

A subspace W such that $V \oplus W = U$ must have dimension 1. Let us denote by w its basis. The vector w must satisfy the following requirements: $w \notin V$ and the vectors u_1, v, w must span U. It is then easy to see that $w = u_3$ satisfies the required conditions. It is equally easy to see that the choice $w = u_2$ also satisfies the required conditions; therefore, it can be deduced that the requested subspace W exists but is not unique.

▶ **Exercise 1.1.8.** Let us consider the following vector subspaces of $\mathbb{R}^4$,

$$U = L\{(1,0,1,0),(0,0,0,1)\}$$
$$V = L\{(1,0,2,0),(0,0,1,1)\},$$

Find a vector subspace $W \subset \mathbb{R}^4$ such that $U \oplus W = V \oplus W = \mathbb{R}^4$. Is such a subspace W unique?

Solution. We set $u_1 = (1,0,1,0)$, $u_2 = (0,0,0,1)$, $v_1 = (1,0,2,0)$, $v_2 = (0,0,1,1)$. The subspaces U and V have dimension 2, hence W must have dimension 2: $W = L\{w_1, w_2\}$. Therefore we must find two vectors w_1 and w_2 such that $\{u_1, u_2, w_1, w_2\}$ and $\{v_1, v_2, w_1, w_2\}$ are bases of $\mathbb{R}^4$. It is easy to check that the vectors $w_1 = (1,0,0,0)$ and $w_2 = (0,1,0,0)$ have the required properties, hence the subspace W exists. However also the choice $w_1 = (1,0,0,0)$ and $w_2 = (0,1,1,0)$ give another subspace W that satisfy the required conditions. It follows that such a subspace W is not unique.

1.2 Sets of generators and bases

▶ **Exercise 1.2.1.** Let V be the vector space of polynomials in x of degree ≤ 3, with real coefficients. Verify that the following sets are bases of V:

(a) $\{1, x, x^2, x^3\}$

(b) $\{1, 1-x, x-x^2, x^2-x^3\}$

(c) $\{1, 1+x, 1+x+x^2, 1+x+x^2+x^3\}$

Solution. (a) Every polynomial $f(x) \in V$ is of the form

$$f(x) = a_0 \cdot 1 + a_1 x + a_2 x^2 + a_3 x^3.$$

This means that the polynomials $1, x, x^2, x^3$ are a set of generators of V. Let us check if they are linearly independent: if $a_0 \cdot 1 + a_1 x + a_2 x^2 + a_3 x^3 = \mathbf{0}$ ($\mathbf{0}$ denotes the zero vector of V, i.e., the zero polynomial), then, by the identity principle of polynomials, we must have $a_0 = a_1 = a_2 = a_3 = 0$. Therefore $\{1, x, x^2, x^3\}$ is a basis of V, hence V has dimension 4.

(b) Also in this case it would be necessary to prove that the polynomials in question are a set of generators and that they are linearly independent. However, knowing that $\dim V = 4$, which is also the number of polynomials of the hypothetical basis, it is sufficient to carry out only one of the two checks. We then verify that these are a set of generators, i.e., that each polynomial $f(x) = a_0 + a_1 x + a_2 x^2 + a_3 x^3$ can be written in the form $\lambda_0 \cdot 1 + \lambda_1(1 - x) + \lambda_2(x - x^2) + \lambda_3(x^2 - x^3)$. Developing the computations and equating the coefficients of the successive powers of x, we obtain the system

$$\begin{cases} \lambda_0 + \lambda_1 = a_0 \\ \lambda_2 - \lambda_1 = a_1 \\ \lambda_3 - \lambda_2 = a_2 \\ -\lambda_3 = a_3 \end{cases}$$

whose solution is

$$\begin{cases} \lambda_3 = -a_3 \\ \lambda_2 = -a_2 - a_3 \\ \lambda_1 = -a_1 - a_2 - a_3 \\ \lambda_0 = a_0 + a_1 + a_2 + a_3 \end{cases}$$

We conclude that the polynomials in question are a set of generators, and therefore, being in the right number, they are a basis of V.

(c) Based on the remarks made in point (b) it is sufficient to prove that the polynomials in question are linearly independent (or that they are a set of generators): let

$$\lambda_0 \cdot 1 + \lambda_1(1 + x) + \lambda_2(1 + x + x^2) + \lambda_3(1 + x + x^2 + x^3) = \mathbf{0}.$$

Developing the computations and equating the coefficients to zero, we obtain

the system

$$\begin{cases} \lambda_0 + \lambda_1 + \lambda_2 + \lambda_3 = 0 \\ \lambda_1 + \lambda_2 + \lambda_3 = 0 \\ \lambda_2 + \lambda_3 = 0 \\ \lambda_3 = 0 \end{cases}$$

whose unique solution is $\lambda_0 = \lambda_1 = \lambda_2 = \lambda_3 = 0$. We thus conclude that the polynomials in question are linearly independent, and therefore, being in the right number, they are a basis of V.

► **Exercise 1.2.2.** In the vector space V of polynomials of degree ≤ 2 consider the polynomials

$$\begin{aligned} p_1(x) &= x^2 + x(1-x) + (1-x)^2 \\ p_2(x) &= x^2 + (1-x)^2 \\ p_3(x) &= x^2 + 1 + (1-x)^2 \\ p_4(x) &= x(1-x). \end{aligned}$$

Is it possible to extract from $\{p_1(x), p_2(x), p_3(x), p_4(x)\}$ a basis of V? If this is possible, find all possible bases.

Solution. The vector space V of polynomials of degree ≤ 2 has dimension 3. In fact a basis is formed by the polynomials $\{1, x, x^2\}$ (because every polynomial of degree ≤ 2 can be uniquely written as a linear combination $a_0 1 + a_1 x + a_2 x^2$ of these three polynomials). It is easy to check that $p_3(x) = 2p_1(x)$, therefore, if a basis can be extracted from those four polynomials, this basis must be given by $\{p_1(x), p_2(x), p_4(x)\}$ (or, equivalently, by $\{p_2(x), p_3(x), p_4(x)\}$). However we can verify that $p_4(x) = p_2(x) - p_1(x)$, hence the polynomials $p_1(x)$, $p_2(x)$ and $p_4(x)$ are linearly dependent and therefore do not form a basis of V. This means that from the four given polynomials it is impossible to extract a basis of V.

► **Exercise 1.2.3.** In the vector space of continuous functions from $\mathbb{R}$ to $\mathbb{R}$ consider the functions $f_1(x) = \sin x$, $f_2(x) = \sin 2x$ and $f_3(x) = \sin 3x$. Are these functions linearly independent?

Solution. Let us consider a linear combination $a_1 f_1 + a_2 f_2 + a_3 f_3$ of the functions f_1, f_2 and f_3. We suppose that such linear combinations is the zero function. Therefore we have:

$$a_1 \sin x + a_2 \sin 2x + a_3 \sin 3x = 0, \quad \forall x \in \mathbb{R}.$$

Since the previous expression must be zero for each value of x, by assigning particular values to x we obtain (for example):

$$
\begin{aligned}
x = \pi/2: &\quad a_1 - a_3 = 0, \\
x = \pi/3: &\quad \frac{\sqrt{3}}{2}\,a_1 + \frac{\sqrt{3}}{2}\,a_2 = 0, \\
x = \pi/4: &\quad \frac{\sqrt{2}}{2}\,a_1 + a_2 + \frac{\sqrt{2}}{2}\,a_3 = 0.
\end{aligned}
$$

The only solution of these equations is: $a_1 = a_2 = a_3 = 0$. This proves that the given functions are linearly independent.

► **Exercise 1.2.4.** In the following cases check if the vectors v_1, v_2, v_3 are a basis of $\mathbb{R}^3$. If not, describe the subspace they span.

(a) $v_1 = (1, 1, 1)$, $v_2 = (3, 0, -1)$, $v_3 = (-1, 1, -1)$;

(b) $v_1 = (1, 2, 3)$, $v_2 = (3, 0, -1)$, $v_3 = (1, 8, 13)$;

(c) $v_1 = (1, 2, -3)$, $v_2 = (1, 0, -1)$, $v_3 = (1, 10, -11)$.

Solution. Let us perform the computations only for case (a). Given that the vectors are three, and that is the right number for a basis of $\mathbb{R}^3$, it will be sufficient to check if they are linearly independent (or if they are a set of generators). By setting $\lambda_1 v_1 + \lambda_2 v_2 + \lambda_3 v_3 = \mathbf{0}$, we obtain the system

$$
\begin{cases}
\lambda_1 + 3\lambda_2 - \lambda_3 = 0 \\
\lambda_1 + \lambda_3 = 0 \\
\lambda_1 - \lambda_2 - \lambda_3 = 0
\end{cases}
$$

whose solution is $\lambda_1 = \lambda_2 = \lambda_3 = 0$. This shows that these three vectors are linearly independent and, consequently, are a basis of $\mathbb{R}^3$.

► **Exercise 1.2.5.** In $\mathbb{R}^4$ the following vectors are:
(1) a free set (i.e., a set of linearly independent vectors)? If yes, complete this set to obtain a basis of $\mathbb{R}^4$, otherwise find the linear dependence relations of the vectors and extract a free set from this set of vectors.
(2) a set of generators? If yes, extract a basis of $\mathbb{R}^4$, otherwise find the dimension of the subspace they span.

(a) $v_1 = (1, 1, 1, 1)$, $v_2 = (0, 1, 2, -1)$, $v_3 = (1, 0, -2, 3)$,
$v_4 = (2, 1, 0, -1)$, $v_5 = (4, 3, 2, 1)$;

(b) $v_1 = (1, 2, 3, 4)$, $v_2 = (0, 1, 2, -1)$, $v_3 = (3, 4, 5, 16)$;

(c) $v_1 = (1, 2, 3, 4)$, $v_2 = (0, 1, 2, -1)$, $v_3 = (2, 1, 0, 11)$, $v_4 = (3, 4, 5, 14)$.

Solution. (a) The given five vectors are surely linearly dependent, since the maximum number of linearly independent vectors in $\mathbb{R}^4$ is 4. In fact it is easy to check that $v_5 = 2v_1 + v_4$. It is also easy to check that the vectors v_1, v_2, v_3 e v_4 are linearly independent, and therefore are a basis of $\mathbb{R}^4$.

(b) The three given vectors are definitely not generators of $\mathbb{R}^4$ (whose dimension is 4). However, it turns out that they are linearly independent. If we introduce a fourth vector $v_4 = (0, 0, 1, 0)$ (for example), it is easy to verify that v_1, v_2, v_3 and v_4 are linearly independent, and therefore form a basis of $\mathbb{R}^4$.

(c) It is easy to verify that the four given vectors are linearly dependent, in fact $v_4 = v_1 + v_2 + v_3$. Therefore they do not form a basis of $\mathbb{R}^4$. It then turns out that v_1 is also a linear combination of v_2 and v_3, in fact we have $v_1 = \frac{3}{2}v_2 + \frac{1}{2}v_3$. Therefore the vectors v_1 and v_4 belong to the subspace spanned by v_2 and v_3. Finally, it is easy to check that v_2 and v_3 are linearly independent. It follows that the four given vectors generate a subspace of dimension 2 of $\mathbb{R}^4$, whose basis is formed by v_2 and v_3. To obtain a basis of $\mathbb{R}^4$ one can consider the vectors v_2, v_3, $(0, 0, 1, 0)$ and $(0, 0, 0, 1)$, which are, as is easily verified, linearly independent.

► **Exercise 1.2.6.** In $\mathbb{R}^4$ let $v_1 = (1, 2, 3, 4)$ and $v_2 = (1, -2, 3, -4)$. Is it possible to find two real numbers x and y such that $(x, 1, y, 1) \in L\{v_1, v_2\}$? (Recall that $L\{v_1, v_2\}$ denotes the subspace spanned by the vectors v_1 and v_2)

Solution. The vector $(x, 1, y, 1)$ belongs to the subspace generated by v_1 and v_2 if and only if it can be expressed as a linear combination of v_1 and v_2:

$$(x, 1, y, 1) = \lambda_1 v_1 + \lambda_2 v_2.$$

The second and fourth equations of this system are, respectively: $2\lambda_1 - 2\lambda_2 = 1$ and $4\lambda_1 - 4\lambda_2 = 1$, which have no common solutions. Therefore $(x, 1, y, 1) \notin L\{v_1, v_2\}$.

▶ **Exercise 1.2.7.** Find a basis of the vector subspace V of $\mathbb{R}^5$ formed by vectors $(x_1, \dots, x_5)$ that are solutions of the following system:

$$\begin{cases} x_1 - 3x_2 + x_4 = 0 \\ x_2 + 3x_3 - x_5 = 0 \\ x_1 + 2x_2 + x_3 - x_4 = 0. \end{cases}$$

Solution. Solving the system we obtain (for example by computing x_1, x_4 and x_5 as a function of x_2 and x_3):

$$\begin{cases} x_1 = \frac{1}{2}x_2 - \frac{1}{2}x_3 \\ x_5 = x_2 + 3x_3 \\ x_4 = \frac{5}{2}x_2 + \frac{1}{2}x_3 \end{cases}$$

This system therefore has infinite solutions, depending on two parameters. In other words, the solution space V has dimension 2. To find a basis of V it is then sufficient to find two vectors, linearly independent, which are solutions of the previous system. These vectors are found simply by attributing 'any' values to the free variables x_2 and x_3 (making sure that the vectors found are linearly independent!). In practice, it will be sufficient to assign the values 0 and 1 alternately to the free variables; we will thus be sure of obtaining linearly independent vectors. By setting $x_2 = 1$ and $x_3 = 0$ we obtain the vector $(\frac{1}{2}, 1, 0, \frac{5}{2}, 1)$, and setting $x_2 = 0$ and $x_3 = 1$ we get $(-\frac{1}{2}, 0, 1, \frac{1}{2}, 3)$. These two vectors are a basis of V.

▶ **Exercise 1.2.8.** Let V be a vector space. Are the following assertions true?

(a) If the vectors v_1, v_2, v_3 are pairwise non-proportional then the set $\{v_1, v_2, v_3\}$ is free.

(b) If none of the vectors $v_1, \dots, v_r$ is a linear combination of the remaining ones then the set $\{v_1, \dots, v_r\}$ is free.

Solution. (a) This assertion is false. To prove this, it suffices to provide a counterexample: let $V = \mathbb{R}^2$, $v_1 = (1, 0)$, $v_2 = (0, 1)$, $v_3 = (1, 1)$. It is clear that these vectors are pairwise non-proportional, but they cannot be linearly independent, given that the dimension of $\mathbb{R}^2$ is 2.

(b) This assertion is true. Indeed if it were false, i.e., if the vectors $v_1, \ldots, v_r$ were linearly dependent, we would have

$$\lambda_1 v_1 + \lambda_2 v_2 + \cdots + \lambda_r v_r = \mathbf{0},$$

with the λ_i not all zero. Suppose then that $\lambda_j \neq 0$. We have

$$v_j = -\frac{1}{\lambda_j} \sum_{i \neq j} \lambda_i v_i,$$

hence the vector v_j would be a linear combination of the other vectors, contradicting the hypothesis.

► **Exercise 1.2.9.** In $\mathbb{R}^4$ let us consider the following vectors: $v_1 = (0, 1, -2, 1)$, $v_2 = (1, 0, 2, -1)$, $v_3 = (3, 2, 2, -1)$, $v_4 = (0, 0, 1, 0)$, $v_5 = (0, 0, 0, 1)$. Check whether the following statements are true.

(a) $L\{v_1, v_2, v_3\} = L\{(1, 1, 0, 0), (-1, 1, -4, 2)\}$;

(b) $(1, 1, 0, 0) \in L\{v_1, v_2\} \cap L\{v_2, v_3, v_4\}$;

(c) $\dim(L\{v_1, v_2\} \cap L\{v_2, v_3, v_4\}) = 1$;

(d) $L\{v_1, v_2\} + L\{v_2, v_3, v_4\} = \mathbb{R}^4$;

(e) $L\{v_1, v_2, v_3\} + L\{v_4, v_5\} = \mathbb{R}^4$.

Solution. To simplify the notation let us set $u_1 = (1, 1, 0, 0)$ and $u_2 = (-1, 1, -4, 2)$.

(a) We have: $v_1 = \frac{1}{2}(u_1 + u_2)$, $v_2 = \frac{1}{2}(u_1 - u_2)$, $v_3 = \frac{1}{2}(5u_1 - u_2)$, hence $L\{v_1, v_2, v_3\} \subset L\{u_1, u_2\}$. Conversely, we also have: $u_1 = v_1 + v_2$ and $u_2 = v_1 - v_2$, hence $L\{u_1, u_2\} \subset L\{v_1, v_2, v_3\}$. It follows that $L\{v_1, v_2, v_3\} = L\{u_1, u_2\}$.

(b) We have seen in (a) that $u_1 \in L\{v_1, v_2\}$. Since $u_1 = \frac{1}{2}(v_3 - v_2)$, we have $u_1 \in L\{v_2, v_3, v_4\}$, hence $u_1 \in L\{v_1, v_2\} \cap L\{v_2, v_3, v_4\}$.

(c) Since $v_2 \in L\{v_1, v_2\} \cap L\{v_2, v_3, v_4\}$, we have

$$\dim(L\{v_1, v_2\} \cap L\{v_2, v_3, v_4\}) \geq 1.$$

In (a) we have seen that $u_1 \in L\{v_1, v_2\} \cap L\{v_2, v_3, v_4\}$, and the vectors u_1, v_2 are linearly independent, therefore

$$\dim(L\{v_1, v_2\} \cap L\{v_2, v_3, v_4\}) \geq 2.$$

On the other hand $L\{v_1, v_2\} \cap L\{v_2, v_3, v_4\} \subset L\{v_1, v_2\}$, hence

$$\dim(L\{v_1, v_2\} \cap L\{v_2, v_3, v_4\}) \leq 2.$$

It follows that $\dim(L\{v_1, v_2\} \cap L\{v_2, v_3, v_4\}) = 2$.

(d) $L\{v_1, v_2\} + L\{v_2, v_3, v_4\} = \mathbb{R}^4$ if and only if $\{v_1, v_2, v_3, v_4\}$ is a basis of $\mathbb{R}^4$, but we have $v_3 = 2v_1 + 3v_2$, hence the vectors v_1, v_2, v_3, v_4 are not linearly independent.

(e) $L\{v_1, v_2, v_3\} + L\{v_4, v_5\} = \mathbb{R}^4$ if and only if the vectors v_1, v_2, v_3, v_4 and v_5 span $\mathbb{R}^4$. It is easy to verify that v_1, v_2, v_4, v_5 are linearly independent, hence they are a basis of $\mathbb{R}^4$. It follows that the vectors v_1, v_2, v_3, v_4 e v_5 generate $\mathbb{R}^4$.

► **Exercise 1.2.10.** Check the linear independence of the following vectors, and find a basis of the subspace they span.

(a) $(1, 0, 1)$, $(0, 2, 2)$, $(3, 7, 1)$, in $\mathbb{R}^3$;

(b) $(1, 0, 0)$, $(0, 1, 1)$, $(1, 1, 1)$, in $\mathbb{R}^3$;

(c) $(1, 2, 1, 2, 1)$, $(2, 1, 2, 1, 2)$, $(1, 0, 1, 1, 0)$, $(0, 1, 0, 0, 1)$, in $\mathbb{R}^5$.

Solution. (a) It is easy to verify that the given vectors are linearly independent, and therefore are a basis of $\mathbb{R}^3$.

(b) Since the third vector is the sum of the previous two, the three given vectors are not a basis of $\mathbb{R}^3$. A basis of the subspace generated by them is formed by any two vectors chosen among the three assigned vectors.

(c) The first three vectors are linearly independent, while the fourth is a linear combination of the first three. The subspace generated by them therefore has dimension 3, and a basis is formed, for example, by the first three vectors.

► **Exercise 1.2.11.** Let V be the vector space of polynomials in x of degree $\leq n$, with coefficients in $\mathbb{R}$. Prove that, for every $a \in \mathbb{R}$, the set

$$\{\, 1, x - a, (x - a)^2, \ldots, (x - a)^n \,\}$$

is a basis of V. Let $f(x) \in V$; write $f(x)$ as a linear combination of the polynomials above. Who are the coefficients of this linear combination?

Solution. V has dimension $n+1$ (a basis is $\{1, x, x^2, \dots, x^n\}$). To prove that $\{1, x-a, (x-a)^2, \dots, (x-a)^n\}$ is a basis of V it suffices to prove that these polynomials are linearly independent. Let

$$\lambda_0 \cdot 1 + \lambda_1(x-a) + \lambda_2(x-a)^2 + \cdots + \lambda_n(x-a)^n = \mathbf{0}.$$

We observe that in the previous expression the term of higher degree is $\lambda_n x^n$, hence it must be $\lambda_n = 0$. Then we get

$$\lambda_0 \cdot 1 + \lambda_1(x-a) + \lambda_2(x-a)^2 + \cdots + \lambda_{n-1}(x-a)^{n-1} = \mathbf{0}.$$

In this expression the term of higher degree is $\lambda_{n-1}x^{n-1}$, hence it must be $\lambda_{n-1} = 0$.

Continuing in this way it is shown that all the coefficients λ_i are zero, and therefore the vectors are linearly independent, hence they are a basis of V. Let now $f(x) \in V$ and set

$$f(x) = \lambda_0 + \lambda_1(x-a) + \lambda_2(x-a)^2 + \cdots + \lambda_n(x-a)^n.$$

Evaluating the above expression for $x = a$, we obtain $f(a) = \lambda_0$. Then, to determine λ_1 it suffices to derive this expression

$$f'(x) = \lambda_1 + 2\lambda_2(x-a) + 3\lambda_3(x-a)^2 + \cdots + n\lambda_n(x-a)^{n-1},$$

and evaluate it for $x = a$, obtaining $\lambda_1 = f'(a)$. Derivating again and evaluating for $x = a$, we get $\lambda_2 = f''(a)/2$. Continuing in this way it is shown that $\lambda_i = f^{(i)}(a)/i!$, for $0 \le i \le n$. We have thus proved that, using the basis $\{1, x-a, (x-a)^2, \dots, (x-a)^n\}$, every polynomial of degree $\le n$ can be written as

$$f(x) = \sum_{i=0}^{n} \frac{f^{(i)}(a)}{i!}(x-a)^i.$$

Note that this is Taylor's formula!

► **Exercise 1.2.12.** Let $U_t = L\{u_1, u_2\}$ and $V_t = L\{v_1, v_2\}$ be two subspaces of $\mathbb{R}^4$, with $u_1 = (1, t, 2t, 0)$, $u_2 = (t, t, t, t)$, $v_1 = (t-2, -t, -3t, t)$ and $v_2 = (2, t, 2t, 0)$.

(a) Find if there exists $t \in \mathbb{R}$ such that $U_t + V_t = \mathbb{R}^4$.

(b) For which $t \in \mathbb{R}$ we have $\dim(U_t \cap V_t) = 1$?

(c) Find a basis of $U_1 \cap V_1$ and extend it to a basis of $\mathbb{R}^4$.

Solution. (a) First of all, we have

$$U_t + V_t = L\{u_1, u_2, v_1, v_2\}.$$

Note that $u_2 - 2u_1 = v_1$, this proves that the vectors u_1, u_2, v_1, v_2 are linearly dependent, hence they are not a basis of $\mathbb{R}^4$. It follows that $U_t + V_t$ is never equal to $\mathbb{R}^4$.

(b) By what we have seen in (a), we have $v_1 \in U_t \cap V_t$, for any t. It follows that $\dim(U_t \cap V_t) \geq 1$, for any t. However, we also have $\dim(U_t \cap V_t) \leq 2$, because U_t and V_t have at most dimension 2, moreover $\dim(U_t \cap V_t) = 2$ if and only if $U_t = V_t$ and $\dim U_t = 2$. Let us therefore look for which values of t we have $U_t = V_t$. Since $v_1 \in U_t \cap V_t$, for any t, it suffices to check for which values of t the vector v_2 can be written as a linear combination of u_1 and u_2: $v_2 = \lambda_1 u_1 + \lambda_2 u_2$. We obtain the linear system

$$\begin{cases} \lambda_1 + t\lambda_2 = 2 \\ t\lambda_1 + t\lambda_2 = t \\ 2t\lambda_1 + t\lambda_2 = 2t \\ t\lambda_2 = 0 \end{cases}$$

which has solution if and only if $t = 0$. Therefore, for $t = 0$ we have $U_t = V_t$, but in this case $\dim U_t = \dim V_t = 1$, hence the dimension of $U_t \cap V_t$ is 1. It follows that $\dim(U_t \cap V_t) = 1$ for any $t \in \mathbb{R}$.

(c) As seen above, v_1 is a basis of $U_t \cap V_t$, for any t. By setting $t = 1$ we get $v_1 = (-1, -1, -3, 1)$. To complete this basis to a basis of $\mathbb{R}^4$ it is sufficient to find three other vectors w_1, w_2 and w_3 such that v_1, w_1, w_2, w_3 are linearly independent. It is easy to check that $w_1 = (1, 0, 0, 0)$, $w_2 = (0, 1, 0, 0)$ and $w_3 = (0, 0, 1, 0)$ satisfy this requirement.

► **Exercise 1.2.13.** In $\mathbb{R}^4$ we consider the subspaces $U = L\{v_1, v_2, v_3\}$ and $V = L\{v_4, v_5\}$, where $v_1 = (1, 2, 3, 4)$, $v_2 = (2, 2, 2, 6)$, $v_3 = (0, 2, 4, 4)$, $v_4 = (1, 0, -1, 2)$ and $v_5 = (2, 3, 0, 1)$. Find bases of the subspaces $U \cap V$, U, V and $U + V$.

Solution. A vector $w \in U \cap V$ can be written as

$$w = \lambda_1 v_1 + \lambda_2 v_2 + \lambda_3 v_3 = \lambda_4 v_4 + \lambda_5 v_5.$$

By solving this system of equations we find:

$$\begin{cases} \lambda_1 = -\lambda_4 \\ \lambda_2 = \lambda_4 \\ \lambda_3 = 0 \\ \lambda_5 = 0. \end{cases}$$

This means that the vector space $U \cap V$ has dimension 1 and its basis is given by the vector w obtained by setting, for example, $\lambda_4 = 1$ in the previous solution: $w = -v_1 + v_2 = v_4 = (1, 0, -1, 2)$.

A basis of U is formed by the vectors v_1, v_2 and v_3, because they are linearly independent (this is easy to check).

The vectors v_4 and v_5 are also linearly independent, so they are a basis of V. Finally, it is easy to see that the vectors v_1, v_2, v_3 and v_5 are linearly independent (while v_4 is a linear combination of v_1 and v_2). Therefore the vector space $U + V$ coincides with $\mathbb{R}^4$ and its bases is given by $\{v_1, v_2, v_3, v_5\}$ (or we can take the canonical basis of $\mathbb{R}^4$).

► **Exercise 1.2.14.** In the vector space V of polynomials of degree ≤ 5 with real coefficients consider the following subspaces:

$$\begin{aligned} U_1 &= \{p(x) \in V \mid p(0) = 0\} \\ U_2 &= \{p(x) \in V \mid p'(1) = 0\} \\ U_3 &= \{p(x) \in V \mid x^2 + 1 \text{ divides } p(x)\} \\ U_4 &= \{p(x) \in V \mid p(-x) = p(x), \forall x\} \\ U_5 &= \{p(x) \in V \mid p(x) = xp'(x), \forall x\} \end{aligned}$$

(a) Find bases of the following subspaces: U_1, U_2, U_3, U_4, U_5, $U_1 \cap U_2$, $U_1 \cap U_3$, $U_1 \cap U_2 \cap U_3$, $U_1 \cap U_2 \cap U_3 \cap U_4$.

(b) Find subspaces W_1, W_2 of V such that $W_1 \oplus U_4 = W_2 \oplus (U_1 \cap U_3) = V$.

Solution. (a) Every polynomial of degree ≤ 5 is written uniquely as follows:

$$p(x) = a_0 + a_1 x + a_2 x^2 + a_3 x^3 + a_4 x^4 + a_5 x^5.$$

From this we deduce that the vector space V has dimension 6, and a basis is given by 1, x, x^2, x^3, x^4, x^5.

The equality $p(0) = 0$ is equivalent to $a_0 = 0$. The subspace U_1 therefore consists of the polynomials of the form

$$p(x) = a_1 x + a_2 x^2 + a_3 x^3 + a_4 x^4 + a_5 x^5.$$

We deduce that U_1 has dimension 5 and a basis is $\{x, x^2, x^3, x^4, x^5\}$.

The equality $p'(1) = 0$ is equivalent to $a_1 = -2a_2 - 3a_3 - 4a_4 - 5a_5$. The subspace U_2 therefore consists of the polynomials of the form

$$p(x) = a_0 + (-2a_2 - 3a_3 - 4a_4 - 5a_5)x + a_2 x^2 + a_3 x^3 + a_4 x^4 + a_5 x^5.$$

It follows that U_2 has dimension 5 and a basis is $\{1, x^2 - 2x, x^3 - 3x, x^4 - 4x, x^5 - 5x\}$.

The requirement that $p(x)$ be divisible by $x^2 + 1$ means that $p(x)$ can be written as follows:

$$p(x) = (x^2 + 1)(b_0 + b_1 x + b_2 x^2 + b_3 x^3).$$

It follows that U_3 has dimension 4 and a basis is $\{(x^2 + 1), (x^2 + 1)x, (x^2 + 1)x^2, (x^2 + 1)x^3\}$.

The equality $p(-x) = p(x)$ is equivalent to $a_1 = a_3 = a_5 = 0$. The subspace U_4 therefore consists of the polynomials of the form

$$p(x) = a_0 + a_2 x^2 + a_4 x^4.$$

We deduce that U_4 has dimension 3 and a basis is $\{1, x^2, x^4\}$.

The equality $p(x) = xp'(x)$ is equivalent to $a_0 = a_2 = a_3 = a_4 = a_5 = 0$. The subspace U_5 therefore consists of the polynomials of the form

$$p(x) = a_1 x.$$

It follows that U_5 has dimension 1 and a basis is $\{x\}$.

The polynomials in $U_1 \cap U_2$ must simultaneously satisfy the conditions $a_0 = 0$ and $a_1 = -2a_2 - 3a_3 - 4a_4 - 5a_5$. The subspace $U_1 \cap U_2$ therefore consists of the polynomials of the form

$$p(x) = (-2a_2 - 3a_3 - 4a_4 - 5a_5)x + a_2 x^2 + a_3 x^3 + a_4 x^4 + a_5 x^5.$$

We deduce that $U_1 \cap U_2$ has dimension 4 and a basis is $\{x^2 - 2x, x^3 - 3x, x^4 - 4x, x^5 - 5x\}$.

$U_1 \cap U_3$ consists of the polynomials of the form

$$p(x) = (x^2 + 1)(b_0 + b_1 x + b_2 x^2 + b_3 x^3),$$

such that $p(0) = 0$, that is equivalent to $b_0 = 0$. We deduce that $U_1 \cap U_3$ has dimension 3 and a basis is $\{(x^2+1)x, (x^2+1)x^2, (x^2+1)x^3\}$.

By requiring the polynomials in $U_1 \cap U_3$ to satisfy the condition defining U_2, it turns out that it must be $b_1 = -\frac{3}{2}\, b_2 - 2\, b_3$. Therefore, the subspace $U_1 \cap U_2 \cap U_3$ has dimension 2 and a basis is $\{(x^2+1)(2x^2 - 3x), (x^2+1)(x^3 - 2x)\}$.

Finally, it is now easy to see that $U_1 \cap U_2 \cap U_3 \cap U_4 = \{\mathbf{0}\}$.

(b) Recalling that the basis of U_4 found previously is $\{1, x^2, x^4\}$, we deduce that a subspace W_1 satisfying the required conditions is, for example, $W_1 = L\{x, x^3, x^5\}$.

Similarly, observing that the subspace $U_1 \cap U_3$ has dimension 3 and does not contain polynomials of degree ≤ 2, we deduce that the subspace W_2 generated by the polynomials 1, x and x^2 satisfies the required conditions.

1.3 Recap exercises

▶ **Exercise 1.3.1.** Let U be the vector subspace of $\mathbb{R}^4$ generated by the vectors $u_1 = (0, 2, 0, -1)$ and $u_2 = (1, 1, 1, 0)$. Let V be the subspace of $\mathbb{R}^4$ consisting of the solutions of the following system:

$$\begin{cases} x_1 + x_3 = x_2 + x_4 \\ x_3 + x_4 = 0. \end{cases}$$

(a) Find the dimension and a basis of $U \cap V$.

(b) Find the dimension and a basis of $U + V$.

Solution. It is easy to verify that the vectors u_1 and u_2 are linearly independent, hence they are a basis of U. It follows that $\dim U = 2$.

From the system of equations that define the subspace V we obtain

$$\begin{cases} x_1 = x_2 + 2x_4 \\ x_3 = -x_4. \end{cases}$$

There are two unknowns which may have any value (x_2 and x_4). This means that V has dimension 2. By setting $x_2 = 1$ and $x_4 = 0$ we get $x_1 = 1$ and

$x_3 = 0$; let us denote by $v_1 = (1, 1, 0, 0)$ the corresponding vector. Setting $x_2 = 0$ and $x_4 = 1$ we obtain $x_1 = 2$ and $x_3 = -1$. This gives the vector $v_2 = (2, 0, -1, 1)$. A basis of V is given by v_1 and v_2.

We now look for the vectors that belong to $U \cap V$. Let w be such a vector: we must have $w = \alpha_1 u_1 + \alpha_2 u_2$, because $w \in U$, but also $w = \beta_1 v_1 + \beta_2 v_2$, because $w \in V$. We obtain the equality $\alpha_1 u_1 + \alpha_2 u_2 = \beta_1 v_1 + \beta_2 v_2$, that is equivalent to the following system:

$$\begin{cases} \alpha_2 = \beta_1 + 2\beta_2 \\ 2\alpha_1 + \alpha_2 = \beta_1 \\ \alpha_2 = -\beta_2 \\ -\alpha_1 = \beta_2 \end{cases}$$

from which it follows

$$\begin{cases} \alpha_1 = \alpha_2 = -\beta_2 \\ \beta_1 = -3\beta_2. \end{cases}$$

This system therefore has infinite solutions, for each value of β_2 (this means that the set of solutions, and therefore also $U \cap V$, has dimension 1). Setting $\beta_2 = -1$ we get $\beta_1 = 3$, $\alpha_1 = 1$ and $\alpha_2 = 1$. The corresponding vector $w \in U \cap V$ is $w = u_1 + u_2$ or, equivalently, $w = 3v_1 - v_2$. We get $w = (1, 3, 1, -1)$. It follows that $U \cap V$ has dimension 1 and is generated by w.

As regards the subspace $U + V$, it is certainly generated by the four vectors u_1, u_2, v_1, v_2 (because U is generated by u_1 and u_2, and V is generated by v_1 and v_2). From Grassmann's formula we get

$$\dim(U + V) = \dim U + \dim V - \dim(U \cap V) = 2 + 2 - 1 = 3,$$

therefore the four vectors u_1, u_2, v_1, v_2 certainly cannot be a basis of $U + V$ (one of these vectors must be a linear combination of the others). We arbitrarily choose to eliminate the vector v_2. Now we need to check if the three vectors u_1, u_2, v_1 are linearly independent. Let us consider a linear combination $\lambda_1 u_1 + \lambda_2 u_2 + \lambda_3 v_1 = \mathbf{0}$. We obtain the following system

$$\begin{cases} \lambda_2 + \lambda_3 = 0 \\ 2\lambda_1 + \lambda_2 + \lambda_3 = 0 \\ \lambda_2 = 0 \\ -\lambda_1 = 0 \end{cases}$$

whose solution is $\lambda_1 = \lambda_2 = \lambda_3 = 0$. We thus conclude that the three vectors u_1, u_2, v_1 are linearly independent and are therefore a basis of $U + V$ (because $U + V$ has dimension 3).

► **Exercise 1.3.2.** In $\mathbb{R}^4$ we consider the vector subspace $U = \langle u_1, u_2 \rangle$, where $u_1 = (-2, 1, 1, 3)$ and $u_2 = (0, -1, 2, 1)$.

(a) Find a basis of a subspace W such that $U \oplus W = \mathbb{R}^4$. Is W unique?

(b) Let $v_t = (t, 3, t-1, 1)$. Determine for which value of t we have $v_t \in U$.

(c) Let V be the vector subspace defined by the equations $x_1+x_2+x_3+x_4 = 0$ and $x_1 + 2x_2 - 2x_4 = 0$. Find a basis of V and a basis of $U \cap V$.

Solution. It is easy to verify that u_1 and u_2 are linearly independent, hence the subspace U has dimension 2. From Grassmann's formula it follows that the dimension of a subspace W such that $U \oplus W = \mathbb{R}^4$ must be 2. Therefore, to find a basis of such a subspace W, it suffices to find two vectors w_1 and w_2 such that $\{u_1, u_2, w_1, w_2\}$ is a basis of $\mathbb{R}^4$. There are endless choices for the vectors w_1 and w_2; for example we can take $w_1 = (1, 0, 0, 0)$ and $w_2 = (0, 1, 0, 0)$ (it is easy to verify that with these choices the vectors $\{u_1, u_2, w_1, w_2\}$ are linearly independent and therefore are a basis of $\mathbb{R}^4$). However one could also take $w_1 = (0, 0, 1, 0)$ and $w_2 = (0, 0, 0, 1)$, which proves that such a subspace W is certainly not unique.

Now let us consider the vector $v_t = (t, 3, t-1, 1)$. Requiring that v_t belongs to U is equivalent to requiring that v_t can be written as a linear combination of the vectors u_1 and u_2 (which are a basis of U):

$$v_t = \alpha_1 u_1 + \alpha_2 u_2.$$

This gives us the following system of linear equations

$$\begin{cases} t = -2\alpha_1 \\ 3 = \alpha_1 - \alpha_2 \\ t - 1 = \alpha_1 + 2\alpha_2 \\ 1 = 3\alpha_1 + \alpha_2 \end{cases}$$

whose solution is

$$\begin{cases} \alpha_1 = 1 \\ \alpha_2 = -2 \\ t = -2. \end{cases}$$

It follows that $v_t \in U$ if and only if $t = -2$.

The subspace V is the set of solutions of the following system of linear equations:

$$V : \begin{cases} x_1 + x_2 + x_3 + x_4 = 0 \\ x_1 + 2x_2 - 2x_4 = 0. \end{cases}$$

Solving the system we find that it admits infinite solutions, depending on two parameters, given by

$$\begin{cases} x_1 = -2x_2 + 2x_4 \\ x_3 = x_2 - 3x_4. \end{cases}$$

Setting $x_2 = 1$ and $x_4 = 0$ we obtain the vector $v_1 = (-2, 1, 1, 0)$, and setting $x_2 = 0$ and $x_4 = 1$ we get $v_2 = (2, 0, -3, 1)$. The vectors v_1 and v_2 are a basis of V. Finally we look for a basis of $U \cap V$.

Every vector $w \in U \cap V$ can be written in the form

$$w = \alpha_1 u_1 + \alpha_2 u_2, \qquad w = \beta_1 v_1 + \beta_2 v_2.$$

The equality $\alpha_1 u_1 + \alpha_2 u_2 = \beta_1 v_1 + \beta_2 v_2$ results in the following system:

$$\begin{cases} -2\alpha_1 = -2\beta_1 + 2\beta_2 \\ \alpha_1 - \alpha_2 = \beta_1 \\ \alpha_1 + 2\alpha_2 = \beta_1 - 3\beta_2 \\ 3\alpha_1 + \alpha_2 = \beta_2 \end{cases}$$

which has infinitely many solutions (depending on a parameter) given by

$$\begin{cases} \beta_1 = \alpha_1 - \alpha_2 \\ \beta_2 = 3\alpha_1 + \alpha_2 \\ \alpha_1 = -\dfrac{2}{3}\alpha_2 \end{cases}$$

A particular solution is given, for example, by $\alpha_1 = -2$ and $\alpha_2 = 3$ (as we will see now, it is not important to compute the values of β_1 and β_2). We can therefore conclude that $U \cap V$ has dimension 1 and a basis is given by the vector

$$w = \alpha_1 u_1 + \alpha_2 u_2 = -2u_1 + 3u_2 = (4, -5, 4, -3).$$

► **Exercise 1.3.3.** Let U be the vector subspace of $\mathbb{R}^4$ spanned by the vectors

$$u_1 = (2, 0, 1, -1), \quad u_2 = (1, 1, 2, 0), \quad u_3 = (4, -2, -1, -3),$$

and W be the vector subspace of $\mathbb{R}^4$ of equation $x + y + 2z = 0$.

(a) Find the dimension and a basis of U.

(b) Determine for which value of t the vector $v = (1+t, 1, 3, -1)$ belongs to U.

(c) Determine the Cartesian equations of U.

(d) Find the dimension and a basis of $U \cap W$.

(e) Find the dimension and a basis of $U + W$.

Solution. To determine the dimension of U we need to find out how many of the vectors u_1, u_2 and u_3 are linearly independent. To do this, we compute the rank of the following matrix

$$\begin{pmatrix} 1 & 1 & 2 & 0 \\ 2 & 0 & 1 & -1 \\ 4 & -2 & -1 & -3 \end{pmatrix}$$

We carry out the following elementary operations on the rows: from the second row we subtract twice the first and from the third row we subtract the first row multiplied by 4. The following matrix is obtained:

$$\begin{pmatrix} 1 & 1 & 2 & 0 \\ 0 & -2 & -3 & -1 \\ 0 & -6 & -9 & -3 \end{pmatrix}$$

Now we subtract three times the second from the third row, obtaining

$$\begin{pmatrix} 1 & 1 & 2 & 0 \\ 0 & -2 & -3 & -1 \\ 0 & 0 & 0 & 0 \end{pmatrix}$$

The rank of this matrix is 2, which means that the vectors u_1 and u_2 are linearly independent, and u_3 is a linear combination of u_1 and u_2. We conclude that the subspace U has dimension 2 and a basis is given by u_1 and u_2.

The vector $v = (1+t, 1, 3, -1)$ belongs to U if and only if it can be written as a linear combination of u_1 and u_2. The equation

$$v = \lambda_1 u_1 + \lambda_2 u_2$$

gives the system

$$\begin{cases} 1 + t = 2\lambda_1 + \lambda_2 \\ 1 = \lambda_2 \\ 3 = \lambda_1 + 2\lambda_2 \\ -1 = -\lambda_1, \end{cases}$$

whose solution is

$$\begin{cases} t = 2 \\ \lambda_1 = 1 \\ \lambda_2 = 1. \end{cases}$$

Thus we find that $v \in U$ if and only if $t = 2$.

To find the Cartesian equations of U we observe that a generic vector $(x, y, z, w) \in \mathbb{R}^4$ belongs to U if and only if it is a linear combination of the vectors u_1 and u_2, which form a basis of U. The equality

$$(x, y, z, w) = \lambda_1 u_1 + \lambda_2 u_2$$

gives the system

$$\begin{cases} x = 2\lambda_1 + \lambda_2 \\ y = \lambda_2 \\ z = \lambda_1 + 2\lambda_2 \\ w = -\lambda_1. \end{cases}$$

To find the Cartesian equations of U it suffices to eliminate the two parameters λ_1 and λ_2 from the previous equations. We have $\lambda_2 = y$ (from the second equation) and $\lambda_1 = -w$ (from the fourth); substituting into the two remaining equations, we find the system

$$\begin{cases} x - y + 2w = 0 \\ 2y - z - w = 0. \end{cases}$$

These are the Cartesian equations of U.

Since we now know the equations of U, to determine $U \cap W$ it is enough to combine the equations of U with those of W:

$$\begin{cases} x - y + 2w = 0 \\ 2y - z - w = 0 \\ x + y + 2z = 0. \end{cases}$$

The solutions of this system are given by

$$\begin{cases} x = -\frac{4}{3}\,w \\ y = \frac{2}{3}\,w \\ z = \frac{1}{3}\,w \\ w \text{ any value.} \end{cases}$$

This means that $U \cap W$ has dimension 1 and a basis is formed by the vector $u = (-4, 2, 1, 3)$, obtained by setting $w = 3$ in the previous equations.

Noting that W is a subspace of $\mathbb{R}^4$ defined by a single linear equation, we conclude that $\dim W = 3$ and therefore, using Grassmann's formula, we have

$$\dim(U + W) = \dim U + \dim W - \dim(U \cap W) = 2 + 3 - 1 = 4.$$

Since $U + W$ is a subspace of $\mathbb{R}^4$ of dimension 4, it must necessarily be $U + W = \mathbb{R}^4$. Looking for a basis of $U + W$ is therefore equivalent to looking for a basis of $\mathbb{R}^4$, hence we can take the canonical basis $\{e_1, e_2, e_3, e_4\}$.

► **Exercise 1.3.4.** In the vector space $\mathbb{R}^4$ let U be the subspace generated by the vectors $u_1 = (3, 4, -1, 1)$, $u_2 = (1, -2, -1, -3)$ and W the subspace defined by the equations $x_1 - 2x_2 + x_3 - x_4 = 0$ and $2x_1 + 3x_3 + x_4 = 0$.

(a) Find whether the sum of U and W is direct and find bases of $U + W$ and $U \cap W$.

(b) Find a basis of a subspace $L \subset \mathbb{R}^4$ such that $U \oplus L = W \oplus L = \mathbb{R}^4$.

(c) Given the vector $v = (0, 1, -1, 3)$, determine the equation of the smallest subspace of $\mathbb{R}^4$ containing U and v.

(d) Given $\bar{v} = (2, -1, 0, 3)$, let us consider the set $S = \bar{v} + U = \{\bar{v} + u \mid u \in U\}$. Write a linear system whose set solutions is S.

Solution. We begin by determining a basis of W. The vectors of W are the solutions of the following linear system

$$W : \begin{cases} x_1 - 2x_2 + x_3 - x_4 = 0 \\ 2x_1 + 3x_3 + x_4 = 0 \end{cases}$$

Computing x_2 and x_4 as a function of x_1 and x_3, we find that the previous system is equivalent to

$$W : \begin{cases} x_2 = \dfrac{3}{2}x_1 + 2x_3 \\ x_4 = -2x_1 - 3x_3 \end{cases}$$

therefore W has dimension 2 and a basis is formed by the vectors $w_1 = (2, 3, 0, -4)$ and $w_2 = (0, 2, 1, -3)$.

We now look for a basis of $U \cap W$. If $v \in U \cap W$, it must be

$$v = a_1 u_1 + a_2 u_2 = b_1 w_1 + b_2 w_2.$$

The equality $a_1 u_1 + a_2 u_2 = b_1 w_1 + b_2 w_2$ gives the system

$$\begin{cases} 3a_1 + a_2 = 2b_1 \\ 4a_1 - 2a_2 = 3b_1 + 2b_2 \\ -a_1 - a_2 = b_2 \\ a_1 - 3a_2 = -4b_1 - 3b_2 \end{cases}$$

whose solutions are

$$\begin{cases} a_1 = a_2 \\ b_1 = 2a_2 \\ b_2 = -2a_2 \\ a_2 \text{ any value.} \end{cases}$$

It follows that $U \cap W$ has dimension 1. To find a basis we can set $a_2 = 1$, obtaining $a_1 = 1$, $b_1 = 2$, $b_2 = -2$. It follows that $v = u_1 + u_2 = 2w_1 - 2w_2$, hence $v = (4, 2, -2, -2)$. This vector is a basis of $U \cap W$. Since $U \cap W \neq \{\mathbf{0}\}$, the sum of U and W is not direct. From Grassmann's formula we can deduce that

$$\dim(U + W) = \dim U + \dim W - \dim(U \cap W) = 3.$$

Since $U + W$ is generated by the four vectors u_1, u_2, w_1 and w_2, but has dimension 3, to find a basis we need to eliminate one of these four vectors. We choose (arbitrarily) to eliminate u_2 and, after verifying that the remaining vectors u_1, w_1 and w_2 are linearly independent, we can state that they are a basis of $U + W$.

Now we have to determine a basis of a subspace $L \subset \mathbb{R}^4$ such that $U \oplus L = W \oplus L = \mathbb{R}^4$. Since $\dim U = \dim W = 2$, from Grassmann's formula we deduce that L must have dimension 2. To find a basis of L we need to find two vectors ℓ_1 and ℓ_2 such that $\{u_1, u_2, \ell_1, \ell_2\}$ and $\{w_1, w_2, \ell_1, \ell_2\}$ are bases of $\mathbb{R}^4$. We can set $\ell_1 = (1, 0, 0, 0)$ and $\ell_2 = (0, 1, 0, 0)$: it is easy to verify that both the vectors $\{u_1, u_2, \ell_1, \ell_2\}$ and $\{w_1, w_2, \ell_1, \ell_2\}$ are linearly independent, so the vectors ℓ_1 and ℓ_2 are a basis of the subspace L we are looking for.

We note that other choices of ℓ_1 and ℓ_2, such as $\ell_1 = (0, 0, 1, 0)$ and $\ell_2 = (0, 0, 0, 1)$ would do (indeed there are infinitely many possible choices), which means that the subspace L is not unique.

The smallest subspace of $\mathbb{R}^4$ containing U and $v = (0, 1, -1, 3)$ is the subspace generated by the vectors u_1, u_2 and v. Denoting by (x_1, x_2, x_3, x_4)

the generic vector of this subspace, we must therefore have $(x_1, x_2, x_3, x_4) = a_1u_1 + a_2u_2 + a_3v$. This equation is equivalent to the following system

$$\begin{cases} x_1 = 3a_1 + a_2 \\ x_2 = 4a_1 - 2a_2 + a_3 \\ x_3 = -a_1 - a_2 - a_3 \\ x_4 = a_1 - 3a_2 + 3a_3 \end{cases}$$

Computing a_1, a_2 and a_3, we obtain the equivalent system

$$\begin{cases} a_2 = x_1 - 3a_1 \\ a_3 = 2a_1 - x_1 - x_3 \\ a_1 = (3x_1 + x_2 + x_3)/12 \\ 6x_1 - 4x_2 + 5x_3 + 3x_4 = 0. \end{cases}$$

The last equation does not contain the parameters a_1, a_2 and a_3; it is therefore the equation we are looking for. We conclude that the smallest subspace of $\mathbb{R}^4$ containing U and v is given by the equation

$$6x_1 - 4x_2 + 5x_3 + 3x_4 = 0.$$

The last point is similar to what we have just seen. The elements of the set $S = \bar{v} + U$ are written in the form $\bar{v} + a_1u_1 + a_2u_2$, with $a_1, a_2 \in \mathbb{R}$. If we denote by (x_1, x_2, x_3, x_4) the generic vector of S, we have $(x_1, x_2, x_3, x_4) = \bar{v} + a_1u_1 + a_2u_2$, which gives the system

$$\begin{cases} x_1 = 2 + 3a_1 + a_2 \\ x_2 = -1 + 4a_1 - 2a_2 \\ x_3 = -a_1 - a_2 \\ x_4 = 3 + a_1 - 3a_2 \end{cases}$$

Computing a_1 and a_2, we obtain the equivalent system

$$\begin{cases} a_1 = -x_3 - a_2 \\ a_2 = (2 - x_1 - 3x_3)/2 \\ 2x_1 + 5x_3 - x_4 = 1 \\ 3x_1 - x_2 + 5x_3 = 7. \end{cases}$$

The last two equations do not contain the parameters a_1 and a_2; therefore they form a system of linear equations having S as the set of solutions

$$S : \begin{cases} 2x_1 + 5x_3 - x_4 = 1 \\ 3x_1 - x_2 + 5x_3 = 7. \end{cases}$$

▶ **Exercise 1.3.5.** In $\mathbb{R}^4$ let U be the subspace spanned by the vector $u = (12, 3, -2, 0)$ and W be the subspace of equation $x_1 - 2x_2 + 3x_3 - 4x_4 = 0$.

(a) Prove that $U \subset W$ and complete the basis of U to a basis of W.

(b) Let $V \subset \mathbb{R}^4$ be the subspace spanned by the vectors $v_1 = (1, 2, 3, 0)$ and $v_2 = (2, 3, 4, -1)$. Find a basis of $V \cap W$ and a basis of $V + W$.

(c) Let $v_t = (t, 0, 1, 2)$. Determine the value of t for which the vectors v_1, v_2, v_t are linearly dependent.

(d) Is there a linear function $f: \mathbb{R}^4 \to \mathbb{R}^4$ such that $W = \operatorname{Ker} f$ and $U = \operatorname{Im} f$?

Solution. To prove that $U \subset W$ it suffices to observe that $u \in W$ (in fact the coordinates of u satisfy the equation of W).

The subspace W has dimension 3, being defined by an equation in $\mathbb{R}^4$. Therefore to complete the basis of U to a basis of W it is sufficient to find two vectors $w_1, w_2 \in W$ such that the vectors u, w_1 and w_2 are linearly independent.

From the equation of W we obtain

$$x_1 = 2x_2 - 3x_3 + 4x_4.$$

Setting $w_1 = (2, 1, 0, 0)$ and $w_2 = (-3, 0, 1, 0)$ we find $u = 3w_1 - 2w_2$, hence the vectors u, w_1, w_2 are linearly dependent, therefore we must change our choice. So let's try to set $w_2 = (4, 0, 0, 1)$; now it is easy to verify that u, w_1, w_2 are linearly independent and therefore are a basis of W.

To determine a basis of $V \cap W$ we can start by determining the equations of V. A generic vector $(x_1, x_2, x_3, x_4) \in V$ is a linear combination of the vectors v_1 and v_2 which generate V, therefore we have

$$(x_1, x_2, x_3, x_4) = a_1 v_1 + a_2 v_2,$$

that gives the system

$$\begin{cases} x_1 = a_1 + 2a_2 \\ x_2 = 2a_1 + 3a_2 \\ x_3 = 3a_1 + 4a_2 \\ x_4 = -a_2 \end{cases}$$

Computing a_1 and a_2 we obtain the system

$$\begin{cases} a_2 = -x_4 \\ a_1 = x_1 + 2x_4 \\ 2x_1 - x_2 + x_4 = 0 \\ 3x_1 - x_3 + 2x_4 = 0 \end{cases}$$

from which it follows that the Cartesian equations of V are as follows:

$$V : \begin{cases} 2x_1 - x_2 + x_4 = 0 \\ 3x_1 - x_3 + 2x_4 = 0. \end{cases}$$

The subspace $V \cap W$ is therefore described by the following system:

$$V \cap W : \begin{cases} 2x_1 - x_2 + x_4 = 0 \\ 3x_1 - x_3 + 2x_4 = 0 \\ x_1 - 2x_2 + 3x_3 - 4x_4 = 0 \end{cases}$$

whose solutions are

$$V \cap W : \begin{cases} x_1 = 0 \\ x_2 = x_4 \\ x_3 = 2x_4 \\ x_4 \text{ any value.} \end{cases}$$

It follows that $V \cap W$ has dimension 1 and a basis is given by the vector $(0, 1, 2, 1)$. From Grassmann's formula it follows that

$$\dim(V + W) = \dim V + \dim W - \dim(V \cap W) = 2 + 3 - 1 = 4,$$

hence we have $V + W = \mathbb{R}^4$. As the basis of $V + W$ one can therefore take any basis of $\mathbb{R}^4$ (for example, the canonical basis).

Let us now find for which value of t the vectors v_1, v_2, v_t are linearly dependent. The linear combination

$$a_1 v_1 + a_2 v_2 + a_3 v_t = \mathbf{0}$$

gives the system

$$\begin{cases} a_1 + 2a_2 + a_3 t = 0 \\ 2a_1 + 3a_2 = 0 \\ 3a_1 + 4a_2 + a_3 = 0 \\ -a_2 + 2a_3 = 0. \end{cases}$$

Solving this system we find that, if $t \neq -1$ then we have $a_1 = a_2 = a_3 = 0$ and therefore the vectors v_1, v_2, v_t are linearly independent, otherwise, if $t = -1$, they are linearly dependent.

To find out if there exists a linear function $f\colon \mathbb{R}^4 \to \mathbb{R}^4$ such that $W = \operatorname{Ker} f$ and $U = \operatorname{Im} f$ we recall that, for any linear function $f\colon \mathbb{R}^4 \to \mathbb{R}^4$ we must have

$$\dim \operatorname{Ker}(f) + \dim \operatorname{Im}(f) = 4.$$

Since $\dim W = 3$ and $\dim U = 1$, this equality is verified. Of course this is not enough to say that such an f exists!

However we recall that to define a linear function $f\colon \mathbb{R}^4 \to \mathbb{R}^4$ it is sufficient to assign the images via f of the vectors of a basis of $\mathbb{R}^4$. Let us then consider a basis $\{w_1, w_2, w_3\}$ of W and let $v \in \mathbb{R}^4$ be a vector such that $\{w_1, w_2, w_3, v\}$ is a basis of $\mathbb{R}^4$ (such a vector v always exists). We define a linear function $f\colon \mathbb{R}^4 \to \mathbb{R}^4$ by setting $f(w_1) = f(w_2) = f(w_3) = \mathbf{0}$ and $f(v) = u$. For such a f we have $\operatorname{Ker}(f) = W$ and $\operatorname{Im}(f) = U$, as required.

► **Exercise 1.3.6.** Let $T \subset \mathbb{R}^4$ be the vector subspace generated by the vectors $v_1 = (1, 1, 1, 2)$, $v_2 = (3, 0, 0, -1)$. Let S be the subspace given by the solutions of the equations

$$\begin{cases} -x + 2y + 2z = 0 \\ -x + 2z + 2w = 0 \end{cases}$$

(a) Find $S \cap T$ and a basis of $S + T$.

(b) Find a subspace L of $\mathbb{R}^4$ such that $(T + S) \oplus L = \mathbb{R}^4$.

(c) Find another subspace L_1, $L_1 \neq L$ such that $(T + S) \oplus L_1 = \mathbb{R}^4$ and $L \oplus L_1$. Is it possible that $L \oplus L_1 = \mathbb{R}^4$?

(d) Find a subspace M of $\mathbb{R}^4$, of dimension 2 and different from S, such that $T + S = T + M$.

Solution. The generic vector of T is a linear combination of the vectors v_1 and v_2; so it is written as follows:

$$w = a_1 v_1 + a_2 v_2 = (a_1 + 3a_2, a_1, a_1, 2a_1 - a_2).$$

For w to belong to $S \cap T$, its components must satisfy the equations of S. Solving the system thus obtained, we find $a_1 = a_2$, from which we deduce that $S \cap T$ has dimension 1 and a basis is given by the vector $w = (4, 1, 1, 1)$ (obtained by setting $a_1 = a_2 = 1$).

From the equations of S

$$S : \begin{cases} -x + 2y + 2z = 0 \\ -x + 2z + 2w = 0 \end{cases}$$

we obtain

$$S : \begin{cases} x = 2z + 2w \\ y = w \end{cases}$$

It follows that S has dimension 2 and a basis is given by the vectors $s_1 = (2, 0, 1, 0)$ and $s_2 = (2, 1, 0, 1)$.

From Grassmann's formula we have

$$\dim(S + T) = \dim S + \dim T - \dim(S \cap T) = 3$$

and, since the vectors v_1, v_2, s_1 are linearly independent (as is easily verified) they are necessarily a basis of $S + T$.

A subspace L of $\mathbb{R}^4$ such that $(T+S) \oplus L = \mathbb{R}^4$ must have dimension 1. To determine L it is therefore sufficient to find a vector ℓ such that the vectors v_1, v_2, s_1, ℓ are linearly independent. There are infinite possible choices for ℓ; for example the vector $\ell = (1, 0, 0, 0)$ satisfies the requests (verify it).

Similarly, L_1 must also have dimension 1 and a basis must consist of a vector ℓ_1 such that the vectors v_1, v_2, s_1, ℓ_1 are linearly independent. If we also require that L and L_1 be in direct sum, this means that $L \cap L_1 = \{\mathbf{0}\}$ and therefore that the vectors ℓ and ℓ_1 must be linearly independent. Therefore, it is sufficient to take $\ell_1 = (0, 0, 0, 1)$ and verify that v_1, v_2, s_1, ℓ_1 are linearly independent. Of course it can't be $L \oplus L_1 = \mathbb{R}^4$ because $\dim L + \dim L_1 = 2$.

Finally, to determine a subspace M of $\mathbb{R}^4$ of dimension 2 and different from S, such that $T + S = T + M$, we recall that $T + S$ is the subspace generated by the vectors v_1, v_2, s_1. A basis of M must therefore be formed by two vectors chosen in such a way as to guarantee that $T+S = T+M$ and, at the same time, that $M \neq S$. Since S is generated by the vectors s_1 and s_2, we can take $M = \langle s_1, v_2 \rangle$. It is easy to verify that $M \neq S$ (because M does not contain s_2) and that $T + M$ has as basis $\{v_1, v_2, s_1\}$, and therefore is equal to $T + S$.

► **Exercise 1.3.7.** Let $U \subset \mathbb{R}^4$ be the subspace of equations $3x_2 + x_3 - x_4 = 0$ and $x_2 - 2x_3 + x_4 = 0$, and let $W \subset \mathbb{R}^4$ be the smallest subspace which contains the vector $(1, 1, 0, 0)$ and the vectors $(0, t, t+1, t+2)$, for each $t \in \mathbb{Z}$.

(a) Find the dimension and a basis of U.

(b) Find the Cartesian equations and a basis of W.

(c) Find bases of $U \cap W$ and $U + W$.

(d) Determine if there exists a nonzero subspace $L \subset \mathbb{R}^4$ that is in direct sum with both U and W. If such a L exists, find a basis. Also determine (if it exists) a subspace $L' \subset \mathbb{R}^4$ such that $(U \oplus L) \oplus L' = \mathbb{R}^4$.

Solution. The subspace U is given by the set of solutions of the following system:

$$U : \begin{cases} 3x_2 + x_3 - x_4 = 0 \\ x_2 - 2x_3 + x_4 = 0 \end{cases}$$

This system is equivalent to

$$U : \begin{cases} x_3 = 4x_2 \\ x_4 = 7x_2 \end{cases}$$

Since there are two free parameters (x_2 and x_1), U has dimension 2 and a basis is given by the vectors $u_1 = (1, 0, 0, 0)$ and $u_2 = (0, 1, 4, 7)$.

From the definition of W we deduce that it is the vector subspace of $\mathbb{R}^4$ generated by the infinite vectors $(1, 1, 0, 0)$ and $(0, t, t + 1, t + 2)$, for any integer t. However, we observe that, for each t, we have

$$(0, t, t + 1, t + 2) = (0, 0, 1, 2) + t(0, 1, 1, 1),$$

so the three vectors $w_1 = (1, 1, 0, 0)$, $w_2 = (0, 0, 1, 2)$ and $w_3 = (0, 1, 1, 1)$ are enough to generate W. Since, as can be easily seen, these vectors are linearly independent, they are a basis of W, which therefore has dimension 3.

To find the Cartesian equations of W (actually there will be only one equation, given that $\dim W = 3$) we write the generic vector of W as a linear combination of the three basis vectors:

$$(x_1, x_2, x_3, x_4) = \alpha_1 w_1 + \alpha_2 w_2 + \alpha_3 w_3.$$

We obtain the system

$$\begin{cases} x_1 = \alpha_1 \\ x_2 = \alpha_1 + \alpha_3 \\ x_3 = \alpha_2 + \alpha_3 \\ x_4 = 2\alpha_2 + \alpha_3 \end{cases}$$

from which, eliminating the parameters α_1, α_2 and α_3, we obtain the equation $x_1 - x_2 + 2x_3 - x_4 = 0$. This is the Cartesian equation of W.

To find the intersection of U and W we need to solve the system formed by the equations of U and those of W:

$$U \cap W : \begin{cases} 3x_2 + x_3 - x_4 = 0 \\ x_2 - 2x_3 + x_4 = 0 \\ x_1 - x_2 + 2x_3 - x_4 = 0 \end{cases}$$

The solutions of this system are given by

$$U \cap W : \begin{cases} x_1 = 0 \\ x_3 = 4x_2 \\ x_4 = 7x_2 \end{cases}$$

from which we deduce that $\dim(U \cap W) = 1$ and a basis of $U \cap W$ is given by the vector $(0, 1, 4, 7)$. Then we have $\dim(U + W) = 2 + 3 - 1 = 4$, which implies that $U + W = \mathbb{R}^4$. As the basis of $U + W$ we can therefore take the canonical basis of $\mathbb{R}^4$.

Recalling that $\dim U = 2$ and $\dim W = 3$, a nonzero subspace $L \subset \mathbb{R}^4$ that is in direct sum with both U and W must necessarily have dimension 1. We therefore have to find a subspace L, generated by a vector ℓ, such that $U \cap L = \{\mathbf{0}\}$ and $W \cap L = \{\mathbf{0}\}$ (*attention*: the text of this exercise does not ask for $U + L = \mathbb{R}^4$ and $W + L = \mathbb{R}^4$, it only asks that the subspaces U and L, and W and L, be in direct sum, i.e., that $U \cap L = \{\mathbf{0}\}$ and $W \cap L = \{\mathbf{0}\}$). We therefore have to find a vector ℓ such that $\ell \notin U$ and $\ell \notin W$; for example, the vector $\ell = (0, 0, 0, 1)$ satisfies these requirements.

Therefore we conclude that such a subspace L exists and a basis is, for example, $\ell = (0, 0, 0, 1)$.

Now we have $\dim(U \oplus L) = 3$, so there certainly exists a subspace L', of dimension 1, such that $(U \oplus L) \oplus L' = \mathbb{R}^4$. Such a subspace will be generated by a vector ℓ' chosen such that the vectors $\{u_1, u_2, \ell, \ell'\}$ are a basis of $\mathbb{R}^4$. A possible choice is given by the vector $\ell' = (0, 0, 1, 0)$ (also in this case this choice is certainly not unique).

2

Linear Functions and Matrices

2.1 Linear functions

► **Exercise 2.1.1.** Determine if the following functions are linear:

(a) $f\colon \mathbb{R}^2 \to \mathbb{R}^3$, $(x, y) \mapsto (x - y, x + y + 1, 0)$

(b) $f\colon \mathbb{R}^2 \to \mathbb{R}^2$, $(x, y) \mapsto (2x, x + y)$

(c) $f\colon \mathbb{R}^2 \to \mathbb{R}$, $(x, y) \mapsto \sin(x - y)$.

Solution. Recalling the definition of linear function, it is only a matter of checking if $f(v + w) = f(v) + f(w)$ and $f(\lambda v) = \lambda f(v)$, for any choice of vectors v and w and for any scalar λ. Thus we find that only the function at point (b) is linear.

► **Exercise 2.1.2.** Find for which values of $t \in \mathbb{R}$ the following function is linear:

$$f\colon \mathbb{R}^3 \to \mathbb{R}^2, \quad (x, y, z) \mapsto (x + ty, tyz).$$

Solution. Reasoning as in the previous exercise it turns out that the only value of t that makes this function linear is $t = 0$.

► **Exercise 2.1.3.** Let $f\colon \mathbb{C}^2 \to \mathbb{C}$ be the function given by $f(x, y) = x + \bar{y}$, where $\bar{y}$ denotes the complex conjugate of the complex number y. Determine if f is $\mathbb{C}$-linear.

Solution. It is easy to check that

$$f((x_1, y_1) + (x_2, y_2)) = f(x_1, y_1) + f(x_2, y_2),$$

hence f is additive. However, we have

$$f(\lambda(x, y)) = \lambda x + \overline{\lambda y} = \lambda x + \bar{\lambda}\bar{y},$$

while

$$\lambda f(x, y) = \lambda x + \lambda \bar{y},$$

and, in general, $\lambda \neq \bar{\lambda}$ for $\lambda \in \mathbb{C}$. This proves that f is not linear.

► **Exercise 2.1.4.** Let $f : V \to W$ be a function between two vector spaces. Prove that f is linear if and only if its graph is a vector subspace of $V \times W$.

Solution. Recall that the graph of f is the subset Γ_f of $V \times W$ defined by

$$\Gamma_f = \{ (v, w) \mid w = f(v) \}.$$

To state that Γ_f is a vector subspace is equivalent to saying that it is closed under the operations of addition and product for a scalar, i.e., for every (v_1, w_1), $(v_2, w_2) \in \Gamma_f$ and every scalar λ, we have $(v_1, w_1) + (v_2, w_2) \in \Gamma_f$ and $\lambda(v_1, w_1) \in \Gamma_f$. The sum $(v_1, w_1) + (v_2, w_2) = (v_1 + v_2, w_1 + w_2)$ belongs to Γ_f if and only if $w_1 + w_2 = f(v_1 + v_2)$, i.e., if and only if $f(v_1) + f(v_2) = f(v_1 + v_2)$, since, by definition, we have $w_1 = f(v_1)$ and $w_2 = f(v_2)$. Similarly, we find that $\lambda(v_1, w_1) \in \Gamma_f$ if and only if $\lambda w_1 = f(\lambda v_1)$, i.e., if and only if $\lambda f(v_1) = f(\lambda v_1)$.

2.2 Linear functions and matrices

► **Exercise 2.2.1.** Let $f : V \to W$ be a linear function between two vector spaces. Let $\{v_1, v_2, v_3\}$ be a basis of V and $\{w_1, w_2, w_3, w_4\}$ be a basis of W, and suppose that f is defined by $f(v_1) = 2w_1 - 3w_2 + w_4$, $f(v_2) = w_2 - 2w_3 + 3w_4$ and $f(v_3) = w_1 + w_2 + w_3 - 3w_4$. Write the matrix of f in the given bases.

Solution. Let's denote by A the matrix of f. We recall that, by definition, the *columns* of A consists of the coordinates of the vectors $f(v_i)$ with respect to the basis of W. Therefore, from the equality $f(v_1) = 2w_1 - 3w_2 + w_4$ we deduce that the coefficients 2, -3, 0 (because w_3 is missing) and 1 are the elements of the first column of A. Similarly, from $f(v_2) = w_2 - 2w_3 + 3w_4$ it follows that the second column of A consists of the numbers 0 (because w_1 is missing), 1, -2 and 3, while the equality $f(v_3) = w_1 + w_2 + w_3 - 3w_4$ gives us the elements of the third column, which are 1, 1, 1 and -3. Therefore we have

$$A = \begin{pmatrix} 2 & 0 & 1 \\ -3 & 1 & 1 \\ 0 & -2 & 1 \\ 1 & 3 & -3 \end{pmatrix}$$

► **Exercise 2.2.2.** Let V and W be two vector spaces with bases $\{v_1, v_2, v_3\}$ and $\{w_1, w_2\}$, respectively. Let $f\colon V \to W$ be the linear function whose matrix (in the given bases) is

$$A = \begin{pmatrix} 2 & -1 & 1 \\ 3 & 2 & -3 \end{pmatrix}$$

(a) We choose for V the new basis $v_1' = v_2 + v_3$, $v_2' = v_1 + v_3$, $v_3' = v_1 + v_2$. Write the matrix A' of f with respect to the bases $\{v_1', v_2', v_3'\}$ and $\{w_1, w_2\}$.

(b) We choose for W the new basis $w_1' = \frac{1}{2}(w_1 + w_2)$ and $w_2' = \frac{1}{2}(w_1 - w_2)$. Write the matrix A'' of f with respect to the bases $\{v_1', v_2', v_3'\}$ and $\{w_1', w_2'\}$.

Solution. From the hypotheses we know that $f(v_1) = 2w_1 + 3w_2$, $f(v_2) = -w_1 + 2w_2$, $f(v_3) = w_1 - 3w_2$.

(a) We know that the columns of the matrix A' contain the coordinates, with respect to the basis $\{w_1, w_2\}$, of the images of the vectors of the basis $\{v_1', v_2', v_3'\}$. Therefore it sufficies to compute $f(v_1')$, $f(v_2')$ and $f(v_3')$. Since f is linear, we have:

$$\begin{aligned} f(v_1') &= f(v_2) + f(v_3) = -w_2, \\ f(v_2') &= f(v_1) + f(v_3) = 3w_1, \\ f(v_3') &= f(v_1) + f(v_2) = w_1 + 5w_2, \end{aligned}$$

hence

$$A' = \begin{pmatrix} 0 & 3 & 1 \\ -1 & 0 & 5 \end{pmatrix}.$$

(b) Similarly, the columns of the matrix A'' contain the coordinates, with respect to the basis $\{w_1', w_2'\}$, of the images of the vectors of the basis $\{v_1', v_2', v_3'\}$. We therefore have to express the vectors $f(v_1')$, $f(v_2')$ and $f(v_3')$ in terms of the vectors w_1' and w_2'. The vectors w_1 and w_2 must therefore be expressed as a linear combination of the vectors of the new basis. Solving the system

$$\begin{cases} \frac{1}{2}(w_1 + w_2) = w_1' \\ \frac{1}{2}(w_1 - w_2) = w_2' \end{cases}$$

we obtain

$$\begin{cases} w_1 = w_1' + w_2' \\ w_2 = w_1' - w_2' \end{cases}$$

from which it follows

$$\begin{aligned} f(v_1') &= -w_2 = -w_1' + w_2', \\ f(v_2') &= 3w_1 = 3w_1' + 3w_2', \\ f(v_3') &= w_1 + 5w_2 = 6w_1' - 4w_2'. \end{aligned}$$

The matrix A'' is

$$A'' = \begin{pmatrix} -1 & 3 & 6 \\ 1 & 3 & -4 \end{pmatrix}.$$

► **Exercise 2.2.3.** Let V be the subspace of $C^\infty(\mathbb{R})$ generated by the functions $f_1(x) = e^{2x} + \cos x$, $f_2(x) = \cos x + \sin x$ and $f_3(x) = \sin x$. Prove that f_1, f_2 and f_3 are linearly independent and write the matrix (with respect to the basis $\{f_1, f_2, f_3\}$) of the endomorphism of V that sends a function to its derivative.

Solution. Let's suppose that $\lambda_1 f_1 + \lambda_2 f_2 + \lambda_3 f_3 = \mathbf{0}$. This is equivalent to saying that the function

$$\lambda_1 e^{2x} + (\lambda_1 + \lambda_2)\cos x + (\lambda_2 + \lambda_3)\sin x$$

vanishes identically. Taking the limit for $x \to +\infty$, and remembering that $\sin x$ and $\cos x$ are two bounded functions, we deduce that it must be $\lambda_1 = 0$. Therefore we are left with the function

$$\lambda_2 \cos x + (\lambda_2 + \lambda_3)\sin x,$$

that vanishes identically. For example, if we set $x = 0$ we obtain $\lambda_2 = 0$, from which it follows that also $\lambda_3 = 0$, since the function $\lambda_3 \sin x$ must be identically zero. In conclusion, the three functions are linearly independent and therefore are a basis of the subspace V.

Now let $D\colon V \to V$ be the endomorphism given by $D(f) = f'$. We have:

$$\begin{aligned} D(f_1) &= 2e^{2x} - \sin x = 2f_1 - 2f_2 + f_3, \\ D(f_2) &= -\sin x + \cos x = f_2 - 2f_3, \\ D(f_3) &= \cos x = f_2 - f_3. \end{aligned}$$

It follows that the matrix of D with respect to the basis $\{f_1, f_2, f_3\}$ is

$$A = \begin{pmatrix} 2 & 0 & 0 \\ -2 & 1 & 1 \\ 1 & -2 & -1 \end{pmatrix}$$

► **Exercise 2.2.4.** Find the matrices, with respect to the canonical bases, of all the linear functions $f\colon \mathbb{R}^3 \to \mathbb{R}^4$ such that $f(1,2,-1) = (0,1,0,1)$, $f(3,-1,2) = (1,2,0,-1)$ and $f(-1,5,-4) = (2,0,3,2)$.

Solution. We start by setting

$$\begin{aligned} v_1 &= (1,2,-1),\ v_2 = (3,-1,2),\ v_3 = (-1,5,-4), \\ w_1 &= (0,1,0,1),\ w_2 = (1,2,0,-1),\ w_3 = (2,0,3,2). \end{aligned}$$

Given that a linear function is uniquely determined by the images of the vectors of a basis, we deduce that, if the three vectors v_1, v_2 and v_3 are a basis of $\mathbb{R}^3$ then such a function exists and is unique. Let us then check if these three vectors are linearly independent:

$$\lambda_1 v_1 + \lambda_2 v_2 + \lambda_3 v_3 = \mathbf{0}$$

is equivalent to

$$\begin{cases} \lambda_1 + 3\lambda_2 - \lambda_3 = 0 \\ 2\lambda_1 - \lambda_2 + 5\lambda_3 = 0 \\ -\lambda_1 + 2\lambda_2 - 4\lambda_3 = 0 \end{cases}$$

from which we obtain

$$\begin{cases} \lambda_1 = -3\lambda_2 + \lambda_3 \\ \lambda_2 = \lambda_3 \\ 0 = 0 \end{cases}$$

It turns out that there are non-zero solutions and, consequently, the three vectors are linearly dependent. By setting, for example, $\lambda_3 = 1$, we obtain

$\lambda_2 = 1$ and $\lambda_1 = -2$, and therefore a linear dependence relation between the three vectors is as follows:

$$-2v_1 + v_2 + v_3 = \mathbf{0}.$$

If such a f exists, by linearity we must have

$$\mathbf{0} = f(-2v_1 + v_2 + v_3) = -2f(v_1) + f(v_2) + f(v_3) = -2w_1 + w_2 + w_3,$$

but this equality is not verified by the vectors w_1, w_2 and w_3. Thus we conclude that there is no linear function with the required properties.

► **Exercise 2.2.5.** Find the matrices, with respect to the canonical bases, of all the linear functions $f\colon \mathbb{R}^3 \to \mathbb{R}^2$ such that $f(0,-2,1) = (3,-1)$, $f(1,1,-2) = (1,2)$ and $f(2,-4,-1) = (11,1)$.

Solution. We start by setting

$$v_1 = (0,-2,1),\ v_2 = (1,1,-2),\ v_3 = (2,-4,-1),$$
$$w_1 = (3,-1),\ w_2 = (1,2),\ w_3 = (11,1).$$

Given that a linear function is uniquely determined by the images of the vectors of a basis, we deduce that, if the three vectors v_1, v_2 and v_3 are a basis of $\mathbb{R}^3$ then such a function exists and is unique. Let us then check if these three vectors are linearly independent:

$$\lambda_1 v_1 + \lambda_2 v_2 + \lambda_3 v_3 = \mathbf{0}$$

is equivalent to

$$\begin{cases} \lambda_2 + 2\lambda_3 = 0 \\ -2\lambda_1 + \lambda_2 - 4\lambda_3 = 0 \\ \lambda_1 - 2\lambda_2 - \lambda_3 = 0 \end{cases}$$

from which we obtain

$$\begin{cases} \lambda_2 = -2\lambda_3 \\ \lambda_1 = -3\lambda_3 \\ 0 = 0 \end{cases}$$

It turns out that there are non-zero solutions and, consequently, the three vectors are linearly dependent. By setting, for example, $\lambda_3 = 1$, we obtain

$\lambda_1 = -3$ and $\lambda_2 = -2$, and therefore a linear dependence relation between the three vectors is as follows:

$$-3v_1 - 2v_2 + v_3 = \mathbf{0}.$$

If such a f exists, by linearity we must have

$$\mathbf{0} = f(-3v_1 - 2v_2 + v_3) = -3f(v_1) - 2f(v_2) + f(v_3) = -3w_1 - 2w_2 + w_3,$$

and this relation is verified.

Let us then consider v_1 and v_2 and complete them to a basis of $\mathbb{R}^3$ by adding, for example, the vector $e_3 = (0, 0, 1)$ (check that v_1, v_2 and e_3 are linearly independent). Since f is determined by the images of the vectors of a basis, it suffices to specify the images of v_1, v_2 and e_3. While the images of v_1 and v_2 are assigned, the image of e_3 is not subject to any conditions and is therefore arbitrary, so let's set $f(e_3) = (a, b)$. From what has been said, it can be deduced that there exist infinite linear functions which satisfy the required conditions, and furthermore that these linear functions depend on two real parameters (which we have denoted by a and b). To determine the matrices of these functions with respect to the canonical bases we need to know the images of the basis vectors e_1, e_2 and e_3. We therefore have to write the vectors of the canonical basis in terms of the vectors of the basis $\{v_1, v_2, e_3\}$. From the equations

$$\begin{cases} v_1 = -2e_2 + e_3 \\ v_2 = e_1 + e_2 - 2e_3 \end{cases}$$

we obtain

$$\begin{cases} e_1 = \frac{1}{2}\, v_1 + v_2 + \frac{3}{2}\, e_3 \\ e_2 = -\frac{1}{2}\, v_1 + \frac{1}{2}\, e_3 \end{cases}$$

hence we have

$$\begin{aligned} f(e_1) &= \tfrac{1}{2} f(v_1) + f(v_2) + \tfrac{3}{2} f(e_3) = \left(\tfrac{5}{2} + \tfrac{3}{2}\, a, \tfrac{3}{2} + \tfrac{3}{2}\, b\right), \\ f(e_2) &= -\tfrac{1}{2} f(v_1) + \tfrac{1}{2} f(e_3) = \left(-\tfrac{3}{2} + \tfrac{1}{2}\, a, \tfrac{1}{2} + \tfrac{1}{2}\, b\right). \end{aligned}$$

Therefore the matrices we are looking for are

$$A = \begin{pmatrix} \frac{5}{2} + \frac{3}{2}a & -\frac{3}{2} + \frac{1}{2}a & a \\ \frac{3}{2} + \frac{3}{2}b & \frac{1}{2} + \frac{1}{2}b & b \end{pmatrix}$$

for $a, b \in \mathbb{R}$.

Remark: another way to solve this exercise is to consider an unknown matrix

$$A = \begin{pmatrix} a_{11} & a_{12} & a_{13} \\ a_{21} & a_{22} & a_{23} \end{pmatrix}$$

and impose the conditions $Av_1 = w_1$ and $Av_2 = w_2$ (note that the condition $Av_3 = w_3$ is then automatically satisfied). We thus obtain 4 equations in the 6 unknowns a_{ij} which, once solved, will provide all the required matrices.

► **Exercise 2.2.6.** Let V be the set of all polynomial functions of degree ≤ 4 with real coefficients that vanish in 0 and 1, and W be the set of all polynomial functions of degree ≤ 3 with real coefficients whose integral between 0 and 1 is zero.

(a) Prove that V and W are vector spaces and write bases of V and W.

(b) Let $D\colon V \to W$ be the linear function that sends a function to its derivative. Prove that D is well defined and write the matrix of D with respect to the bases previously found.

Solution. (a) The elements of V are the polynomials of the form $f(x) = a_0 + a_1x + a_2x^2 + a_3x^3 + a_4x^4$ such that $f(0) = f(1) = 0$, which is equivalent to $a_0 = 0$ and $a_0 + a_1 + a_2 + a_3 + a_4 = 0$. We obtain

$$V = \{-(a_2 + a_3 + a_4)x + a_2x^2 + a_3x^3 + a_4x^4 \mid a_2, a_3, a_4 \in \mathbb{R}\},$$

which is easily verified to be a vector space (i.e., closed under the operations of addition and product for a scalar). Furthermore, by successively assigning to a_2, a_3, a_4 the values 1, 0, 0, then 0, 1, 0, etc., we obtain the polynomials $f_1(x) = x^2 - x$, $f_2(x) = x^3 - x$, $f_3(x) = x^4 - x$, which are therefore a basis of V.

The elements of W are the polynomials of the form $g(x) = b_0 + b_1x + b_2x^2 + b_3x^3$ such that

$$\int_0^1 (b_0 + b_1x + b_2x^2 + b_3x^3)\, dx = 0,$$

which is equivalent to

$$b_0 + \frac{1}{2}b_1 + \frac{1}{3}b_2 + \frac{1}{4}b_3 = 0.$$

Therefore we have

$$W = \left\{ -\left(\frac{1}{2}b_1 + \frac{1}{3}b_2 + \frac{1}{4}b_3\right) + b_1x + b_2x^2 + b_3x^3 \,\middle|\, b_1, b_2, b_3 \in \mathbb{R} \right\},$$

which is easily verified to be a vector space. Furthermore, by successively assigning to b_1, b_2, b_3 the values 1, 0, 0, then 0, 1, 0, etc., we obtain the polynomials $g_1(x) = x - \frac{1}{2}$, $g_2(x) = x^2 - \frac{1}{3}$, $g_3(x) = x^3 - \frac{1}{4}$, which are therefore a basis of W.

(b) If $f(x) \in V$, we have

$$\int_0^1 D(f(x))\,dx = f(1) - f(0) = 0,$$

therefore $D(f(x)) \in W$, hence D is well defined. To determine the matrix of D it suffices to compute the images of the vectors of the basis of V:

$$\begin{aligned} D(f_1(x)) &= 2x - 1 = 2g_1(x), \\ D(f_2(x)) &= 3x^2 - 1 = 3g_2(x), \\ D(f_3(x)) &= 4x^3 - 1 = 4g_3(x), \end{aligned}$$

from which we deduce that the matrix is

$$A = \begin{pmatrix} 2 & 0 & 0 \\ 0 & 3 & 0 \\ 0 & 0 & 4 \end{pmatrix}$$

▶ **Exercise 2.2.7.** Let $\phi_\lambda \colon \mathbb{R}^3 \to \mathbb{R}^4$ be the homomorphism whose matrix (with respect to the canonical bases) is

$$A_\lambda = \begin{pmatrix} 1 & \lambda & 0 \\ 0 & \lambda & 0 \\ 1 & 0 & 1 \\ 0 & 0 & 0 \end{pmatrix}$$

(a) Is it true that, for every $\lambda \in \mathbb{R}$, there exists a homomorphism $\psi \colon \mathbb{R}^4 \to \mathbb{R}^3$ such that $\psi \circ \phi_\lambda$ is surjective?

(b) Find the values of λ for which there exist $x, y, z \in \mathbb{R}$ such that $BA_\lambda = \mathbf{1}$, where

$$B = \begin{pmatrix} 1 & x & 0 & 0 \\ 0 & y & 0 & 0 \\ -1 & z & 1 & 0 \end{pmatrix}$$

Solution. (a) If $\psi \circ \phi_\lambda \colon \mathbb{R}^3 \to \mathbb{R}^3$ is surjective then it is also injective, but this implies that ϕ_λ is injective too, and this must hold for every λ. However, we observe that the image of ϕ_λ is generated by the vectors $(1, 0, 1, 0)$, $(\lambda, \lambda, 0, 0)$ and $(0, 0, 1, 0)$, and therefore has dimension 2 if $\lambda = 0$. Consequently, at least in the case $\lambda = 0$, ϕ_λ is not injective, so it follows that there is no homomorphism ψ with the required characteristics.

(b) Computing the product, we get:

$$BA_\lambda = \begin{pmatrix} 1 & \lambda(1+x) & 0 \\ 0 & \lambda y & 0 \\ 0 & \lambda(z-1) & 1 \end{pmatrix}$$

which is the identity matrix if and only if

$$\begin{cases} \lambda(1+x) = 0 \\ \lambda y = 1 \\ \lambda(z-1) = 0 \end{cases}$$

This system has solutions if and only if $\lambda \neq 0$. (Note that B is the matrix of a homomorphism $\psi \colon \mathbb{R}^4 \to \mathbb{R}^3$ and that BA_λ is the matrix of the composition $\psi \circ \phi_\lambda$). From what we have seen, we deduce that, if $\lambda \neq 0$, there exists ψ such that $\psi \circ \phi_\lambda$ is the identity.

► **Exercise 2.2.8.** Let f be the endomorphism of $\mathbb{R}^3$ whose matrix, with respect to the canonical bases of the domain and codomain, is

$$A = \begin{pmatrix} 2 & 1 & 0 \\ -1 & 3 & 2 \\ 0 & 1 & 1 \end{pmatrix}$$

(a) Find a basis $\mathbf{v} = \{v_1, v_2, v_3\}$ of the domain such that the matrix of f with respect to the basis $\mathbf{v}$ of the domain and the canonical basis of the codomain is

$$B = \begin{pmatrix} 3 & 1 & 4 \\ 4 & 1 & 0 \\ 2 & 0 & 1 \end{pmatrix}$$

(b) Find a basis $\mathbf{w} = \{w_1, w_2, w_3\}$ of the codomain such that the matrix of f with respect to the canonical basis of the domain and the basis $\mathbf{w}$ of the codomain is

$$C = \begin{pmatrix} -5 & 0 & 1 \\ -7 & -1 & 1 \\ 9 & 3 & 0 \end{pmatrix}$$

Solution. (a) Requiring that B is the matrix of f with respect to the basis $\{v_1, v_2, v_3\}$ of the domain and the basis $\{e_1, e_2, e_3\}$ of the codomain means requiring that we have $f(v_1) = 3e_1 + 4e_2 + 2e_3$, $f(v_2) = e_1 + e_2$, $f(v_3) = 4e_1 + e_3$.

Let's start by finding the vector v_1. If we set $v_1 = (x_1, x_2, x_3)$, we have

$$f(v_1) = A\begin{pmatrix} x_1 \\ x_2 \\ x_3 \end{pmatrix} = \begin{pmatrix} 2x_1 + x_2 \\ -x_1 + 3x_2 + 2x_3 \\ x_2 + x_3 \end{pmatrix}$$

The equation $f(v_1) = 3e_1 + 4e_2 + 2e_3 = (3, 4, 2)$ gives the system

$$\begin{cases} 2x_1 + x_2 = 3 \\ -x_1 + 3x_2 + 2x_3 = 4 \\ x_2 + x_3 = 2 \end{cases}$$

whose solution is

$$\begin{cases} x_1 = 1 \\ x_2 = 1 \\ x_3 = 1 \end{cases}$$

Therefore the first vector of the basis is $v_1 = (1, 1, 1)$.

Similarly, setting $v_2 = (x_1, x_2, x_3)$, the equation $f(v_2) = e_1 + e_2 = (1, 1, 0)$ gives the system

$$\begin{cases} 2x_1 + x_2 = 1 \\ -x_1 + 3x_2 + 2x_3 = 1 \\ x_2 + x_3 = 0 \end{cases}$$

whose solution is

$$\begin{cases} x_1 = 0 \\ x_2 = 1 \\ x_3 = -1 \end{cases}$$

Therefore the second vector of the basis is $v_2 = (0, 1, -1)$.

Finally, setting $v_3 = (x_1, x_2, x_3)$, we have $f(v_3) = 4e_1 + e_3 = (4, 0, 1)$ which gives the system

$$\begin{cases} 2x_1 + x_2 = 4 \\ -x_1 + 3x_2 + 2x_3 = 0 \\ x_2 + x_3 = 1 \end{cases}$$

whose solution is

$$\begin{cases} x_1 = 2 \\ x_2 = 0 \\ x_3 = 1 \end{cases}$$

Therefore the third vector of the basis is $v_3 = (2, 0, 1)$.

Note that if we denote by P the matrix whose columns contain the components of the vectors v_1, v_2 and v_3,

$$P = \begin{pmatrix} 1 & 0 & 2 \\ 1 & 1 & 0 \\ 1 & -1 & 1 \end{pmatrix}$$

we have $B = AP$, as can easily be verified by a direct computation.

(b) Requiring that C is the matrix of f with respect to the basis $\{e_1, e_2, e_3\}$ of the domain and the basis $\{w_1, w_2, w_3\}$ of the codomain means requiring that we have $f(e_1) = -5w_1 - 7w_2 + 9w_3$, $f(e_2) = -w_2 + 3w_3$, $f(e_3) = w_1 + w_2$. Now we need to compute the vectors w_1, w_2, w_3 from these equations.

From the last equality we deduce that $w_1 = f(e_3) - w_2$, while from the second one we get $w_3 = \frac{1}{3}\big(f(e_2) + w_2\big)$. Substituting into the first one, we find $f(e_1) = -5f(e_3) + 3f(e_2) + w_2$, from which we obtain

$$w_2 = f(e_1) - 3f(e_2) + 5f(e_3) = (-1, 0, 2).$$

Having determined the vector w_2, we have

$$w_1 = f(e_3) - w_2 = (1, 2, -1), \qquad w_3 = \tfrac{1}{3}\big(f(e_2) + w_2\big) = (0, 1, 1).$$

Note that if we denote by Q the matrix whose columns contain the components of the vectors w_1, w_2, w_3,

$$Q = \begin{pmatrix} 1 & -1 & 0 \\ 2 & 0 & 1 \\ -1 & 2 & 1 \end{pmatrix}$$

we have $A = QC$, as can easily be verified by a direct computation.

2.3 Kernel and image of a linear function

► **Exercise 2.3.1.** Let V and W be two vector spaces with bases $\{v_1, v_2, v_3, v_4\}$ and $\{w_1, w_2, w_3\}$, respectively. Write the matrix (with respect to these bases) of the linear function $\phi\colon V \to W$ defined by $\phi(v_1) = w_1 - w_2$, $\phi(v_2) = 2w_2 - 6w_3$, $\phi(v_3) = -2w_1 + 2w_2$, $\phi(v_4) = w_2 - 3w_3$. Compute the dimensions of $\operatorname{Ker}\phi$ and $\operatorname{Im}\phi$ and write bases of these subspaces. Check whether $w_1 + w_2 + w_3 \in \operatorname{Im}\phi$.

Solution. Recalling that the columns of the matrix of ϕ consist of the coordinates, with respect to the basis $\{w_1, w_2, w_3\}$, of the images of the vectors of the basis $\{v_1, v_2, v_3, v_4\}$, we deduce that the matrix is

$$A = \begin{pmatrix} 1 & 0 & -2 & 0 \\ -1 & 2 & 2 & 1 \\ 0 & -6 & 0 & -3 \end{pmatrix}$$

$\operatorname{Im}\phi$ is the subspace of W generated by the vectors $\phi(v_1)$, $\phi(v_2)$, $\phi(v_3)$ and $\phi(v_4)$. Therefore we will find a basis among these vectors. Indeed, it suffices to note that $\phi(v_3) = -2\phi(v_1)$ and $\phi(v_2) = 2\phi(v_4)$, while $\phi(v_1)$ and $\phi(v_4)$ are linearly independent, to conclude that a basis of $\operatorname{Im}\phi$ consists, for example, of the vectors $\phi(v_1)$ and $\phi(v_4)$. Therefore we have $\dim(\operatorname{Im}\phi) = 2$ and, consequently, $\dim(\operatorname{Ker}\phi) = \dim(V) - \dim(\operatorname{Im}\phi) = 2$. To find a basis of $\operatorname{Ker}\phi$ then it is sufficient to find two linearly independent vectors which belong to the kernel of ϕ. From the linear relations between the vectors $\phi(v_i)$ found previously, and from the linearity of ϕ, we deduce that $\phi(2v_1 + v_3) = \mathbf{0}$ and $\phi(v_2 - 2v_4) = \mathbf{0}$, hence the two vectors $2v_1 + v_3$ and $v_2 - 2v_4$ belong to $\operatorname{Ker}\phi$, hence they are a basis (because they are linearly independent). Finally, the vector $w_1 + w_2 + w_3$ belongs to the image of ϕ if and only if it is a linear combination of the vectors that form a basis of $\operatorname{Im}\phi$. We then have to see if there exist two scalars λ_1 and λ_2 such that

$$w_1 + w_2 + w_3 = \lambda_1\phi(v_1) + \lambda_2\phi(v_4) = \lambda_1(w_1 - w_2) + \lambda_2(w_2 - 3w_3).$$

We obtain the system

$$\begin{cases} \lambda_1 = 1 \\ -\lambda_1 + \lambda_2 = 1 \\ -3\lambda_2 = 1 \end{cases}$$

that has no solutions. It can therefore be concluded that $w_1 + w_2 + w_3 \notin \operatorname{Im}\phi$.

▶ **Exercise 2.3.2.** Find out whether the endomorphism of $\mathbb{R}^3$ defined by

$$f(x, y, z) = (x + 2y, y + z, 2z - x)$$

is injective or surjective. Write bases of $\operatorname{Ker} f$ and $\operatorname{Im} f$ and determine if the sum of the kernel and the image of f is direct.

Solution. Note that f is an endomorphism of $\mathbb{R}^3$, i.e., a linear function from $\mathbb{R}^3$ to $\mathbb{R}^3$. From this it follows that f is injective if and only if it is surjective, i.e., for example, if and only if $\operatorname{Ker} f = \{\mathbf{0}\}$. We therefore determine the kernel of f: $f(x,y,z) = \mathbf{0}$ if and only if

$$\begin{cases} x + 2y = 0 \\ y + z = 0 \\ 2z - x = 0 \end{cases}$$

from which we obtain

$$\begin{cases} x = -2y \\ z = -y. \end{cases}$$

It thus turns out that the kernel of f contains non-zero vectors, that it has dimension 1 (in the previous system there is only one free variable) and that the vector $(-2,1,-1)$ (corresponding to the choice $y = 1$) is a basis (f is therefore neither injective nor surjective). The dimension of the image of f is then

$$\dim(\operatorname{Im} f) = \dim(\mathbb{R}^3) - \dim(\operatorname{Ker} f) = 2,$$

and given that $\operatorname{Im} f$ is generated from the image of the vectors of a basis of $\mathbb{R}^3$, for example the canonical basis, to find a basis of $\operatorname{Im} f$ it is sufficient to find, among these, two linearly independent vectors. Computing $f(1,0,0) = (1,0,-1)$ and $f(0,1,0) = (2,1,0)$ we realize that these are linearly independent; therefore we can conclude that the vectors $(1,0,-1)$ and $(2,1,0)$ are a basis of $\operatorname{Im} f$. Finally, we recall that requiring the sum of two subspaces to be direct is equivalent to requiring their intersection to be zero. Then it suffices to check whether the vector $(-2,1,-1)$, which generates $\operatorname{Ker} f$, can be written as a linear combination of the vectors which form a basis of $\operatorname{Im} f$:

$$(-2,1,-1) = \lambda_1(1,0,-1) + \lambda_2(2,1,0).$$

We get a linear system that has no solutions. Consequently the sum of $\operatorname{Ker} f$ and $\operatorname{Im} f$ is direct and, from the computation of the dimensions, it can be deduced that $\operatorname{Ker} f \oplus \operatorname{Im} f = \mathbb{R}^3$.

▶ **Exercise 2.3.3.** Let $f\colon \mathbb{R}^3 \to \mathbb{R}^3$ be the endomorphism defined by

$$\begin{aligned} f(1,0,0) &= (2,-1,0) \\ f(0,1,0) &= (1,-1,1) \\ f(0,1,-1) &= (0,2,2). \end{aligned}$$

Write the matrix of f with respect to the canonical basis of $\mathbb{R}^3$. Compute the dimensions of the kernel and the image of f and write bases of these subspaces.

Solution. The images of the vectors $(1,0,0)$ and $(0,1,0)$ respectively constitute the first and second column of the matrix. The third column is given by the image of the vector $(0,0,1)$. Since $(0,0,1) = (0,1,0)-(0,1,-1)$, it follows from the linearity of f that $f(0,0,1) = (1,-1,1) - (0,2,2) = (1,-3,-1)$. Therefore the matrix is

$$A = \begin{pmatrix} 2 & 1 & 1 \\ -1 & -1 & -3 \\ 0 & 1 & -1 \end{pmatrix}$$

It is now easy to see that $\operatorname{Ker} f = \{\mathbf{0}\}$, and therefore $\operatorname{Im} f = \mathbb{R}^3$. As the basis of $\operatorname{Im} f$ one can then take the canonical basis of $\mathbb{R}^3$.

▶ **Exercise 2.3.4.** Let $f\colon \mathbb{R}^3 \to \mathbb{R}^3$ be the endomorphism given by the matrix (with respect to the canonical basis)

$$A = \begin{pmatrix} 1 & 2 & 3 \\ 1 & 1 & 1 \\ 1 & 1 & 1 \end{pmatrix}$$

Compute the rank of f and write bases of $\operatorname{Ker} f$ and $\operatorname{Im} f$.

Solution. Observing the matrix A we immediately see that $\operatorname{rk}(f) = \dim \operatorname{Im}(f) = \operatorname{rk}(A) = 2$. Consequently, one has $\dim \operatorname{Ker}(f) = 1$. A basis of the kernel of f is then given by the vector $(-1,2,-1)$, while as the basis of the image of f one can take, for example, the vectors constituting the first two columns of the matrix A.

▶ **Exercise 2.3.5.** Let $f\colon \mathbb{R}^3 \to \mathbb{R}^2$ be the linear function whose matrix (with respect to the canonical bases) is

$$A = \begin{pmatrix} 2 & -1 & 1 \\ 1 & 3 & 0 \end{pmatrix}$$

(a) Find a basis of the kernel of f and determine if f is surjective.

(b) Let $U \subset \mathbb{R}^3$ be the vector subspace generated by $u_1 = (-2,2,3)$, $u_2 = (1,1,-4)$, and consider the restriction of f to the subspace U:

$$f|_U : U \to \mathbb{R}^2, \qquad u \mapsto f(u).$$

Determine if the function $f|_U$ is surjective and write its matrix with respect to the basis $\{u_1, u_2\}$ of U and the canonical basis of $\mathbb{R}^2$.

Solution. The kernel of f contains the vectors which are solutions of the system

$$\begin{pmatrix} 2 & -1 & 1 \\ 1 & 3 & 0 \end{pmatrix} \begin{pmatrix} x_1 \\ x_2 \\ x_3 \end{pmatrix} = \begin{pmatrix} 0 \\ 0 \end{pmatrix}$$

that is

$$\begin{cases} 2x_1 - x_2 + x_3 = 0 \\ x_1 + 3x_2 = 0 \end{cases}$$

This system has infinitely many solutions, given by

$$\begin{cases} x_1 = -3x_2 \\ x_3 = 7x_2 \end{cases}$$

from which we deduce that the kernel of f has dimension 1 and a basis is given by the vector $v = (-3,1,7)$. Since $\dim \operatorname{Ker} f = 1$, we have $\dim \operatorname{Im} f = 2$ and, therefore, the function f is surjective.

To find out if $f|_U : U \to \mathbb{R}^2$ is surjective we can try to determine its kernel. We have already seen that the kernel of f is generated by the vector $v = (-3,1,7)$, so we wonder if this vector belongs to the subspace U or not. Checking if v is a linear combination of the vectors u_1 and u_2 we discover that $v = u_1 - u_2$, therefore $v \in U$ and, consequently, $\operatorname{Ker} f|_U = \operatorname{Ker} f = \langle v \rangle$. Since $\dim U = 2$ and $\dim \operatorname{Ker} f|_U = 1$, we conclude that $\dim \operatorname{Im} f|_U = 1$ and therefore $f|_U : U \to \mathbb{R}^2$ is not surjective.

Finally, we write the matrix of $f|_U$ with respect to the basis $\{u_1, u_2\}$ of U and the canonical basis of $\mathbb{R}^2$. We have:

$$f|_U(u_1) = f(u_1) = (-3,4) = -3e_1 + 4e_2$$
$$f|_U(u_2) = f(u_2) = (-3,4) = -3e_1 + 4e_2$$

from which it follows that the matrix is

$$\begin{pmatrix} -3 & -3 \\ 4 & 4 \end{pmatrix}$$

The fact that this matrix has rank 1 confirms that $\dim \operatorname{Im} f|_U = 1$.

► **Exercise 2.3.6.** Let V be a finite dimensional vector space. State under which conditions on V there exists an endomorphism $\phi\colon V \to V$ such that $\operatorname{Ker}\phi = \operatorname{Im}\phi$.

Solution. Recall that the dimensions of the kernel and the image of a linear function are related by the formula

$$\dim V = \dim(\operatorname{Ker}\phi) + \dim(\operatorname{Im}\phi).$$

Therefore, if there exists $\phi\colon V \to V$ such that $\operatorname{Ker}\phi = \operatorname{Im}\phi$ one must necessarily have $\dim V = 2\dim(\operatorname{Ker}\phi)$, i.e., the dimension of V must be even.

Conversely, we may ask whether the fact that V has even dimension is also sufficient to guarantee the existence of an endomorphism ϕ with the required characteristics. So let V be a vector space of dimension $2n$ and choose a basis $\{v_1, v_2, \ldots, v_{2n}\}$. Let us now try to construct a linear function ϕ such that $\operatorname{Ker}\phi = \operatorname{Im}\phi$. Recall that a linear function is uniquely determined by the images of the elements of a basis. Therefore, to define ϕ, we just need to specify $\phi(v_1), \phi(v_2), \ldots, \phi(v_{2n})$. We then set $\phi(v_1) = \phi(v_2) = \cdots = \phi(v_n) = \mathbf{0}$, so as to obtain that the kernel is the subspace of V generated by the first n vectors of the basis. Now that we know the kernel, we just need to make sure that the image of ϕ is also generated by the vectors $v_1, v_2, \ldots, v_n$. Given that the image of ϕ is generated by the images of the vectors of the basis of V, and given that the image of the first n vectors is zero, then it suffices to set $\phi(v_{n+1}) = v_1, \phi(v_{n+2}) = v_2, \ldots, \phi(v_{2n}) = v_n$. This proves that such a ϕ exists and therefore that the condition that V has even dimension is also sufficient.

2.4 Gaussian elimination, computation of the rank of a matrix

► **Exercise 2.4.1.** Compute the rank of the matrix

$$A = \begin{pmatrix} 0 & 1 & 2 & 1 \\ 1 & 1 & 1 & 0 \\ 0 & -1 & 1 & 1 \\ 1 & 1 & 4 & 2 \end{pmatrix}$$

Solution. Recalling that the rank of a matrix is the number of linearly independent rows (or columns), it is clear that by adding to a row (or column) a linear combination of the remaining rows (columns), or by exchanging two rows (or columns) , the rank does not change. By performing this type of operations (called elementary operations) on the rows (or columns), we can reduce the matrix A to a simpler form, in which the determination of the rank is immediate. In this specific case, the operations to be performed are, for example, the following: we swap the first with the second row of A, obtaining

$$\begin{pmatrix} 1 & 1 & 1 & 0 \\ 0 & 1 & 2 & 1 \\ 0 & -1 & 1 & 1 \\ 1 & 1 & 4 & 2 \end{pmatrix}$$

Now we subtract the first from the fourth row, obtaining

$$\begin{pmatrix} 1 & 1 & 1 & 0 \\ 0 & 1 & 2 & 1 \\ 0 & -1 & 1 & 1 \\ 0 & 0 & 3 & 2 \end{pmatrix}$$

Now we add the second to the third row, obtaining

$$\begin{pmatrix} 1 & 1 & 1 & 0 \\ 0 & 1 & 2 & 1 \\ 0 & 0 & 3 & 2 \\ 0 & 0 & 3 & 2 \end{pmatrix}$$

Finally we subtract the third from the fourth row, obtaining

$$\begin{pmatrix} 1 & 1 & 1 & 0 \\ 0 & 1 & 2 & 1 \\ 0 & 0 & 3 & 2 \\ 0 & 0 & 0 & 0 \end{pmatrix}$$

The matrix is now in the so-called "echelon form." Now it is immediate to count how many linearly independent rows there are: in our case 3. So this is the rank of A.

► **Exercise 2.4.2.** Determine, as $a \in \mathbb{R}$ varies, the rank of the matrix

$$A = \begin{pmatrix} 0 & 1 & 2 & 1 & 0 \\ 1 & 2 & 2 & 1 & 1 \\ 1 & 1 & a & 0 & 1 \\ 0 & a & 2a & a^2 & 0 \end{pmatrix}$$

Solution. With elementary operations on the rows (see the previous exercise) we obtain the matrix

$$\begin{pmatrix} 1 & 2 & 2 & 1 & 1 \\ 0 & 1 & 2 & 1 & 0 \\ 0 & 0 & a & 0 & 0 \\ 0 & 0 & 0 & a^2-a & 0 \end{pmatrix}$$

Then we see that, if $a = 0$ we have $\mathrm{rk}(A) = 2$, if $a = 1$ we have $\mathrm{rk}(A) = 3$, while for all the other values of a we have $\mathrm{rk}(A) = 4$.

► **Exercise 2.4.3.** Find a basis of the space of linear relations between the following vectors of $\mathbb{R}^4$:

$$v_1 = (2, 0, -1, 3),\ v_2 = (1, -1, 2, -1),\ v_3 = (-3, 2, 1, -2),\ v_4 = (2, 1, 1, 3).$$

Solution. We solve this exercise using a method derived from the Gauss elimination procedure. The matrix A, whose rows are the given vectors, is

$$A = \begin{pmatrix} 2 & 0 & -1 & 3 \\ 1 & -1 & 2 & -1 \\ -3 & 2 & 1 & -2 \\ 2 & 1 & 1 & 3 \end{pmatrix}$$

We write the identity matrix next to A, obtaining

$$\left(\begin{array}{cccc|cccc} 2 & 0 & -1 & 3 & 1 & 0 & 0 & 0 \\ 1 & -1 & 2 & -1 & 0 & 1 & 0 & 0 \\ -3 & 2 & 1 & -2 & 0 & 0 & 1 & 0 \\ 2 & 1 & 1 & 3 & 0 & 0 & 0 & 1 \end{array}\right)$$

Now we perform elementary row operations to transform A into an echelon matrix S. We can start, for example, by swapping the first row with the second:

$$\left(\begin{array}{cccc|cccc} 1 & -1 & 2 & -1 & 0 & 1 & 0 & 0 \\ 2 & 0 & -1 & 3 & 1 & 0 & 0 & 0 \\ -3 & 2 & 1 & -2 & 0 & 0 & 1 & 0 \\ 2 & 1 & 1 & 3 & 0 & 0 & 0 & 1 \end{array}\right)$$

Now we subtract twice the first from the second row, add three times the first to the third row and subtract the first row multiplied by 2 from the fourth row, thus obtaining the matrix

$$\left(\begin{array}{cccc|cccc} 1 & -1 & 2 & -1 & 0 & 1 & 0 & 0 \\ 0 & 2 & -5 & 5 & 1 & -2 & 0 & 0 \\ 0 & -1 & 7 & -5 & 0 & 3 & 1 & 0 \\ 0 & 3 & -3 & 5 & 0 & -2 & 0 & 1 \end{array}\right)$$

Let's swap the second line with the third:

$$\left(\begin{array}{cccc|cccc} 1 & -1 & 2 & -1 & 0 & 1 & 0 & 0 \\ 0 & -1 & 7 & -5 & 0 & 3 & 1 & 0 \\ 0 & 2 & -5 & 5 & 1 & -2 & 0 & 0 \\ 0 & 3 & -3 & 5 & 0 & -2 & 0 & 1 \end{array}\right)$$

Now we add to the third row twice the second and to the fourth row three times the second:

$$\left(\begin{array}{cccc|cccc} 1 & -1 & 2 & -1 & 0 & 1 & 0 & 0 \\ 0 & -1 & 7 & -5 & 0 & 3 & 1 & 0 \\ 0 & 0 & 9 & -5 & 1 & 4 & 2 & 0 \\ 0 & 0 & 18 & -10 & 0 & 7 & 3 & 1 \end{array}\right)$$

Finally, we subtract twice the third from the fourth row:

$$\left(\begin{array}{cccc|cccc} 1 & -1 & 2 & -1 & 0 & 1 & 0 & 0 \\ 0 & -1 & 7 & -5 & 0 & 3 & 1 & 0 \\ 0 & 0 & 9 & -5 & 1 & 4 & 2 & 0 \\ 0 & 0 & 0 & 0 & -2 & -1 & -1 & 1 \end{array}\right)$$

The matrix A has thus been transformed into the following echelon form

$$S = \begin{pmatrix} 1 & -1 & 2 & -1 \\ 0 & -1 & 7 & -5 \\ 0 & 0 & 9 & -5 \\ 0 & 0 & 0 & 0 \end{pmatrix}$$

The matrix that appeared next to S, in place of the identity matrix,

$$B = \begin{pmatrix} 0 & 1 & 0 & 0 \\ 0 & 3 & 1 & 0 \\ 1 & 4 & 2 & 0 \\ -2 & -1 & -1 & 1 \end{pmatrix}$$

encodes all the elementary operations we performed to transform the initial matrix A into the echelon matrix S; it has the property that $BA = S$ (as can be verified with a direct computation).

Since the last row of S is zero, the matrix A has rank 3, so the subspace of $\mathbb{R}^4$ generated by the vectors v_1, v_2, v_3, v_4 has dimension 3. From this we deduce that the space R of the linear relations between the given vectors has dimension 1 and a basis is given by the last row of the matrix B. Therefore we have

$$R = \langle(-2, -1, -1, 1)\rangle.$$

This means that the following linear dependence relation exists between the vectors v_1, v_2, v_3, v_4:

$$-2v_1 - v_2 - v_3 + v_4 = \mathbf{0}.$$

2.5 Linear systems

► **Exercise 2.5.1.** Solve the following system according to the value of $m \in \mathbb{R}$:

$$\begin{cases} x + (m+1)y = m+2 \\ mx + (m+4)y = 3. \end{cases}$$

Solution. Recall that a linear system of the type $AX = B$ admits solutions if and only if the rank of the (incomplete) matrix A is equal to the rank of the complete matrix $(A|B)$. In this case we have

$$(A|B) = \left(\begin{array}{cc|c} 1 & m+1 & m+2 \\ m & m+4 & 3 \end{array}\right)$$

To compute the rank of this matrix we subtract the first row multiplied by m from the second row, obtaining

$$\left(\begin{array}{cc|c} 1 & m+1 & m+2 \\ 0 & 4-m^2 & 3-m^2-2m \end{array}\right)$$

We therefore see that, if $m \neq \pm 2$ we have $\mathrm{rk}(A) = \mathrm{rk}(A|B)$, and therefore the system has a unique solution, while for $m = 2$ or $m = -2$ we have $\mathrm{rk}(A) = 1$ but $\mathrm{rk}(A|B) = 2$, therefore the system has no solutions.

However, note that this exercise requires explicitly solving the system; in cases like this we can then avoid studying the rank of the matrices and we can instead solve the system directly; along the way, we will discover when the solution exists (and whether it is unique or not) and when it does not exist. To solve the system we use the Gauss elimination method (which is exactly the same method we used to determine the rank of a matrix): we subtract the first equation multiplied by m from the second equation, obtaining

$$\begin{cases} x + (m+1)y = m+2 \\ (4-m^2)y = 3-m^2-2m \end{cases}$$

from which we deduce that, if $m \neq \pm 2$, we have

$$\begin{cases} y = \dfrac{m^2+2m-3}{m^2-4} \\ x = -\dfrac{m^2+3m+5}{m^2-4} \end{cases}$$

while, if $m = 2$ or $m = -2$, the system has no solutions.

► **Exercise 2.5.2.** Solve the following system according to the value of $m \in \mathbb{R}$:

$$\begin{cases} mx + (m-1)y = m+2 \\ (m+1)x - my = 5m+3. \end{cases}$$

Solution. We solve the system directly using the Gaussian elimination method. We note first of all that, if $m = 0$, the system has a unique solution given by

$$\begin{cases} y = -2 \\ x = 3 \end{cases}$$

If $m \neq 0$ we replace the second equation with the sum of the second, multiplied by m, and the first, multiplied by $-(m+1)$, obtaining

$$\begin{cases} mx + (m-1)y = m+2 \\ (1-2m^2)y = 4m^2-2 \end{cases}$$

Now, if $m^2 = \frac{1}{2}$, in the second equation we obtain $0 = 0$, therefore the system has infinite solutions (depending on a parameter) which are obtained from the first equation by writing, for example, x as a function of y:

$$x = \frac{1-m}{m}y + \frac{m+2}{m},$$

(and obviously substituting for m the values $-1/\sqrt{2}$ and $1/\sqrt{2}$, respectively), while, if $m^2 \neq \frac{1}{2}$, we obtain

$$\begin{cases} y = -2 \\ x = 3 \end{cases}$$

In conclusion, the system has a unique solution (which does not depend on m) if $m^2 \neq \frac{1}{2}$, otherwise it has infinitely many solutions.

► **Exercise 2.5.3.** Solve the following system according to the value of $a \in \mathbb{R}$:

$$\begin{cases} x + (a-1)y + (2-a)z = a + 5 \\ x + ay + 2z = 4 \\ x + (a-2)y + (2-2a^2)z = 6. \end{cases}$$

Solution. We recall the Rouché–Capelli theorem: a linear system $AX = B$ has solutions if and only if the rank of the incomplete matrix A is equal to the rank of the complete matrix $(A|B)$. In this case we have

$$(A|B) = \left(\begin{array}{ccc|c} 1 & a-1 & 2-a & a+5 \\ 1 & a & 2 & 4 \\ 1 & a-2 & 2-2a^2 & 6 \end{array}\right)$$

To compute the rank of this matrix we will use elementary transformations on the rows to reduce it to an echelon form. Let's start by exchanging the first and second rows, obtaining:

$$\left(\begin{array}{ccc|c} 1 & a & 2 & 4 \\ 1 & a-1 & 2-a & a+5 \\ 1 & a-2 & 2-2a^2 & 6 \end{array}\right)$$

Now we replace the second row with the difference between the first and second row, and subtract the first from the third row:

$$\left(\begin{array}{ccc|c} 1 & a & 2 & 4 \\ 0 & 1 & a & -a-1 \\ 0 & -2 & -2a^2 & 2 \end{array}\right)$$

Now we add the second row multiplied by 2 to the third row:

$$\left(\begin{array}{ccc|c} 1 & a & 2 & 4 \\ 0 & 1 & a & -a-1 \\ 0 & 0 & 2(a-a^2) & -2a \end{array}\right)$$

Now we see that, for $a = 0$ we have $\text{rk}(A) = \text{rk}(A|B) = 2$, therefore the system is compatible and has infinitely many solutions, depending on $3-2 = 1$ parameter. The solutions are found by solving the system corresponding to the first two rows of the last matrix (obviously for $a = 0$), i.e.:

$$\begin{cases} x + 2z = 4 \\ y = -1 \end{cases}$$

from which we obtain

$$\begin{cases} x = 4 - 2t \\ y = -1 \\ z = t \end{cases}$$

for every $t \in \mathbb{R}$.

For $a = 1$ instead we have $\text{rk}(A) = 2$ but $\text{rk}(A|B) = 3$, therefore the system is incompatible, i.e., it has no solutions.

Finally, if $a \neq 0, 1$, we have $\text{rk}(A) = \text{rk}(A|B) = 3$, therefore the system has a unique solution, which is found by solving the system corresponding to last matrix:

$$\begin{cases} x + ay + 2z = 4 \\ y + az = -a - 1 \\ 2(a - a^2)z = -2a \end{cases}$$

This system is easily solved by substitution, obtaining z from the last equation, substituting the value found in the previous equations, then obtaining y from the penultimate one, etc. We get:

$$\begin{cases} z = \dfrac{1}{a-1} \\ y = \dfrac{-a^2 - a + 1}{a - 1} \\ x = \dfrac{a^3 + a^2 + 3a - 6}{a - 1} \end{cases}$$

► **Exercise 2.5.4.** Determine for which values of $\lambda \in \mathbb{R}$ the following system is solvable, and solve it when it has more than one solution.

$$\begin{cases} x + 2y + 2w + z = 1 \\ y + 2w + z = 0 \\ x + y + \lambda w = 0 \\ \lambda y + 2\lambda w + \lambda^2 z = 0. \end{cases}$$

Solution. In this case it is not required to always solve the system, so we can limit ourselves, for now, to studying the rank of the system's matrices:

$$(A|B) = \left(\begin{array}{cccc|c} 1 & 2 & 2 & 1 & 1 \\ 0 & 1 & 2 & 1 & 0 \\ 1 & 1 & \lambda & 0 & 0 \\ 0 & \lambda & 2\lambda & \lambda^2 & 0 \end{array}\right)$$

With elementary operations on the rows, this matrix can be transformed to obtain the following matrix:

$$\left(\begin{array}{cccc|c} 1 & 2 & 2 & 1 & 1 \\ 0 & 1 & 2 & 1 & 0 \\ 0 & 0 & \lambda & 0 & -1 \\ 0 & 0 & 0 & \lambda^2-\lambda & 0 \end{array}\right)$$

It then turns out that, if $\lambda = 0$ we have $\mathrm{rk}(A) = 2$ and $\mathrm{rk}(A|B) = 3$, therefore the system has no solution. If $\lambda = 1$ we have $\mathrm{rk}(A) = 3 = \mathrm{rk}(A|B)$, then the system has solutions and, since the rank of A is not 4, the solutions are infinitely many (and depend on $4-3 = 1$ parameter). Finally, if $\lambda \neq 0, 1$, the matrix A has maximum rank ($= 4$), therefore it is invertible, and the system has a unique solution.

Now we have to solve the system for $\lambda = 1$. In this case the system becomes

$$\begin{cases} x + 2y + 2w + z = 1 \\ y + 2w + z = 0 \\ x + y + w = 0 \\ y + 2w + z = 0 \end{cases}$$

which, by Gaussian elimination, is equivalent to the system

$$\begin{cases} x + 2y + 2w + z = 1 \\ y + 2w + z = 0 \\ w = -1 \\ 0 = 0 \end{cases}$$

from which we get

$$\begin{cases} x = z - 1 \\ y = 2 - z \\ w = -1. \end{cases}$$

► **Exercise 2.5.5.** Solve the following system according to the values of the parameters $b_1, \dots, b_4 \in \mathbb{R}$:

$$\begin{cases} x + 3y + 5z + 3t = b_1 \\ x + 4y + 7z + 3t = b_2 \\ y + 2z = b_3 \\ x + 2y + 3z + 2t = b_4. \end{cases}$$

Solution. Using the Gauss elimination method we get the system

$$\begin{cases} x + 3y + 5z + 3t = b_1 \\ y + 2z = b_2 - b_1 \\ 0 = b_3 - b_2 + b_1 \\ t = 2b_1 - b_2 - b_4 \end{cases}$$

We therefore find that, if $b_3 - b_2 + b_1 \neq 0$, the system has no solution, while if $b_3 - b_2 + b_1 = 0$ there are infinite solutions, given (for example) by

$$\begin{cases} t = 2b_1 - b_2 - b_4 \\ y = b_2 - b_1 - 2z \\ x = -2b_1 + 3b_4 + z. \end{cases}$$

► **Exercise 2.5.6.** As $\lambda \in \mathbb{Q}$ varies, determine how many solutions there are in $\mathbb{Q}^4$ for the following system of linear equations:

$$\begin{cases} (\lambda - 1)x_1 + 2x_2 + 3x_4 = 0 \\ \lambda x_2 + (\lambda + 1)x_4 = 1 \\ x_1 + \lambda x_3 + x_4 = 0 \\ (\lambda - 1)x_1 + x_4 = 0. \end{cases}$$

Solution. In this case it is not required to solve the system, so it is enough to study the rank of the complete and incomplete matrices.

$$(A|B) = \left(\begin{array}{cccc|c} \lambda - 1 & 2 & 0 & 3 & 0 \\ 0 & \lambda & 0 & \lambda + 1 & 1 \\ 1 & 0 & \lambda & 1 & 0 \\ \lambda - 1 & 0 & 0 & 1 & 0 \end{array}\right)$$

With elementary operations on the rows it is possible to transform this matrix to obtain

$$\left(\begin{array}{cccc|c} 1 & 0 & \lambda & 1 & 0 \\ 0 & 2 & \lambda-\lambda^2 & 4-\lambda & 0 \\ 0 & 0 & \lambda-\lambda^2 & 2-\lambda & 0 \\ 0 & 0 & 0 & 1 & 1 \end{array}\right)$$

Now it turns out that if $\lambda = 0$ or $\lambda = 1$, we have $\text{rk}(A) = 3$ and $\text{rk}(A|B) = 4$, so the system has no solutions. If instead $\lambda \neq 0, 1$ the matrix A has maximum rank, therefore it is invertible and therefore the solution exists and is unique. The fact that this solution belongs to $\mathbb{Q}^4$ is obvious, since all the operations necessary to solve the system involve only sums and products of rational numbers.

▶ **Exercise 2.5.7.** Given the linear system

$$\begin{cases} \lambda x - \mu y - \mu z = \mu \\ \mu x - \lambda y = \lambda \\ x - y - z = 0 \end{cases}$$

determine for which pairs $(\lambda, \mu) \in \mathbb{R}^2$ the system is solvable and for which the solution is unique.

Solution. We compute the rank of the complete and incomplete matrices:

$$(A|B) = \left(\begin{array}{ccc|c} \lambda & -\mu & -\mu & \mu \\ \mu & -\lambda & 0 & \lambda \\ 1 & -1 & -1 & 0 \end{array}\right)$$

With elementary operations on the rows we obtain the matrix

$$\left(\begin{array}{ccc|c} 1 & -1 & -1 & 0 \\ 0 & \lambda-\mu & \lambda-\mu & \mu \\ 0 & 0 & \lambda & \lambda+\mu \end{array}\right)$$

We must therefore distinguish the following cases:

(1) $\lambda = 0$, $\mu = 0$: we have $\text{rk}(A) = \text{rk}(A|B) = 1$, so there are infinitely many solutions (depending on $3 - 1 = 2$ parameters);

(2) $\lambda = 0$, $\mu \neq 0$: we have $\text{rk}(A) = 2$, $\text{rk}(A|B) = 3$, so there are no solutions;

(3) $\lambda \neq 0$, $\mu = \lambda$: we have $\mathrm{rk}(A) = 2$, $\mathrm{rk}(A|B) = 3$, so there are no solutions;

(4) $\lambda \neq 0$, $\mu \neq \lambda$: we have $\mathrm{rk}(A) = \mathrm{rk}(A|B) = 3$, so the system has a unique solution.

▶ **Exercise 2.5.8.** Determine the intersection of the sets of solutions of the following linear systems:

$$\begin{cases} 2x - y = 1 \\ x + y - z + w = 0 \end{cases} \qquad \begin{cases} 2y + z - w = -1 \\ x + 2z - 2w = 0. \end{cases}$$

Solution. The intersection of the sets of solutions of the two given linear systems consists of the values of x, y, z and w which simultaneously satisfy the equations of the first system and the equations of the second system. So we are dealing with the solutions of the following system of equations:

$$\begin{cases} 2x - y = 1 \\ x + y - z + w = 0 \\ 2y + z - w = -1 \\ x + 2z - 2w = 0 \end{cases}$$

By solving this system, we get:

$$\begin{cases} x = 2/7 \\ y = -3/7 \\ z = w - 1/7 \\ w \text{ any value.} \end{cases}$$

▶ **Exercise 2.5.9.** Solve the following linear system as the parameter $a \in \mathbb{R}$ varies:

$$\begin{cases} x_1 - x_2 + ax_3 + 2x_4 = 0 \\ -x_1 + x_2 - 2ax_3 + ax_4 = 1 \\ 2x_2 + (a-2)x_3 + (1-2a)x_4 = -1 \\ x_1 - x_2 - ax_3 + 4x_4 = 2. \end{cases}$$

Solution. The complete matrix of the system is:

$$\left(\begin{array}{cccc|c} 1 & -1 & a & 2 & 0 \\ -1 & 1 & -2a & a & 1 \\ 0 & 2 & a-2 & 1-2a & -1 \\ 1 & -1 & -a & 4 & 2 \end{array}\right)$$

We use Gaussian elimination, using only elementary row operations. If we add the first to the second row and subtract the first from the fourth row, we get the matrix

$$\left(\begin{array}{cccc|c} 1 & -1 & a & 2 & 0 \\ 0 & 0 & -a & a+2 & 1 \\ 0 & 2 & a-2 & 1-2a & -1 \\ 0 & 0 & -2a & 2 & 2 \end{array}\right)$$

Now let's swap the second row with the third:

$$\left(\begin{array}{cccc|c} 1 & -1 & a & 2 & 0 \\ 0 & 2 & a-2 & 1-2a & -1 \\ 0 & 0 & -a & a+2 & 1 \\ 0 & 0 & -2a & 2 & 2 \end{array}\right)$$

Finally, we subtract twice the third row from the fourth row, obtaining the following matrix:

$$\left(\begin{array}{cccc|c} 1 & -1 & a & 2 & 0 \\ 0 & 2 & a-2 & 1-2a & -1 \\ 0 & 0 & -a & a+2 & 1 \\ 0 & 0 & 0 & -2a-2 & 0 \end{array}\right)$$

We can now observe that, if $a \neq -1, 0$, the rank of the incomplete matrix of the system is maximum (equal to 4). This matrix is therefore invertible, hence the system admits a unique solution. The system corresponding to the last matrix is

$$\begin{cases} x_1 - x_2 + ax_3 + 2x_4 = 0 \\ 2x_2 + (a-2)x_3 + (1-2a)x_4 = -1 \\ -ax_3 + (a+2)x_4 = 1 \\ (-2a-2)x_4 = 0 \end{cases}$$

which can be easily solved by a "back substitution," yielding the following solution:

$$\begin{cases} x_1 = \frac{a-1}{a} \\ x_2 = -\frac{1}{a} \\ x_3 = -\frac{1}{a} \\ x_4 = 0. \end{cases}$$

Let us now consider the case $a = -1$. The previous matrix now becomes

$$\left(\begin{array}{cccc|c} 1 & -1 & -1 & 2 & 0 \\ 0 & 2 & -3 & 3 & -1 \\ 0 & 0 & 1 & 1 & 1 \\ 0 & 0 & 0 & 0 & 0 \end{array}\right)$$

In this case the two matrices, complete and incomplete, have the same rank equal to 3. By Rouché–Capelli's Theorem this means that the system has infinitely many solutions, depending on a parameter. In fact, the system corresponding to the last matrix is the following:

$$\begin{cases} x_1 - x_2 - x_3 + 2x_4 = 0 \\ 2x_2 - 3x_3 + 3x_4 = -1 \\ x_3 + x_4 = 1 \\ 0 = 0 \end{cases}$$

which has infinitely many solutions given by

$$\begin{cases} x_1 = 2 - 6x_4 \\ x_2 = 1 - 3x_4 \\ x_3 = 1 - x_4 \\ x_4 \text{ any value.} \end{cases}$$

To finish we just have to consider the case $a = 0$. The matrix of the system becomes

$$\left(\begin{array}{cccc|c} 1 & -1 & 0 & 2 & 0 \\ 0 & 2 & -2 & 1 & -1 \\ 0 & 0 & 0 & 2 & 1 \\ 0 & 0 & 0 & -2 & 0 \end{array}\right)$$

This matrix is not in echelon form. We therefore add the third to the fourth row, obtaining the following matrix:

$$\left(\begin{array}{cccc|c} 1 & -1 & 0 & 2 & 0 \\ 0 & 2 & -2 & 1 & -1 \\ 0 & 0 & 0 & 2 & 1 \\ 0 & 0 & 0 & 0 & 1 \end{array}\right)$$

It thus turns out that, in this case, the incomplete matrix has rank 3, while the complete matrix has rank 4 and therefore the system has no solutions (which is completely obvious, given that the equation corresponding to the last row of the matrix above would be $0 = 1$).

2.6 Recap exercises

► **Exercise 2.6.1.** Determine for which values of $t \in \mathbb{R}$ there exists a linear function $f\colon \mathbb{R}^3 \to \mathbb{R}^3$ such that $f(0,1,-1) = (3,-1,0)$, $f(-2,1,3) = (-t,-1,t+3)$ and the kernel of f is generated by the vector $(1, t^2+3t, -2)$. For the values of t for which f exists, also determine whether it is unique or not.

Solution. Recall that a linear function between two vector spaces V and W is uniquely determined when the images of the vectors of a basis of V are known. Knowing that the kernel of f is generated by the vector $(1, t^2+3t, -2)$, we deduce that $f(1, t^2+3t, -2) = (0,0,0)$. Therefore we know the images, through f, of the three vectors $v_1 = (0,1,-1)$, $v_2 = (-2,1,3)$ and $v_3 = (1, t^2+3t, -2)$.

The first thing to do is therefore to determine for which values of the parameter t the vectors v_1, v_2 and v_3 form a basis of $\mathbb{R}^3$. We consider a linear combination

$$\lambda_1 v_1 + \lambda_2 v_2 + \lambda_3 v_3 = \mathbf{0}.$$

The following system is obtained:

$$\begin{cases} -2\lambda_2 + \lambda_3 = 0 \\ \lambda_1 + \lambda_2 + (t^2+3t)\lambda_3 = 0 \\ -\lambda_1 + 3\lambda_2 - 2\lambda_3 = 0 \end{cases}$$

from which we get

$$\begin{cases} \lambda_1 = -\lambda_2 \\ \lambda_3 = 2\lambda_2 \\ 2(t^2+3t)\lambda_2 = 0. \end{cases}$$

If $t^2 + 3t \neq 0$, i.e., if $t \neq 0, -3$, the only solution is $\lambda_1 = \lambda_2 = \lambda_3 = 0$. In this case the vectors v_1, v_2, v_3 are linearly independent and are therefore a basis of $\mathbb{R}^3$. We can therefore state that, for $t \neq 0, -3$, the linear function f exists and is unique.

Let us now analyze separately the two special cases $t = 0$ and $t = -3$. If $t = 0$ the previous system becomes

$$\begin{cases} \lambda_1 = -\lambda_2 \\ \lambda_3 = 2\lambda_2 \\ 0 = 0 \end{cases}$$

which has infinitely many solutions (this means that, in this case, the vectors v_1, v_2, v_3 are linearly dependent). A particular solution is given, for example, by $\lambda_1 = -1$, $\lambda_2 = 1$, $\lambda_3 = 2$. This means that, for $t = 0$, the following linear dependence relation holds:

$$-v_1 + v_2 + 2v_3 = \mathbf{0}.$$

The vectors v_1, v_2, v_3 are not a basis of $\mathbb{R}^3$, therefore the images of the vectors of a basis of the domain of the function f are not known. We can conclude that, even if such a function f were to exist, it would not be uniquely determined. However, since f must be linear, we should have

$$f(-v_1 + v_2 + 2v_3) = -f(v_1) + f(v_2) + 2f(v_3) = f(\mathbf{0}) = \mathbf{0}.$$

By replacing $f(v_1)$, $f(v_2)$ and $f(v_3)$ with the values given by the problem, we obtain the equality

$$-(3, -1, 0) + (0, -1, 3) + (0, 0, 0) = (0, 0, 0),$$

which is not verified. We thus conclude that for $t = 0$ there cannot exist a linear function f that satisfies all the requests.

We now pass to the case $t = -3$. As in the previous case the system becomes

$$\begin{cases} \lambda_1 = -\lambda_2 \\ \lambda_3 = 2\lambda_2 \\ 0 = 0 \end{cases}$$

which has infinitely many solutions. A particular solution is given, as before, by $\lambda_1 = -1$, $\lambda_2 = 1$, $\lambda_3 = 2$. This means that, for $t = -3$, the following linear dependence relation holds:

$$-v_1 + v_2 + 2v_3 = \mathbf{0}.$$

Also in this case the vectors v_1, v_2, v_3 are not a basis of $\mathbb{R}^3$, therefore, exactly as before, we can conclude that, even if such a function f exists, it would not be uniquely determined. In this case however, if we apply f to the previous equality, we obtain

$$f(-v_1 + v_2 + 2v_3) = -f(v_1) + f(v_2) + 2f(v_3) = f(\mathbf{0}) = \mathbf{0}$$

and replacing $f(v_1)$, $f(v_2)$ and $f(v_3)$ with the values given by the problem, we obtain the equality

$$-(3, -1, 0) + (3, -1, 0) + (0, 0, 0) = (0, 0, 0),$$

which is now verified. However, to construct a linear function f that satisfies the requirements we should neglect the vector v_3 (which is a linear combination of the vectors v_1 and v_2) and complete the vectors v_1, v_2 to a basis of $\mathbb{R}^3$ by adding some vector w (chosen such that the vectors v_1, v_2, w are linearly independent). The function f would be uniquely determined if we also knew the image of w which however is not assigned and, on the contrary, can be fixed (almost*) arbitrarily. Thus we conclude that, for $t = -3$, there exist infinite linear functions f which satisfy the requirements imposed by the problem [*There is such a function f for every choice of $f(w) \in \mathbb{R}^3$. However, to be precise, $f(w)$ cannot be chosen completely arbitrarily, otherwise one might find a function f having a kernel of dimension 2 and not 1 as required by the problem! To make $\operatorname{Ker} f$ have dimension 1 we must choose $f(w)$ so that the image of f has dimension 2, in other words, $f(w)$ must be linearly independent of $f(v_1)$. However, there are infinitely many choices for $f(w)$].

► **Exercise 2.6.2.** Consider the linear function $f\colon \mathbb{R}^4 \to \mathbb{R}^4$ whose matrix (with respect to the canonical bases) is

$$A = \begin{pmatrix} 2 & 0 & -1 & 1 \\ t & 1 & 0 & 0 \\ 0 & 2 & 3 & 1 \\ 1 & -1 & 1 & -2 \end{pmatrix}$$

Determine for which values of $t \in \mathbb{R}$ the function f *is not* surjective. For such values of t determine a basis of $\operatorname{Im}(f)$.

Solution. Since the domain and codomain of f have the same dimension, the function f is surjective if and only if it is injective (that is, if and only if it is bijective). In terms of the matrix A, f is bijective if and only if $\det A \neq 0$. Thus, requiring f not to be surjective is equivalent to requiring the determinant of A to be equal to zero.

$$\begin{aligned}
\det A &= -t\begin{vmatrix} 0 & -1 & 1 \\ 2 & 3 & 1 \\ -1 & 1 & -2 \end{vmatrix} + \begin{vmatrix} 2 & -1 & 1 \\ 0 & 3 & 1 \\ 1 & 1 & -2 \end{vmatrix} \\
&= -t\left(\begin{vmatrix} 2 & 1 \\ -1 & -2 \end{vmatrix} + \begin{vmatrix} 2 & 3 \\ -1 & 1 \end{vmatrix}\right) + 2\begin{vmatrix} 3 & 1 \\ 1 & -2 \end{vmatrix} + \begin{vmatrix} -1 & 1 \\ 3 & 1 \end{vmatrix} \\
&= -2t - 18.
\end{aligned}$$

From the computation of the determinant of A it turns out that it vanishes for $t = -9$. We therefore conclude that the function f is not surjective for $t = -9$ (while for $t \neq -9$ it is bijective).

For $t = -9$ the matrix A becomes

$$A = \begin{pmatrix} 2 & 0 & -1 & 1 \\ -9 & 1 & 0 & 0 \\ 0 & 2 & 3 & 1 \\ 1 & -1 & 1 & -2 \end{pmatrix}$$

We know that the image of f is generated by the columns of the matrix A. On the other hand, since f is not surjective, the dimension of $\operatorname{Im} f$ is < 4 (the four columns of A are linearly dependent, since $\det A = 0$). We arbitrarily exclude the last column and consider the first three. We check if they are linearly independent:

$$\lambda_1\begin{pmatrix} 2 \\ -9 \\ 0 \\ 1 \end{pmatrix} + \lambda_2\begin{pmatrix} 0 \\ 1 \\ 2 \\ -1 \end{pmatrix} + \lambda_3\begin{pmatrix} -1 \\ 0 \\ 3 \\ 1 \end{pmatrix} = \begin{pmatrix} 0 \\ 0 \\ 0 \\ 0 \end{pmatrix}$$

We get the system

$$\begin{cases} 2\lambda_1 - \lambda_3 = 0 \\ -9\lambda_1 + \lambda_2 = 0 \\ 2\lambda_2 + 3\lambda_3 = 0 \\ \lambda_1 - \lambda_2 + \lambda_3 = 0 \end{cases}$$

whose unique solution is $\lambda_1 = \lambda_2 = \lambda_3 = 0$.

Thus we conclude that the first three columns of the matrix A are linearly independent (hence the rank of A is 3) and are therefore a basis of the image of f (which has dimension 3).

► **Exercise 2.6.3.** In $\mathbb{R}^4$ consider the subspaces $U_1 = \langle u_1, u_2, u_3, u_4 \rangle$, where $u_1 = (1, -1, 0, 2)$, $u_2 = (2, 1, -1, 2)$, $u_3 = (1, -4, 1, 4)$, $u_4 = (-3, -3, 2, -2)$, and U_2 of equations $3x_1 - 4x_2 = 0$ and $5x_1 + 7x_2 = 0$.

(a) Find a basis of U_1 and a basis of U_2. Find a basis $\{v_1, v_2, v_3, v_4\}$ of $\mathbb{R}^4$ (if it exists) such that $U_1 = \langle v_1, v_2 \rangle$ and $U_2 = \langle v_3, v_4 \rangle$.

(b) Let $w_1 = (2, -1, 3)$, $w_2 = (1, 1, 2)$, $w_3 = (5, -4, t)$. Find for which values of t there is a linear function $g \colon \mathbb{R}^4 \to \mathbb{R}^3$ such that $g(u_1) = w_1$, $g(u_2) = w_2$, $g(u_3) = w_3$. Is this function g unique?

Solution. The vectors u_1, u_2, u_3, u_4 generate U_1, so from them we can extract a basis of U_1. To determine the dimension of U_1 we compute the rank of the following matrix:

$$\begin{pmatrix} 1 & -1 & 0 & 2 \\ 2 & 1 & -1 & 2 \\ 1 & -4 & 1 & 4 \\ -3 & -3 & 2 & -2 \end{pmatrix}$$

We use the Gauss elimination method. From the second row we subtract twice the first, from the third row we subtract the first and to the fourth row we add three times the first, thus obtaining the following matrix:

$$\begin{pmatrix} 1 & -1 & 0 & 2 \\ 0 & 3 & -1 & -2 \\ 0 & -3 & 1 & 2 \\ 0 & -6 & 2 & 4 \end{pmatrix}$$

Now we carry out the following elementary operations: to the third row we add the second while to the fourth row we add the second multiplied by 2.

$$\begin{pmatrix} 1 & -1 & 0 & 2 \\ 0 & 3 & -1 & -2 \\ 0 & 0 & 0 & 0 \\ 0 & 0 & 0 & 0 \end{pmatrix}$$

It turns out that the matrix has rank 2. This means that the subspace U_1 has dimension 2 and a basis consists, for example, of the vectors u_1 and u_2 (naturally, as the basis of U_1 one could also consider the vectors u_3 and u_4, or u_1 and u_3, etc.).

The subspace U_2 is the set of solutions of the following system:

$$U_2 : \begin{cases} 3x_1 - 4x_2 = 0 \\ 5x_1 + 7x_2 = 0. \end{cases}$$

This is a system of two equations with *four* unknowns x_1, x_2, x_3, x_4 (remember that U_2 is a subspace of $\mathbb{R}^4$), whose solutions are given by

$$U_2 : \begin{cases} x_1 = 0 \\ x_2 = 0 \\ x_3,\ x_4 \text{ any value.} \end{cases}$$

There are therefore infinitely many solutions, depending on two parameters, hence U_2 has dimension 2 and a basis consists of the vectors $w_1 = (0, 0, 1, 0)$ and $w_2 = (0, 0, 0, 1)$. If we set $v_1 = u_1$, $v_2 = u_2$, $v_3 = w_1$ and $v_4 = w_2$, we have $U_1 = \langle v_1, v_2 \rangle$ and $U_2 = \langle v_3, v_4 \rangle$, as required. We just have to check if the vectors v_1, v_2, v_3, v_4 are a basis of $\mathbb{R}^4$, i.e., if they are linearly independent. To do this we compute the determinant of the matrix whose rows are the given vectors:

$$\det \begin{pmatrix} 1 & -1 & 0 & 2 \\ 2 & 1 & -1 & 2 \\ 0 & 0 & 1 & 0 \\ 0 & 0 & 0 & 1 \end{pmatrix} = \det \begin{pmatrix} 1 & -1 \\ 2 & 1 \end{pmatrix} = 3.$$

Since this determinant is not zero, the vectors v_1, v_2, v_3, v_4 are linearly independent and are therefore a basis of $\mathbb{R}^4$.

Now we consider $w_1 = (2, -1, 3)$, $w_2 = (1, 1, 2)$, $w_3 = (5, -4, t)$. We have already seen that the vectors u_1, u_2 and u_3 are linearly dependent, so there must be a linear dependence relation

$$\lambda_1 u_1 + \lambda_2 u_2 + \lambda_3 u_3 = \mathbf{0},$$

with λ_1, λ_2, λ_3 not all zero. By substituting the components of u_1, u_2 and u_3 we obtain the system

$$\begin{cases} \lambda_1 + 2\lambda_2 + \lambda_3 = 0 \\ -\lambda_1 + \lambda_2 - 4\lambda_3 = 0 \\ -\lambda_2 + \lambda_3 = 0 \\ 2\lambda_1 + 2\lambda_2 + 4\lambda_3 = 0 \end{cases}$$

whose solutions are

$$\begin{cases} \lambda_1 = -3\lambda_3 \\ \lambda_2 = \lambda_3. \end{cases}$$

A particular solution is $\lambda_1 = -3$, $\lambda_2 = 1$, $\lambda_3 = 1$, which gives the following linear dependence relation:

$$-3u_1 + u_2 + u_3 = \mathbf{0}.$$

Therefore we have $u_3 = 3u_1 - u_2$. In order that there exists a linear function $g\colon \mathbb{R}^4 \to \mathbb{R}^3$ such that $g(u_1) = w_1$, $g(u_2) = w_2$, $g(u_3) = w_3$ one must therefore have

$$g(u_3) = g(3u_1 - u_2) = 3g(u_1) - g(u_2),$$

that is $w_3 = 3w_1 - w_2$. Hence, the following equality must hold

$$(5, -4, t) = 3(2, -1, 3) - (1, 1, 2),$$

from which it follows that $t = 7$.

We conclude that a linear function g with the required properties exists if and only if $t = 7$. However, such a linear function is not uniquely determined, given that the images via g of the vectors of a basis of its domain are not known (in other words, for $t = 7$ there exist infinitely many linear functions g with the required properties).

► **Exercise 2.6.4.** Let V and W be two vector spaces, let $\{v_1, v_2, v_3, v_4\}$ be a basis of V and $\{w_1, w_2, w_3\}$ be a basis of W. We denote by $f\colon V \to W$ a linear function such that

$$f(v_1 - v_3) = w_1 - 2w_2 - 2w_3 \qquad f(v_1 - v_2 + v_3) = w_2$$
$$f(v_1 + v_3) = w_1 + 2w_2 \qquad f(v_1 - v_3 + v_4) = 5w_1 - 4w_3$$

(a) Find if f is uniquely determined by the given conditions and write the matrix of f with respect to the given bases.

(b) Find a basis of $\operatorname{Ker}(f)$ and a basis of $\operatorname{Im}(f)$.

(c) Compute $f^{-1}(w_1 + w_3)$.

(d) Determine if there exists a linear function $g\colon W \to V$ such that $f \circ g\colon W \to W$ is the identity (**Attention:** it is not required to determine such a g, but only to say whether it exists or not, justifying the answer).

Solution. To write the matrix of f with respect to the bases $\{v_1, v_2, v_3, v_4\}$ of V and $\{w_1, w_2, w_3\}$ of W we must compute $f(v_1)$, $f(v_2)$, $f(v_3)$ and $f(v_4)$. By the linearity of f, we have:

$$f(v_1 - v_3) + f(v_1 + v_3) = f(2v_1) = 2f(v_1),$$

hence

$$f(v_1) = \frac{1}{2}\big(f(v_1 - v_3) + f(v_1 + v_3)\big) = w_1 - w_3.$$

Similarly we have

$$f(v_1 + v_3) - f(v_1) = f(v_3),$$

hence

$$f(v_3) = f(v_1 + v_3) - f(v_1) = 2w_2 + w_3.$$

Then we have:

$$f(v_1 - v_2 + v_3) - f(v_1 + v_3) = f(-v_2) = -f(v_2),$$

from which it follows

$$f(v_2) = -\big(f(v_1 - v_2 + v_3) - f(v_1 + v_3)\big) = w_1 + w_2.$$

Finally we have

$$f(v_1 - v_3 + v_4) - f(v_1 - v_3) = f(v_4),$$

hence

$$f(v_4) = f(v_1 - v_3 + v_4) - f(v_1 - v_3) = 4w_1 + 2w_2 - 2w_3.$$

The matrix of f with respect to the assigned bases is therefore

$$A = \begin{pmatrix} 1 & 1 & 0 & 4 \\ 0 & 1 & 2 & 2 \\ -1 & 0 & 1 & -2 \end{pmatrix}$$

which proves, among other things, the existence of a unique linear function f which satisfies the requirements of the problem.

Using the matrix A we can easily determine the kernel of f. In fact the vectors $v \in \mathrm{Ker}(f)$ are the vectors $v = x_1v_1 + x_2v_2 + x_3v_3 + x_4v_4$ such that

$$\begin{pmatrix} 1 & 1 & 0 & 4 \\ 0 & 1 & 2 & 2 \\ -1 & 0 & 1 & -2 \end{pmatrix} \begin{pmatrix} x_1 \\ x_2 \\ x_3 \\ x_4 \end{pmatrix} = \begin{pmatrix} 0 \\ 0 \\ 0 \end{pmatrix}$$

We then have to solve the following linear system:

$$\begin{cases} x_1 + x_2 + 4x_4 = 0 \\ x_2 + 2x_3 + 2x_4 = 0 \\ -x_1 + x_3 - 2x_4 = 0 \end{cases}$$

This system has infinitely many solutions, given by

$$\begin{cases} x_1 = -2x_4 \\ x_2 = -2x_4 \\ x_3 = 0 \\ x_4 \text{ any value.} \end{cases}$$

We therefore conclude that the kernel of f has dimension 1 and is generated by the vector $v = 2v_1 + 2v_2 - v_4$ (obtained by setting $x_4 = -1$).

Since $\dim \operatorname{Ker}(f) + \dim \operatorname{Im}(f) = \dim V = 4$, we have $\dim \operatorname{Im}(f) = 3$, which means that f is surjective (indeed $\dim W = 3$). Since $\operatorname{Im}(f) = W$, as the basis of the image of f we can take any basis of W, for example the basis w_1, w_2, w_3 assigned at the beginning.

Recalling the definition of the inverse image, we have

$$f^{-1}(w_1 + w_3) = \{v \in V \mid f(v) = w_1 + w_3\}.$$

In this regard, we point out that, in this case, the function f is not invertible, therefore there is no "inverse function" of f, $f^{-1}\colon W \to V$.

If we set $v = x_1v_1 + x_2v_2 + x_3v_3 + x_4v_4$, the requirement that $v \in f^{-1}(w_1 + w_3)$, i.e., that $f(v) = w_1 + w_3$, is equivalent to requiring that

$$\begin{pmatrix} 1 & 1 & 0 & 4 \\ 0 & 1 & 2 & 2 \\ -1 & 0 & 1 & -2 \end{pmatrix} \begin{pmatrix} x_1 \\ x_2 \\ x_3 \\ x_4 \end{pmatrix} = \begin{pmatrix} 1 \\ 0 \\ 1 \end{pmatrix}$$

We therefore have to solve the following linear system:

$$\begin{cases} x_1 + x_2 + 4x_4 = 1 \\ x_2 + 2x_3 + 2x_4 = 0 \\ -x_1 + x_3 - 2x_4 = 1 \end{cases}$$

This system has infinitely many solutions, given by

$$\begin{cases} x_1 = -3 - 2x_4 \\ x_2 = 4 - 2x_4 \\ x_3 = -2 \\ x_4 \text{ any value.} \end{cases}$$

Setting $x_4 = \alpha$, we can write

$$\begin{aligned} f^{-1}(w_1 + w_3) &= \{(-3 - 2\alpha)v_1 + (4 - 2\alpha)v_2 - 2v_3 + \alpha v_4 \mid \alpha \in \mathbb{R}\} \\ &= -3v_1 + 4v_2 - 2v_3 + \operatorname{Ker}(f). \end{aligned}$$

To answer the last question we recall that the function $f\colon V \to W$ is surjective. This means that for every $w \in W$ there exists a vector $u \in V$ such that $f(u) = w$. In particular, given the vectors w_1, w_2, w_3 of the basis of W, there exist vectors u_1, u_2, $u_3 \in V$ such that $f(u_i) = w_i$, for $i = 1, 2, 3$ (the vectors u_i are not uniquely determined, they are determined only up to the sum of elements of the kernel of f). We also recall that to define a linear function $g\colon W \to V$ it is sufficient to specify the images via g of the vectors of a basis of W. We can therefore define the function g by setting $g(w_i) = u_i$, for $i = 1, 2, 3$. With this definition we have

$$(f \circ g)(w_i) = f\big(g(w_i)\big) = f(u_i) = w_i, \qquad \text{for } i = 1, 2, 3,$$

which proves that the composition $f \circ g\colon W \to W$ is the identity. We therefore conclude that such a function g exists (this is due to the fact that f is surjective), but it is not unique, since its definition depends on the choice of vectors $u_i \in V$ such that $f(u_i) = w_i$.

► **Exercise 2.6.5.** Let $f\colon \mathbb{R}^4 \to \mathbb{R}^3$ be the linear function defined by

$$\begin{aligned} f(1,1,0,0) &= (3,1,2), & f(1,0,1,0) &= (2,0,2), \\ f(0,1,0,0) &= (-1,-2,1), & f(0,0,1,1) &= (1,-1,2). \end{aligned}$$

(a) Write the matrix of f with respect to the canonical bases.

(b) Find a basis of the kernel of f and a basis of the image of f.

(c) Find a basis of $\operatorname{Ker}(f) \cap U$, where U is the vector subspace of $\mathbb{R}^4$ of equation $2x_1 + x_2 + x_3 = 0$.

Solution. We set $v_1 = (1,1,0,0)$, $v_2 = (0,1,0,0)$, $v_3 = (1,0,1,0)$, $v_4 = (0,0,1,1)$, $w_1 = (3,1,2)$, $w_2 = (-1,-2,1)$, $w_3 = (2,0,2)$, $w_4 = (1,-1,2)$; then we have $f(v_i) = w_i$, for $i = 1,\dots,4$.

To write the matrix of f with respect to the canonical bases we have to compute the images of the vectors of the canonical basis of $\mathbb{R}^4$. We have $e_1 = v_1 - v_2$, hence

$$f(e_1) = f(v_1) - f(v_2) = w_1 - w_2 = (4,3,1).$$

Note that $e_2 = v_2$, hence $f(e_2) = w_2 = (-1,-2,1)$. Now we have $e_3 = v_3 - e_1$, hence

$$f(e_3) = f(v_3) - f(e_1) = (-2,-3,1).$$

Finally we have $e_4 = v_4 - e_3$, hence

$$f(e_4) = f(v_4) - f(e_3) = (3,2,1).$$

The matrix of f with respect to the canonical bases is therefore

$$A = \begin{pmatrix} 4 & -1 & -2 & 3 \\ 3 & -2 & -3 & 2 \\ 1 & 1 & 1 & 1 \end{pmatrix}.$$

The kernel of f is the set of solutions of the following linear system:

$$\begin{cases} 4x_1 - x_2 - 2x_3 + 3x_4 = 0 \\ 3x_1 - 2x_2 - 3x_3 + 2x_4 = 0 \\ x_1 + x_2 + x_3 + x_4 = 0 \end{cases}$$

which is equivalent to

$$\begin{cases} x_1 = 4x_2 + 5x_3 \\ x_4 = -5x_2 - 6x_3. \end{cases}$$

There are therefore infinitely many solutions, depending on two parameters (x_2 and x_3); this means that $\dim \operatorname{Ker}(f) = 2$. A basis of the kernel of f consists of the vectors $u_1 = (4,1,0,-5)$ and $u_2 = (5,0,1,-6)$, obtained by setting $x_2 = 1$, $x_3 = 0$ and $x_2 = 0$, $x_3 = 1$, respectively.

As for the image of f, we have

$$\dim \operatorname{Im}(f) = \dim \mathbb{R}^4 - \dim \operatorname{Ker}(f) = 2.$$

Since the image of f is generated by the columns of A, to find a basis of $\operatorname{Im} f$ it is enough to take two linearly independent columns of the matrix

A, for example the columns $(4, 3, 1)$ and $(-1, -2, 1)$ (obviously, the vectors w_1 and w_2 could also be taken as the basis of $\operatorname{Im} f$, since they are linearly independent, or even w_1 and w_3, etc.).

We now look for the vectors that belong to the subspace $\operatorname{Ker}(f) \cap U$. As we have seen previously, the vectors of $\operatorname{Ker} f$ are given by the solutions of the linear system

$$\begin{cases} x_1 = 4x_2 + 5x_3 \\ x_4 = -5x_2 - 6x_3. \end{cases}$$

Since the vectors of U are the solutions of the equation $2x_1 + x_2 + x_3 = 0$, to find the vectors of $\operatorname{Ker}(f) \cap U$ we need to solve the system given by the equations of $\operatorname{Ker} f$ and of U:

$$\begin{cases} x_1 = 4x_2 + 5x_3 \\ x_4 = -5x_2 - 6x_3 \\ 2x_1 + x_2 + x_3 = 0 \end{cases}$$

The solutions of this system are given by

$$\begin{cases} x_1 = \frac{1}{9}\, x_3 \\ x_2 = -\frac{11}{9}\, x_3 \\ x_4 = \frac{1}{9}\, x_3 \\ x_3 \text{ any value.} \end{cases}$$

The presence of a free parameter means that the subspace $\operatorname{Ker}(f) \cap U$ has dimension 1; it is therefore generated by the vector $(1, -11, 9, 1)$, obtained by setting $x_3 = 9$.

► **Exercise 2.6.6.** (a) Given the vectors $v_1 = (1, 0, 2, -1)$, $v_2 = (0, 1, -1, 2) \in \mathbb{R}^4$ and $w_1 = (2, 0, 1)$, $w_2 = (1, -1, 2) \in \mathbb{R}^3$, determine if there is a linear function $f\colon \mathbb{R}^4 \to \mathbb{R}^3$ such that $\operatorname{Ker}(f) = \langle v_1, v_2 \rangle$ and $\operatorname{Im}(f) = \langle w_1, w_2 \rangle$. If such a function exists, determine its rank and whether it is unique. Also write the matrix of such a f with respect to the canonical bases.

(b) Let $w = (t, 2, -1)$. Determine for which value of $t \in \mathbb{R}$ we have $w \in \operatorname{Im}(f)$.

Solution. Requiring that w_1 and w_2 belong to the image of f is equivalent to requiring that there exist vectors $u_1, u_2 \in \mathbb{R}^4$ such that $f(u_1) = w_1$ and $f(u_2) = w_2$. We arbitrarily decide that $u_1 = e_3 = (0, 0, 1, 0)$ and $u_2 = e_4 =$

$(0,0,0,1)$. In this way w_1 and w_2 will be respectively the images of the third and fourth vector of the canonical basis of $\mathbb{R}^4$, i.e., they will be the third and fourth column of the matrix A of f. The matrix of f will therefore be:

$$A = \begin{pmatrix} a & a' & 2 & 1 \\ b & b' & 0 & -1 \\ c & c' & 1 & 2 \end{pmatrix}$$

We point out that our choice is completely arbitrary (which suggests that such a function f is certainly not unique). For example, if we had decided that the vectors w_1 and w_2 had to be the images of the first and second vector of the canonical basis of $\mathbb{R}^4$, i.e., the first and second column of A, the matrix of f with respect to the canonical bases would have the following form:

$$A = \begin{pmatrix} 2 & 1 & a & a' \\ 0 & -1 & b & b' \\ 1 & 2 & c & c' \end{pmatrix}$$

Let's stick with our first choice anyway. We now need to determine the 6 unknown coefficients a, b, c, a', b', c'. To do this we recall that the vectors v_1 and v_2 must belong to the kernel of f. This means that we must have

$$\begin{pmatrix} a & a' & 2 & 1 \\ b & b' & 0 & -1 \\ c & c' & 1 & 2 \end{pmatrix} \begin{pmatrix} 1 \\ 0 \\ 2 \\ -1 \end{pmatrix} = \begin{pmatrix} 0 \\ 0 \\ 0 \end{pmatrix}, \qquad \begin{pmatrix} a & a' & 2 & 1 \\ b & b' & 0 & -1 \\ c & c' & 1 & 2 \end{pmatrix} \begin{pmatrix} 0 \\ 1 \\ -1 \\ 2 \end{pmatrix} = \begin{pmatrix} 0 \\ 0 \\ 0 \end{pmatrix}$$

By solving these systems we find

$$\begin{cases} a = -3 \\ b = -1 \\ c = 0 \end{cases} \qquad \begin{cases} a' = 0 \\ b' = 2 \\ c' = -3 \end{cases}$$

The matrix is therefore

$$A = \begin{pmatrix} -3 & 0 & 2 & 1 \\ -1 & 2 & 0 & -1 \\ 0 & -3 & 1 & 2 \end{pmatrix}$$

It can now be easily verified that the kernel of this matrix has dimension 2 and is generated by the vectors v_1 and v_2, while the image, also of dimension 2, is generated by the vectors w_1 and w_2, as requested. From what we have just seen, we conclude that a function f having the required properties exists but is not unique; its rank is the dimension of the image of f, i.e., it is equal to 2 (since the vectors w_1 and w_2 are linearly independent).

Finally, requiring that the vector $w = (t, 2, -1)$ belongs to the image of f is equivalent to requiring that it be a linear combination of the vectors w_1 and w_2:

$$w = \lambda_1 w_1 + \lambda_2 w_2.$$

Thus we obtain the following linear system

$$\begin{cases} t = 2\lambda_1 + \lambda_2 \\ 2 = -\lambda_2 \\ -1 = \lambda_1 + 2\lambda_2 \end{cases}$$

whose solution is

$$\begin{cases} t = 4 \\ \lambda_1 = 3 \\ \lambda_2 = -2 \end{cases}$$

We conclude that $w \in \operatorname{Im}(f)$ if and only if $t = 4$.

▶ **Exercise 2.6.7.** Let $f\colon \mathbb{R}^3 \to \mathbb{R}^2$ be the linear function with matrix

$$A = \begin{pmatrix} 2 & 0 & 1 \\ 3 & -1 & 2 \end{pmatrix}$$

with respect to the canonical bases. Given the vector $u = (1, -2) \in \mathbb{R}^2$, compute $f^{-1}(u)$. Is the function f invertible?

Solution. A linear function $f\colon \mathbb{R}^3 \to \mathbb{R}^2$ certainly cannot be invertible (otherwise it would be an isomorphism and therefore $\mathbb{R}^3$ would be isomorphic to $\mathbb{R}^2$, which is false). Therefore the notation $f^{-1}(u)$ does not mean the "inverse function of f applied to the vector u" (given that, in this case, the "inverse function of f" does not exist), but rather the *inverse image* of the vector u, defined as follows:

$$f^{-1}(u) = \{\, v \in \mathbb{R}^3 \mid f(v) = u \,\}.$$

This definition makes perfect sense (even if f is not invertible), since the inverse function f^{-1} is never used.

Assuming $v = (x_1, x_2, x_3)$, requiring that $f(v) = u$ is equivalent to requiring that we have

$$\begin{pmatrix} 2 & 0 & 1 \\ 3 & -1 & 2 \end{pmatrix} \begin{pmatrix} x_1 \\ x_2 \\ x_3 \end{pmatrix} = \begin{pmatrix} 1 \\ -2 \end{pmatrix}$$

that is

$$\begin{cases} 2x_1 + x_3 = 1 \\ 3x_1 - x_2 + 2x_3 = -2. \end{cases}$$

Solving these equations we find

$$\begin{cases} x_2 = 4 - x_1 \\ x_3 = 1 - 2x_1 \\ x_1 \text{ any value.} \end{cases}$$

Setting $x_1 = t$, we have

$$f^{-1}(u) = \{(t, 4 - t, 1 - 2t) \,|\, \forall t \in \mathbb{R}\} = (0, 4, 1) + \langle(1, -1, -2)\rangle.$$

Finally, it can be noted that the vector $(1, -1, -2)$ is the generator of the kernel of f.

► **Exercise 2.6.8.** Let W be the vector subspace of $\mathbb{R}^4$ of equation $x_1 + 3x_2 - x_3 + 2x_4 = 0$.

(a) Given the vectors $w_1 = (2, 3, a, -1)$ and $w_2 = (1, 4, -1, b)$, determine the values of a and b such that $w_1, w_2 \in W$ (in all the following questions replace a and b with the values found).

(b) Determine $w_3 \in W$ such that $\{w_1, w_2, w_3\}$ is a basis of W.

(c) Determine if there exists a linear function $f\colon \mathbb{R}^4 \to \mathbb{R}^4$ such that $\operatorname{Im} f = W$ and if such a function is unique. If such a function exists, write its matrix with respect to the canonical bases and compute its kernel. Is it true that such a function must necessarily have an eigenvalue equal to zero?

(d) Given the subspace U generated by the vectors $u_1 = (3, -1, 2, 2)$ and $u_2 = (1, 2, -1, 2)$, find a basis of $U \cap W$.

Solution. The vector $w_1 = (2, 3, a, -1)$ belongs to the subspace W if and only if its components satisfy the equation $x_1 + 3x_2 - x_3 + 2x_4 = 0$ of W. By substituting, we find $a = 9$. Similarly, substituting the components of the vector $w_2 = (1, 4, -1, b)$ into the equation of W, we find $b = -7$. From now on we set $w_1 = (2, 3, 9, -1)$ and $w_2 = (1, 4, -1, -7)$.

To find a vector w_3 such that $\{w_1, w_2, w_3\}$ is a basis of W we observe that the subspace W has dimension 3 (being defined by *one* linear equation with *four* unknowns). Therefore it is sufficient to find a vector w_3 whose components satisfy the equation $x_1 + 3x_2 - x_3 + 2x_4 = 0$ and such that the vectors w_1, w_2 and w_3 are linearly independent. Of course, there are infinitely many possible choices of such a vector; one of the simplest is $w_3 = (1, 0, 1, 0)$ (to verify that the vectors w_1, w_2 and w_3 are linearly independent, we can compute the rank of the matrix whose rows are the three given vectors; using Gaussian elimination it is easy to verify that this matrix has rank 3).

To determine if there exists a linear function $f\colon \mathbb{R}^4 \to \mathbb{R}^4$ such that $\operatorname{Im} f = W$ it suffices to remember that the image of a linear function $f\colon \mathbb{R}^4 \to \mathbb{R}^4$ is generated by the images of the vectors of a basis of $\mathbb{R}^4$. To define such a f it is therefore sufficient to set $f(e_1) = w_1$, $f(e_2) = w_2$, $f(e_3) = w_3$ and $f(e_4) = \mathbf{0}$, where $\{e_1, e_2, e_3, e_4\}$ is the canonical basis of $\mathbb{R}^4$. Naturally infinitely many other choices are possible, for example $f(e_1) = w_1$, $f(e_2) = w_2$, $f(e_3) = w_3$ and $f(e_4) = w_1$, or $f(e_1) = \mathbf{0}$, $f(e_2) = w_1$, $f(e_3) = w_2$ and $f(e_4) = w_3$, etc. The conclusion is therefore that a function f with the required characteristics exists, but it is not uniquely determined (indeed, there are infinitely many of them). So let's choose one of these functions, for example the one defined by setting $f(e_1) = w_1$, $f(e_2) = w_2$, $f(e_3) = w_3$ and $f(e_4) = \mathbf{0}$. The matrix of this function with respect to the canonical bases is

$$\begin{pmatrix} 2 & 1 & 1 & 0 \\ 3 & 4 & 0 & 0 \\ 9 & -1 & 1 & 0 \\ -1 & -7 & 0 & 0 \end{pmatrix}$$

and from the definition we immediately deduce that the kernel of f is generated by the vector e_4 (this is because $f(e_4) = \mathbf{0}$ and $\dim(\operatorname{Ker} f) = 4 - \dim(\operatorname{Im} f) = 4 - 3 = 1$). Finally we observe that every function f with the required characteristics must necessarily have a non-zero kernel (more precisely, it must have a kernel of dimension 1, given that $\dim(\operatorname{Ker} f) + \dim(\operatorname{Im} f) = 4$). For each vector $v \in \operatorname{Ker} f$ we have $f(v) = \mathbf{0} = 0\,v$, which means that v is an eigenvector corresponding to the eigenvalue $\lambda = 0$. We thus conclude that every such linear function f necessarily has an eigenvalue equal to zero.

Let us now consider the subspace U generated by the vectors $u_1 = (3,-1,2,2)$ and $u_2 = (1,2,-1,2)$. The generic vector $u \in U$ is a linear combination of the vectors u_1 and u_2:

$$u = \alpha u_1 + \beta u_2 = (3\alpha + \beta, -\alpha + 2\beta, 2\alpha - \beta, 2\alpha + 2\beta).$$

Requiring that u belongs to the subspace W is equivalent to requiring that its components satisfy the equation of W, i.e., that we have

$$(3\alpha + \beta) + 3(-\alpha + 2\beta) - (2\alpha - \beta) + 2(2\alpha + 2\beta) = 0,$$

which is equivalent to $\alpha = -6\beta$. The fact that there is only one free parameter (in this case β) allows us to conclude that $U \cap W$ has dimension 1. To find a basis we can set $\beta = -1$, which gives $\alpha = 6$. Thus we get the vector

$$u = 6u_1 - u_2 = (17, -8, 13, 10)$$

which is a basis of the subspace $U \cap W$.

► **Exercise 2.6.9.** In the vector space $\mathbb{R}^4$ let U be the subspace generated by the vectors $u_1 = (3,-1,2,2)$, $u_2 = (1,2,-1,3)$, $u_3 = (1,-5,4,-4)$. Let W be the image of the function $f\colon \mathbb{R}^3 \to \mathbb{R}^4$ defined by

$$f(x,y,z) = (2x - y, x + 3z, 3x + y + z, 2y - 3z).$$

(a) Determine the Cartesian equations and a basis of U.

(b) Determine the Cartesian equations and a basis of W.

(c) Find a basis of $U \cap W$ and a basis of $U + W$.

(d) Let $\tilde{v} = (1, 4, t, -1)$. Determine if there exists a value of $t \in \mathbb{R}$ for which we have $\tilde{v} = f(v)$, for some $v \in \mathbb{R}^3$.

(e) Determine if there exist linear functions $g, h\colon \mathbb{R}^4 \to \mathbb{R}^4$ such that $g(U) = W$ and $h(W) = U$.

Solution. It is easy to verify that the vectors u_1, u_2 and u_3 are linearly dependent (in fact we have $u_3 = u_1 - 2u_2$), while the vectors u_1 and u_2 are linearly independent. This allows us to state that a basis of U is given by the vectors u_1 and u_2. A generic vector $(x_1, x_2, x_3, x_4) \in U$ is therefore written as a linear combination of the vectors u_1 and u_2:

$$(x_1, x_2, x_3, x_4) = \alpha(3,-1,2,2) + \beta(1,2,-1,3).$$

This vector equation is equivalent to the system

$$\begin{cases} x_1 = 3\alpha + \beta \\ x_2 = -\alpha + 2\beta \\ x_3 = 2\alpha - \beta \\ x_4 = 2\alpha + 3\beta \end{cases}$$

from which, deriving α and β from two of the four equations and substituting in the remaining two equations, the following system is obtained

$$\begin{cases} 3x_1 - 5x_2 - 7x_3 = 0 \\ 8x_1 - 7x_3 - 5x_4 = 0. \end{cases}$$

So these are the Cartesian equations of the subspace U (as a check, it can easily be verified that the vectors u_1 and u_2 satisfy the equations just found).

From the definition of the function f we deduce that its matrix with respect to the canonical bases is

$$A = \begin{pmatrix} 2 & -1 & 0 \\ 1 & 0 & 3 \\ 3 & 1 & 1 \\ 0 & 2 & -3 \end{pmatrix}$$

This matrix has rank 3 (as can easily be verified), therefore $W = \operatorname{Im} f$ has dimension 3 and a basis of W consists of the columns of the matrix A. To find the Cartesian equation of W we can use a method analogous to the one used to determine the equations of U. A generic vector $(x_1, x_2, x_3, x_4) \in W$ is written as a linear combination of the three columns of A:

$$(x_1, x_2, x_3, x_4) = \alpha(2, 1, 3, 0) + \beta(-1, 0, 1, 2) + \gamma(0, 3, 1, -3).$$

This vector equation is equivalent to the system

$$\begin{cases} x_1 = 2\alpha - \beta \\ x_2 = \alpha + 3\gamma \\ x_3 = 3\alpha + \beta + \gamma \\ x_4 = 2\beta - 3\gamma \end{cases}$$

from which, obtaining α, β and γ from three of the four equations and substituting in the remaining equation, we obtain the following equation

$$13x_1 + 19x_2 - 15x_3 + 14x_4 = 0,$$

which is the Cartesian equation of the subspace W.

Since we have the Cartesian equations of U and W, to find a basis of $U \cap W$ it is enough to solve the following system:

$$U \cap W : \begin{cases} 3x_1 - 5x_2 - 7x_3 = 0 \\ 8x_1 - 7x_3 - 5x_4 = 0 \\ 13x_1 + 19x_2 - 15x_3 + 14x_4 = 0. \end{cases}$$

By solving this system it turns out that it has infinitely many solutions (depending on a parameter) and that the space of solutions is generated by the vector $(17, -8, 13, 9)$. This vector is therefore a basis of $U \cap W$. From Grassmann's formula we get

$$\dim(U + W) = \dim U + \dim W - \dim(U \cap W) = 2 + 3 - 1 = 4,$$

hence $U + W = \mathbb{R}^4$ and as the basis of $U + W$ we can take any basis of $\mathbb{R}^4$ (for example the canonical basis).

Given $\tilde{v} = (1, 4, t, -1)$, requiring that $\tilde{v} = f(v)$, for some $v \in \mathbb{R}^3$, is equivalent to require that $\tilde{v} \in \operatorname{Im} f = W$, which is equivalent to requiring that $\tilde{v}$ can be written as a linear combination of vectors of a basis of W:

$$(1, 4, t, -1) = \alpha(2, 1, 3, 0) + \beta(-1, 0, 1, 2) + \gamma(0, 3, 1, -3).$$

Thus the following system is obtained

$$\begin{cases} 1 = 2\alpha - \beta \\ 4 = \alpha + 3\gamma \\ t = 3\alpha + \beta + \gamma \\ -1 = 2\beta - 3\gamma \end{cases}$$

from whose solution we find $t = 5$.

To answer the last question it suffices to observe that $\dim U = 2$ and $\dim W = 3$. A linear function $g \colon \mathbb{R}^4 \to \mathbb{R}^4$ such that $g(U) = W$ cannot exist: if it did exist, the three vectors of a basis of W would have to be images by g of three linearly independent vectors of U, but this is impossible since U has dimension 2. Conversely, a linear function $h \colon \mathbb{R}^4 \to \mathbb{R}^4$ such that $h(W) = U$ certainly exists (indeed, there are infinitely many). Such a function h can be defined as follows. Let $\{w_1, w_2, w_3\}$ be a basis of W and complete this basis to a basis $\{w_1, w_2, w_3, v\}$ of $\mathbb{R}^4$. Let $\{u_1, u_2\}$ be a basis of U. We define a linear function $h \colon \mathbb{R}^4 \to \mathbb{R}^4$ setting $h(w_1) = u_1$, $h(w_2) = u_2$, $h(w_3) = \mathbf{0}$ and

$h(v) = \mathbf{0}$ (recall that to define a linear function between two vector spaces it is sufficient to specify the images of the vectors of a basis of the domain). It is easy to verify that such a h satisfies the property $h(W) = U$.

► **Exercise 2.6.10.** Let M be the vector space of 2×2 matrices with real entries.

(a) Find the dimension and a basis of M.

(b) Let S be the subspace of M of symmetric matrices. Find the dimension and a basis of S.

(c) Let U be the vector subspace of M generated by the matrices

$$\begin{pmatrix} 0 & -3 \\ -2 & 4 \end{pmatrix} \quad \text{and} \quad \begin{pmatrix} 1 & 2 \\ 4 & 0 \end{pmatrix}$$

Write the Cartesian equations of U and find a basis of $U \cap S$.

(d) Let $f\colon M \to \mathbb{R}$ be the function that associates the trace of the product matrix XA to a matrix $X \in M$, where A is a fixed matrix (remember that the trace of a square matrix is the sum of the elements on the main diagonal). Determine if the function f is linear.

(e) Let $P \subset M$ denote the set of matrices whose entries are ≥ 0. Is P a vector subspace of M? If the answer is No, who is the subspace generated by P?

(f) Who is the vector subspace of M generated by the set of all invertible matrices?

Solution. M is the set of matrices

$$\begin{pmatrix} a & b \\ c & d \end{pmatrix}, \quad \forall\, a, b, c, d \in \mathbb{R}.$$

Since each such matrix is written uniquely in the form

$$\begin{pmatrix} a & b \\ c & d \end{pmatrix} = a \begin{pmatrix} 1 & 0 \\ 0 & 0 \end{pmatrix} + b \begin{pmatrix} 0 & 1 \\ 0 & 0 \end{pmatrix} + c \begin{pmatrix} 0 & 0 \\ 1 & 0 \end{pmatrix} + d \begin{pmatrix} 0 & 0 \\ 0 & 1 \end{pmatrix}$$

we deduce that the matrices

$$E_1 = \begin{pmatrix} 1 & 0 \\ 0 & 0 \end{pmatrix}, \quad E_2 = \begin{pmatrix} 0 & 1 \\ 0 & 0 \end{pmatrix}, \quad E_3 = \begin{pmatrix} 0 & 0 \\ 1 & 0 \end{pmatrix}, \quad E_4 = \begin{pmatrix} 0 & 0 \\ 0 & 1 \end{pmatrix}$$

are a basis of M and therefore $\dim M = 4$.

The set S (of symmetric matrices) is

$$S = \left\{ \begin{pmatrix} a & b \\ b & c \end{pmatrix} : a, b, c \in \mathbb{R} \right\}.$$

Each matrix in S is written uniquely in the form

$$\begin{pmatrix} a & b \\ b & c \end{pmatrix} = a \begin{pmatrix} 1 & 0 \\ 0 & 0 \end{pmatrix} + b \begin{pmatrix} 0 & 1 \\ 1 & 0 \end{pmatrix} + c \begin{pmatrix} 0 & 0 \\ 0 & 1 \end{pmatrix}$$

hence the matrices

$$E_1 = \begin{pmatrix} 1 & 0 \\ 0 & 0 \end{pmatrix}, \quad F = \begin{pmatrix} 0 & 1 \\ 1 & 0 \end{pmatrix}, \quad E_4 = \begin{pmatrix} 0 & 0 \\ 0 & 1 \end{pmatrix}$$

are a basis of S and therefore $\dim S = 3$.

Since U is generated by the matrices

$$\begin{pmatrix} 0 & -3 \\ -2 & 4 \end{pmatrix} \quad \text{and} \quad \begin{pmatrix} 1 & 2 \\ 4 & 0 \end{pmatrix}$$

the generic matrix of U is a linear combination of the two given matrices:

$$\begin{pmatrix} x & y \\ z & w \end{pmatrix} = \alpha \begin{pmatrix} 0 & -3 \\ -2 & 4 \end{pmatrix} + \beta \begin{pmatrix} 1 & 2 \\ 4 & 0 \end{pmatrix}.$$

Thus we get the system

$$\begin{cases} x = \beta \\ y = -3\alpha + 2\beta \\ z = -2\alpha + 4\beta \\ w = 4\alpha \end{cases}$$

from which it follows that $\alpha = \frac{1}{4}w$ and $\beta = x$. By substituting into the remaining two equations, we obtain the system

$$\begin{cases} 8x - 4y - 3w = 0 \\ 8x - 2z - w = 0. \end{cases}$$

These are the Cartesian equations of the subspace U. Requiring that the generic matrix

$$\begin{pmatrix} x & y \\ z & w \end{pmatrix}$$

of U belong to the set S of symmetric matrices is equivalent to requiring that $y = z$. From this it follows that the equations of the subspace $U \cap S$ are

$$\begin{cases} 8x - 4y - 3w = 0 \\ 8x - 2z - w = 0 \\ y = z. \end{cases}$$

This system has infinitely many solutions (depending on a parameter):

$$\begin{cases} y = 8x \\ z = 8x \\ w = -8x \end{cases}$$

from which we deduce that $U \cap S$ has dimension 1 and a basis is the matrix

$$\begin{pmatrix} 1 & 8 \\ 8 & -8 \end{pmatrix}$$

obtained by setting $x = 1$ in the previous system.

Let us now consider the function $f\colon M \to \mathbb{R}$ which associates the trace of the product matrix XA to a matrix $X \in M$, where A is a fixed matrix. If we set

$$X = \begin{pmatrix} x & y \\ z & w \end{pmatrix} \quad \text{and} \quad A = \begin{pmatrix} a & b \\ c & d \end{pmatrix}$$

we have

$$XA = \begin{pmatrix} ax + cy & bx + dy \\ az + cw & bz + dw \end{pmatrix}$$

hence f is given by the following expression

$$f(X) = ax + cy + bz + dw.$$

From this explicit computation we immediately deduce that f is a linear function of X (i.e., of the four variables x, y, z, w).

Let us now denote by $P \subset M$ the set of matrices whose entries are ≥ 0. If $A \in P$ the entries of the matrix $-A$ are ≤ 0 hence, in general, $-A \notin P$; this means that P is not a vector subspace. We note that the four basis matrices E_1, E_2, E_3, E_4 belong to P, therefore they also belong to the vector subspace generated by P. Since these matrices are a basis of M, the smallest vector space containing them is M itself, so the vector space generated by P is M.

Let us denote by I the set of all invertible matrices

$$I = \{ A \in M \mid \det A \neq 0 \}.$$

To determine the vector space generated by I it is sufficient to observe that it is possible to find a basis of M consisting of invertible matrices. For example, the four matrices

$$G_1 = \begin{pmatrix} 1 & 0 \\ 0 & 1 \end{pmatrix}, \quad G_2 = \begin{pmatrix} 1 & 0 \\ 0 & -1 \end{pmatrix}, \quad G_3 = \begin{pmatrix} 0 & 1 \\ 1 & 0 \end{pmatrix}, \quad G_4 = \begin{pmatrix} 0 & 1 \\ -1 & 0 \end{pmatrix}$$

are invertible and are a basis of M, as can be easily verified (just verify that the matrices G_1, G_2, G_3, G_4 are linearly independent). This means that the vector space generated by I coincides with the vector space generated by the matrices G_1, G_2, G_3, G_4, i.e., with M.

► **Exercise 2.6.11.** Let $f\colon \mathbb{R}^4 \to \mathbb{R}^3$ be the linear function whose matrix, with respect to the canonical bases, is

$$A = \begin{pmatrix} 1 & -2 & 1 & 0 \\ 1 & 0 & 2 & -1 \\ 1 & -6 & -1 & 2 \end{pmatrix}$$

(a) Find a basis of the kernel and a basis of the image of f.

(b) Given the vector $u_t = (7, 2, t, 1)$ find t such that $u_t \in \operatorname{Ker}(f)$.

(c) Given $w_t = (2, t, 0)$ find for which value of t we have $f^{-1}(w_t) \neq \varnothing$.

(d) Let U be the subspace of $\mathbb{R}^4$ generated by $u_1 = (3, -1, 2, 2)$ and $u_2 = (1, 1, 1, 3)$. Find a basis of the subspace $f(U)$.

(e) Determine if there exists a linear function $g\colon \mathbb{R}^3 \to \mathbb{R}^4$ such that the composition $f \circ g\colon \mathbb{R}^3 \to \mathbb{R}^3$ is the identity. If such a g exists write its matrix.

Solution. To find a basis of $\operatorname{Ker}(f)$ we have to solve the following system:

$$\begin{cases} x_1 - 2x_2 + x_3 = 0 \\ x_1 + 2x_3 - x_4 = 0 \\ x_1 - 6x_2 - x_3 + 2x_4 = 0. \end{cases}$$

This system has infinitely many solutions (depending on two parameters), given by

$$\begin{cases} x_1 = 2x_2 - x_3 \\ x_4 = 2x_2 + x_3 \end{cases}$$

from which we deduce that the kernel of f has dimension 2 and a basis is given by the vectors $(2, 1, 0, 2)$ and $(-1, 0, 1, 1)$.

Given that $\dim \operatorname{Ker}(f) = 2$ we deduce that $\dim \operatorname{Im}(f) = 2$, therefore a basis of the image of f consists of two linearly independent columns of the matrix A, for example the vectors $(-2, 0, -6)$ and $(0, -1, 2)$.

Note that another way to compute the dimension of the image of f would be to compute the rank of A. In fact, by means of elementary operations on the rows, the matrix A can be reduced to the following echelon form

$$\begin{pmatrix} 1 & -2 & 1 & 0 \\ 0 & 2 & 1 & -1 \\ 0 & 0 & 0 & 0 \end{pmatrix}$$

from which we deduce that the rank of A (i.e., $\dim \operatorname{Im}(f)$) is 2.

Now we consider the vector $u_t = (7, 2, t, 1)$. Requiring that $u_t \in \operatorname{Ker}(f)$ is equivalent to requiring that

$$\begin{pmatrix} 1 & -2 & 1 & 0 \\ 1 & 0 & 2 & -1 \\ 1 & -6 & -1 & 2 \end{pmatrix} \begin{pmatrix} 7 \\ 2 \\ t \\ 1 \end{pmatrix} = \mathbf{0},$$

which is equivalent to

$$\begin{pmatrix} 3+t \\ 6+2t \\ -3-t \end{pmatrix} = \begin{pmatrix} 0 \\ 0 \\ 0 \end{pmatrix},$$

from which we obtain $t = -3$.

For the next point we note that $f^{-1}(w_t) \neq \varnothing$ is equivalent to requiring that there exists a vector $v = (x_1, x_2, x_3, x_4)$ such that $f(v) = w_t$. We obtain the system

$$\begin{cases} x_1 - 2x_2 + x_3 = 2 \\ x_1 + 2x_3 - x_4 = t \\ x_1 - 6x_2 - x_3 + 2x_4 = 0. \end{cases}$$

By solving this system, it turns out that it has solutions if and only if $t = 3$.

Another way to approach the same problem is to observe that $f^{-1}(w_t) \neq \varnothing$ if and only if $w_t \in \mathrm{Im}(f)$, i.e., if and only if w_t is a linear combination of the two basis vectors of $\mathrm{Im}(f)$ previously chosen. Therefore we must have

$$\begin{pmatrix} 2 \\ t \\ 0 \end{pmatrix} = \alpha \begin{pmatrix} -2 \\ 0 \\ -6 \end{pmatrix} + \beta \begin{pmatrix} 0 \\ -1 \\ 2 \end{pmatrix}$$

from which it follows that $t = 3$.

Let us now consider the subspace U generated by u_1 and u_2. Applying the function f, we have that $f(U)$ is generated by the vectors $f(u_1)$ and $f(u_2)$. Of course the fact that u_1 and u_2 are linearly independent does not imply that their images $f(u_1)$ and $f(u_2)$ are also linearly independent. In fact, we have $f(u_1) = (7, 5, 11)$ and $f(u_2) = (0, 0, 0)$. From this it follows that a basis of the subspace $f(U)$ is given by the vector $f(u_1) = (7, 5, 11)$.

We now come to the last point. For every linear function $g\colon \mathbb{R}^3 \to \mathbb{R}^4$ and for every vector $v \in \mathbb{R}^3$, we have

$$(f \circ g)(v) = f(g(v)) \in \mathrm{Im}(f).$$

Therefore we have $\mathrm{Im}(f \circ g) \subseteq \mathrm{Im}(f)$ and since $\dim \mathrm{Im}(f) = 2$, it follows that $\dim \mathrm{Im}(f \circ g) \leq 2$. This means that there cannot exist a function g such that $f \circ g = \mathrm{id}_{\mathbb{R}^3}$, because $\dim \mathrm{Im}(f \circ g) \leq 2$ while $\dim \mathrm{Im}(\mathrm{id}_{\mathbb{R}^3}) = 3$.

► **Exercise 2.6.12.** Let $f\colon \mathbb{R}^3 \to \mathbb{R}^4$ be the linear function defined by setting $f(1,0,0) = (2,1,0,1)$, $f(1,1,0) = (1,4,2,0)$, $f(1,1,1) = (2,4,5,2)$.

(a) Write the matrix A of f with respect to the canonical bases and find the bases of the kernel and of the image of f.

(b) Let $U \subset \mathbb{R}^3$ be the subspace of equation $2x_1 - x_2 + 3x_3 = 0$ and let $W \subset \mathbb{R}^4$ be the subspace defined by the two equations $y_1 + y_2 + 7y_4 = 0$ and $5y_1 + 2y_3 + 8y_4 = 0$. Prove that $f(U) = W$.

(c) Given the vector $v_t = (t, -1, 1, 5) \in \mathbb{R}^4$ find the value of t such that $f^{-1}(v_t) \neq \varnothing$.

Solution. Recall that the columns of A are the images of the vectors of the canonical basis of $\mathbb{R}^3$. $f(1,0,0) = (2,1,0,1)$ is assigned. Then we have

$$\begin{aligned} f(0,1,0) &= f(1,1,0) - f(1,0,0) \\ &= (1,4,2,0) - (2,1,0,1) \\ &= (-1,3,2,-1) \end{aligned}$$

and

$$\begin{aligned} f(0,0,1) &= f(1,1,1) - f(1,1,0) \\ &= (2,4,5,2) - (1,4,2,0) \\ &= (1,0,3,2) \end{aligned}$$

Therefore the matrix A is

$$A = \begin{pmatrix} 2 & -1 & 1 \\ 1 & 3 & 0 \\ 0 & 2 & 3 \\ 1 & -1 & 2 \end{pmatrix}$$

The kernel of f consists of the vectors (x_1, x_2, x_3) such that

$$\begin{pmatrix} 2 & -1 & 1 \\ 1 & 3 & 0 \\ 0 & 2 & 3 \\ 1 & -1 & 2 \end{pmatrix} \begin{pmatrix} x_1 \\ x_2 \\ x_3 \end{pmatrix} = \begin{pmatrix} 0 \\ 0 \\ 0 \\ 0 \end{pmatrix}$$

that is, of the solutions of the system

$$\begin{cases} 2x_1 - x_2 + x_3 = 0 \\ x_1 + 3x_2 = 0 \\ 2x_2 + 3x_3 = 0 \\ x_1 - x_2 + 2x_3 = 0 \end{cases}$$

The only solution of this system is

$$\begin{cases} x_1 = 0 \\ x_2 = 0 \\ x_3 = 0 \end{cases}$$

hence $\mathrm{Ker}(f) = \{\mathbf{0}\}$. From this we deduce that the image of f has dimension 3 and, since $\mathrm{Im}(f)$ is generated by the columns of A, the three columns of A are therefore linearly independent and are a basis of the image of f.

Let us now consider the subspace $U \subset \mathbb{R}^3$ of equation $2x_1 - x_2 + 3x_3 = 0$. We have

$$U : x_2 = 2x_1 + 3x_3,$$

from which it follows that U has dimension 2 and a basis is given by the vectors $u_1 = (1, 2, 0)$ and $u_2 = (0, 3, 1)$.

The subspace $W \subset \mathbb{R}^4$ is defined by the system

$$\begin{cases} y_1 + y_2 + 7y_4 = 0 \\ 5y_1 + 2y_3 + 8y_4 = 0 \end{cases}$$

from which we obtain

$$\begin{cases} y_2 = -y_1 - 7y_4 \\ y_3 = -\dfrac{5}{2}y_1 - 4y_4 \end{cases}$$

This means that W also has dimension 2 (it is not necessary to determine its basis).

Now let's compute the image of the basis vectors of U:

$$\begin{aligned} f(u_1) &= Au_1 = (0, 7, 4, -1) \\ f(u_2) &= Au_2 = (-2, 9, 9, -1). \end{aligned}$$

Substituting the coordinates of the vectors $f(u_1)$ and $f(u_2)$ in the equations of the subspace W, we verify that $f(u_1) \in W$ and $f(u_2) \in W$, from which it follows that $f(U) \subseteq W$. Now it suffices to observe that the vectors $f(u_1)$ and $f(u_2)$ are linearly independent, therefore $f(U)$ has dimension 2. Since W also has dimension 2, from the inclusion $f(U) \subseteq W$ we deduce that the equality $f(U) = W$ must necessarily hold.

Given the vector $v_t = (t, -1, 1, 5)$, we have $f^{-1}(v_t) \neq \varnothing$ if and only if $v_t \in \operatorname{Im}(f)$, that is, if and only if v_t is a linear combination of the columns of A. Therefore we obtain the system

$$\begin{cases} 2a_1 - a_2 + a_3 = t \\ a_1 + 3a_2 = -1 \\ 2a_2 + 3a_3 = 1 \\ a_1 - a_2 + 2a_3 = 5 \end{cases}$$

from which wc obtain, after suitable computations, $t = 6$.

▶ **Exercise 2.6.13.** Let $v_1 = (4, -2, 6)$, $v_2 = (0, 4, 4)$, $v_3 = (-1, 2, 0) \in \mathbb{R}^3$ and consider the linear function $f\colon \mathbb{R}^4 \to \mathbb{R}^3$ defined by $f(1, 0, 1, 0) = v_1$, $f(1, 0, -1, 0) = v_2$ and such that $(0, 1, 0, 0)$ and $(2, 0, -1, -3)$ belong to $f^{-1}(v_3)$.

(a) Write the matrix A of f with respect to the canonical bases and find bases of $\operatorname{Ker} f$ and $\operatorname{Im} f$.

(b) Let $W \subset \mathbb{R}^4$ be the subspace of equation $x_1 + 3x_3 + x_4 = 0$. Find a basis of W and a basis of $f(W)$. Find a basis of $\operatorname{Ker}(f) \cap W$.

(c) Write the matrix of the function induced by f, $f|_W \colon W \to \mathbb{R}^3$, with respect to the basis of W found in (b) and the canonical basis of the codomain.

Solution. Saying that the vectors $(0, 1, 0, 0)$ and $(2, 0, -1, -3)$ belong to $f^{-1}(v_3)$ means that $f(0, 1, 0, 0) = v_3$ and $f(2, 0, -1, -3) = v_3$.

Then we have

$$f(2, 0, 0, 0) = f(1, 0, 1, 0) + f(1, 0, -1, 0) = v_1 + v_2 = (4, 2, 10)$$

hence $f(1, 0, 0, 0) = (2, 1, 5)$.

Now we have

$$f(0, 0, 1, 0) = f(1, 0, 1, 0) - f(1, 0, 0, 0) = (4, -2, 6) - (2, 1, 5) = (2, -3, 1)$$

and finally

$$\begin{aligned} f(0, 0, 0, -3) &= f(2, 0, -1, -3) - 2f(1, 0, 0, 0) + f(0, 0, 1, 0) \\ &= (-1, 2, 0) - (4, 2, 10) + (2, -3, 1) \\ &= (-3, -3, -9) \end{aligned}$$

hence $f(0, 0, 0, 1) = (1, 1, 3)$.

Therefore the matrix of f is

$$A = \begin{pmatrix} 2 & -1 & 2 & 1 \\ 1 & 2 & -3 & 1 \\ 5 & 0 & 1 & 3 \end{pmatrix}$$

The kernel of f is given by the solutions of the following system

$$\begin{cases} 2x_1 - x_2 + 2x_3 + x_4 = 0 \\ x_1 + 2x_2 - 3x_3 + x_4 = 0 \\ 5x_1 + x_3 + 3x_4 = 0 \end{cases}$$

which is equivalent to

$$\begin{cases} x_2 = -8x_1 - 5x_4 \\ x_3 = -5x_1 - 3x_4 \end{cases}$$

The kernel of f therefore has dimension 2 and a basis consists of the vectors $(1, -8, -5, 0)$ and $(0, -5, -3, 1)$.

Since $\dim \text{Ker}(f) + \dim \text{Im}(f) = 4$, we get $\dim \text{Im}(f) = 2$ and therefore we can take two linearly independent columns of A (for example, the first two).

The subspace W has equation $x_1 + 3x_3 + x_4 = 0$, from which we obtain

$$x_1 = -3x_3 - x_4.$$

Therefore W has dimension 3 (not 2, because x_2 is also free) and a basis consists of the vectors

$$w_1 = (0, 1, 0, 0), \quad w_2 = (-3, 0, 1, 0), \quad w_3 = (-1, 0, 0, 1).$$

The image of W is generated by the images of vectors w_1, w_2 and w_3:

$$\begin{aligned} f(w_1) &= Aw_1 = (-1, 2, 0), \\ f(w_2) &= Aw_2 = (-4, -6, -14), \\ f(w_3) &= Aw_3 = (-1, 0, -2). \end{aligned}$$

It is easy to verify that the vectors $f(w_1)$, $f(w_2)$, $f(w_3)$ are linearly dependent, in fact we have $f(w_2) = -3f(w_1) + 7f(w_3)$, so they are not a basis of $f(W)$. For example, by eliminating $f(w_2)$ the remaining vectors $f(w_1)$ and $f(w_3)$ are linearly independent and are now a basis of $f(W)$.

We have already seen that $\text{Ker}(f)$ is given by the solutions of the system

$$\text{Ker}(f) : \begin{cases} x_2 = -8x_1 - 5x_4 \\ x_3 = -5x_1 - 3x_4 \end{cases}$$

By considering also the equation of W we get

$$\text{Ker}(f) \cap W : \begin{cases} x_2 = -8x_1 - 5x_4 \\ x_3 = -5x_1 - 3x_4 \\ x_1 + 3x_3 + x_4 = 0 \end{cases}$$

We obtain

$$\text{Ker}(f) \cap W : \begin{cases} x_2 = \dfrac{3}{4}x_1 \\ x_3 = \dfrac{1}{4}x_1 \\ x_4 = -\dfrac{7}{4}x_1 \end{cases}$$

Therefore we have $\dim(\text{Ker}(f) \cap W) = 1$ and a basis is given by the vector $(4, 3, 1, -7)$.

Let B be the matrix of the function induced by f, $f|_W\colon W \to \mathbb{R}^3$, with respect to the basis of W found previously and the canonical basis of the codomain. The columns of B are therefore formed by the coordinates of the images via f of the vectors w_1, w_2, w_3, with respect to the canonical basis of $\mathbb{R}^3$. Since we have already seen that

$$f(w_1) = \begin{pmatrix} -1 \\ 2 \\ 0 \end{pmatrix}, \quad f(w_2) = \begin{pmatrix} -4 \\ -6 \\ -14 \end{pmatrix}, \quad f(w_3) = \begin{pmatrix} -1 \\ 0 \\ -2 \end{pmatrix},$$

we have

$$B = \begin{pmatrix} -1 & -4 & -1 \\ 2 & -6 & 0 \\ 0 & -14 & -2 \end{pmatrix}.$$

► **Exercise 2.6.14.** (a) Find an endomorphism f of $\mathbb{R}^3$ such that $\operatorname{Ker}(f) = \langle(1,1,0)\rangle$ and $\operatorname{Im}(f) = \langle(0,1,-1),(2,1,2)\rangle$. Write its matrix with respect to the canonical bases.

(b) Is the function f unique? Why?

(c) For the function f found in (a) determine the inverse image of $(1,1,1)$ and of $(2,2,1)$.

Solution. To determine the matrix of a linear function $f\colon \mathbb{R}^3 \to \mathbb{R}^3$ with the required properties, we recall that the image of f is generated by the columns of its matrix A. Since we know that $\operatorname{Im}(f) = \langle(0,1,-1),(2,1,2)\rangle$, we can write a matrix A that has the vectors $(0,1,-1)$ and $(2,1,2)$ as two of its three columns; for example

$$A = \begin{pmatrix} a & 0 & 2 \\ b & 1 & 1 \\ c & -1 & 2 \end{pmatrix}.$$

Now we need to require that the vector $(1,1,0)$ belongs to the kernel of such a matrix, which means that we must have

$$\begin{pmatrix} a & 0 & 2 \\ b & 1 & 1 \\ c & -1 & 2 \end{pmatrix} \begin{pmatrix} 1 \\ 1 \\ 0 \end{pmatrix} = \begin{pmatrix} 0 \\ 0 \\ 0 \end{pmatrix}$$

We obtain the system

$$\begin{cases} a = 0 \\ b + 1 = 0 \\ c - 1 = 0 \end{cases}$$

from which it follows

$$\begin{cases} a = 0 \\ b = -1 \\ c = 1 \end{cases}$$

The matrix A is

$$A = \begin{pmatrix} 0 & 0 & 2 \\ -1 & 1 & 1 \\ 1 & -1 & 2 \end{pmatrix}$$

Note that the first column is the opposite of the second, so the vector subspace generated by the three columns of A coincides with the subspace generated by the second and third columns of A, which is precisely $\mathrm{Im}(f)$.

Of course such a function f is not unique given the arbitrary choices we made to construct a matrix A with the required properties. For example, if we had started from the matrix

$$A' = \begin{pmatrix} a & 2 & 0 \\ b & 1 & 1 \\ c & 2 & -1 \end{pmatrix}$$

by requiring that the vector $(1, 1, 0)$ belongs to the kernel of A' we would have found

$$\begin{cases} a = -2 \\ b = -1 \\ c = -2 \end{cases}$$

hence

$$A' = \begin{pmatrix} -2 & 2 & 0 \\ -1 & 1 & 1 \\ -2 & 2 & -1 \end{pmatrix}$$

which therefore corresponds to a different linear function with the same properties of f.

Now let's go back to the function f whose matrix is

$$A = \begin{pmatrix} 0 & 0 & 2 \\ -1 & 1 & 1 \\ 1 & -1 & 2 \end{pmatrix}$$

The inverse image of the vector $(1,1,1)$ is given by the solutions of the system

$$\begin{pmatrix} 0 & 0 & 2 \\ -1 & 1 & 1 \\ 1 & -1 & 2 \end{pmatrix} \begin{pmatrix} x \\ y \\ z \end{pmatrix} = \begin{pmatrix} 1 \\ 1 \\ 1 \end{pmatrix}$$

which is equivalent to

$$\begin{cases} 2z = 1 \\ -x + y + z = 1 \\ x - y + 2z = 1 \end{cases}$$

Since this system has no solutions, we have $f^{-1}(1,1,1) = \varnothing$.

Finally, the inverse image of the vector $(2,2,1)$ is given by the solutions of the system

$$\begin{pmatrix} 0 & 0 & 2 \\ -1 & 1 & 1 \\ 1 & -1 & 2 \end{pmatrix} \begin{pmatrix} x \\ y \\ z \end{pmatrix} = \begin{pmatrix} 2 \\ 2 \\ 1 \end{pmatrix}$$

which is equivalent to

$$\begin{cases} 2z = 2 \\ -x + y + z = 2 \\ x - y + 2z = 1 \end{cases}$$

Solving this system we find

$$\begin{cases} z = 1 \\ x = y - 1 \\ y \text{ any value.} \end{cases}$$

Setting $y = t$ we have

$$f^{-1}(2,2,1) = \{(t-1, t, 1) \mid \forall t \in \mathbb{R}\},$$

which can also be rewritten in the form

$$f^{-1}(2,2,1) = (-1,0,1) + \langle (1,1,0) \rangle.$$

► **Exercise 2.6.15.** Given a linear function $f \colon \mathbb{R}^2 \to \mathbb{R}^2$, the *graph* of f is the vector subspace $\Gamma_f \subset \mathbb{R}^4$ defined by setting

$$\Gamma_f = \{\, (x_1, x_2, y_1, y_2) \in \mathbb{R}^4 \mid (y_1, y_2) = f(x_1, x_2) \,\}.$$

Let $f, g \colon \mathbb{R}^2 \to \mathbb{R}^2$ be the linear functions whose matrices (with respect to the canonical bases) are, respectively

$$F = \begin{pmatrix} t-1 & 0 \\ -2 & -2t \end{pmatrix} \qquad G = \begin{pmatrix} 2 & 1 \\ 1 & -1 \end{pmatrix}$$

(a) Determine if g is injective and for which values of $t \in \mathbb{R}$ the function f is not injective.

(b) Write a basis of Γ_g and find $a, b \in \mathbb{R}$ such that $(a, b, 4, 5) \in \Gamma_g$.

(c) As t varies in $\mathbb{R}$ determine the dimension and a basis of $\Gamma_f \cap \Gamma_g$. Determine for which values of t we have $\Gamma_f \oplus \Gamma_g = \mathbb{R}^4$.

Solution. To determine if g is injective we compute the determinant of the matrix G: we have $\det G = -3 \neq 0$, hence g is injective.

We proceed in a similar way to study the injectivity of f:

$$\det F = \det \begin{pmatrix} t-1 & 0 \\ -2 & -2t \end{pmatrix} = -2t(t-1)$$

The function f is not injective when $\det F = 0$, i.e., for $t = 0$ or for $t = 1$.

Since $g(x_1, x_2) = (2x_1 + x_2, x_1 - x_2)$, as is easily verified using the matrix G, from the definition of the graph we deduce that

$$\begin{aligned} \Gamma_g &= \{ (x_1, x_2, y_1, y_2) \in \mathbb{R}^4 \mid (y_1, y_2) = g(x_1, x_2) \} \\ &= \{ (x_1, x_2, 2x_1 + x_2, x_1 - x_2) \in \mathbb{R}^4 \mid \forall x_1, x_2 \in \mathbb{R} \} \end{aligned}$$

Therefore Γ_g has dimension 2 and a basis consists of the vectors $(1, 0, 2, 1)$ and $(0, 1, 1, -1)$ (obtained by assigning to the free parameters x_1 and x_2 the values 1, 0 in the first case and 0, 1 in the second).

Requiring that the vector $(a, b, 4, 5)$ belongs to Γ_g is equivalent to requiring that $g(a, b) = (4, 5)$, so a and b must satisfy the following equations:

$$\begin{cases} 2a + b = 4 \\ a - b = 5 \end{cases}$$

The solution of this system is $a = 3$, $b = -2$.

We now look for the vectors that belong to $\Gamma_f \cap \Gamma_g$. Requiring that $(x_1, x_2, y_1, y_2) \in \Gamma_f \cap \Gamma_g$ is equivalent to requiring that $(y_1, y_2) = f(x_1, x_2) = g(x_1, x_2)$ and therefore $g(x_1, x_2) - f(x_1, x_2) = (g - f)(x_1, x_2) = (0, 0)$, so $(x_1, x_2) \in \operatorname{Ker}(g - f)$. The matrix of the linear function $g - f$ is

$$G - F = \begin{pmatrix} 3-t & 1 \\ 3 & 2t-1 \end{pmatrix}$$

hence $\det(G - F) = -2t^2 + 7t - 6$. We know that $\operatorname{Ker}(g - f) \neq \{\mathbf{0}\}$ if and only if $\det(G - F) = 0$, which occurs if and only if $-2t^2 + 7t - 6 = 0$, i.e., for $t = 3/2$

or for $t = 2$. We can therefore conclude that, if $t \neq 3/2, 2$, $\operatorname{Ker}(g - f) = \{\mathbf{0}\}$, then $\Gamma_f \cap \Gamma_g = \{\mathbf{0}\}$, hence $\Gamma_f \oplus \Gamma_g = \mathbb{R}^4$ (because $\dim \Gamma_f = \dim \Gamma_g = 2$).

The cases $t = 3/2$ and $t = 2$ remain to be considered. If $t = 3/2$ we have

$$G - F = \begin{pmatrix} 3/2 & 1 \\ 3 & 2 \end{pmatrix}$$

and the vectors (x_1, x_2) that belong to $\operatorname{Ker}(g - f)$ are given by the solutions of the system

$$\begin{cases} \frac{3}{2}x_1 + x_2 = 0 \\ 3x_1 + 2x_2 = 0 \end{cases}$$

This system has infinitely many solutions depending on a parameter, therefore $\dim \operatorname{Ker}(g - f) = 1$ and $\operatorname{Ker}(g - f)$ is generated by the vector $(x_1, x_2) = (2, -3)$. From this we deduce that $\dim(\Gamma_f \cap \Gamma_g) = 1$ and a basis of $\Gamma_f \cap \Gamma_g$ is given by the vector $(2, -3, 1, 5)$. Finally if $t = 2$ we have

$$G - F = \begin{pmatrix} 1 & 1 \\ 3 & 3 \end{pmatrix}$$

and $\operatorname{Ker}(g - f)$ is the set of solutions of the system

$$\begin{cases} x_1 + x_2 = 0 \\ 3x_1 + 3x_2 = 0 \end{cases}$$

This system has infinitely many solutions depending on a parameter, therefore $\dim \operatorname{Ker}(g - f) = 1$ and $\operatorname{Ker}(g - f)$ is generated by the vector $(x_1, x_2) = (1, -1)$. From this we deduce that $\dim(\Gamma_f \cap \Gamma_g) = 1$ and a basis of $\Gamma_f \cap \Gamma_g$ is given by the vector $(1, -1, 1, 2)$.

▶ **Exercise 2.6.16.** Let $v_1 = (1, 0, 2)$, $v_2 = (2, 1, -1)$, $v_3 = (0, -1, 5) \in \mathbb{R}^3$.

(a) Determine for which value of $t \in \mathbb{R}$ there exist linear functions $f \colon \mathbb{R}^3 \to \mathbb{R}$ such that $f(v_1) = 2$, $f(v_2) = 3$ and $f(v_3) = t$.

(b) For the value of t found in (a), write the matrices (with respect to the canonical bases) of all the linear functions f that satisfy the requirements indicated in point (a).

(c) Determine the intersection of the kernels of all linear functions associated with the matrices found in point (b).

Solution. The vectors v_1, v_2, v_3 are linearly dependent, in fact we have $v_3 = 2v_1 - v_2$. Since the function $f\colon \mathbb{R}^3 \to \mathbb{R}$ must be linear, it must be $f(v_3) = 2f(v_1) - f(v_2)$, i.e., $t = 4 - 3 = 1$. We therefore conclude that the only value of t for which there exist linear functions f with the required properties is $t = 1$ (it is easy to verify that, for $t = 1$, there are infinitely many such functions f).

A very simple method to find the matrices of all linear functions that satisfy the above requirements is the following. If we denote by $A = (a, b, c)$ the matrix of such a function, we must have

$$(a, b, c)\begin{pmatrix}1\\0\\2\end{pmatrix} = 2, \quad (a, b, c)\begin{pmatrix}2\\1\\-1\end{pmatrix} = 3, \quad (a, b, c)\begin{pmatrix}0\\-1\\5\end{pmatrix} = 1$$

The last request is, in fact, useless, since the vector v_3 is a linear combination of v_1 and v_2. The first two equations give the system

$$\begin{cases} a + 2c = 2 \\ 2a + b - c = 3 \end{cases}$$

which has infinitely many solutions, given by

$$\begin{cases} a = 2 - 2c \\ b = 5c - 1 \end{cases}$$

The matrices can then be written in the form

$$A = (2 - 2c, 5c - 1, c),$$

for any real number c.

Let $v = (x, y, z)$ be a vector belonging to the intersection of the kernels of all the matrices just found. We must have

$$(2 - 2c)x + (5c - 1)y + cz = 0, \qquad \forall c \in \mathbb{R}.$$

If we set $c = 0$, the previous equation becomes $2x - y = 0$, which results in $y = 2x$. If instead we set $c = 1$, the previous equation reduces to $4y + z = 0$, from which we obtain $z = -4y = -8x$. This means that the intersection of the kernels of all previous matrices is contained in the subspace

$$\{\, (x, 2x, -8x) \mid \forall x \in \mathbb{R} \,\}.$$

On the other hand, for any value of x, the vectors $(x, 2x, -8x)$ satisfy the equation $(2-2c)x + (5c-1)y + cz = 0$, for any c.

We therefore conclude that the intersection of the kernels of all the linear functions associated with the matrices found in point (b) coincides with the subspace $\{\, (x, 2x, -8x) \mid \forall x \in \mathbb{R} \,\}$, generated by the vector $(1, 2, -8)$.

▶ **Exercise 2.6.17.** Given $u_1 = (1,1,1,1)$ and $u_2 = (1,2,3,4) \in \mathbb{R}^4$, let $f_1, f_2 \colon \mathbb{R}^4 \to \mathbb{R}$ be the linear functions defined by setting $f_1(v) = v \cdot u_1$ and $f_2(v) = v \cdot u_2$, for each $v \in \mathbb{R}^4$.

(a) Determine if $\mathbb{R}^4$ is the sum of the kernels of f_1 and f_2 and find a basis of $\operatorname{Ker}(f_1) \cap \operatorname{Ker}(f_2)$.

(b) Find two subspaces $U_1 \subset \operatorname{Ker}(f_1)$ and $U_2 \subset \operatorname{Ker}(f_2)$ such that $U_1 \oplus U_2 = \mathbb{R}^4$.

(c) Find $f_1^{-1}(1) \cap f_2^{-1}(1)$.

(d) Determine if there exists a linear function $g \colon \mathbb{R} \to \mathbb{R}^4$ such that both the compositions $f_1 \circ g$ and $f_2 \circ g$ are the identity.

Solution. If we set $v = (x_1, x_2, x_3, x_4)$, the functions f_1 and f_2 are written as follows:

$$f_1(v) = x_1 + x_2 + x_3 + x_4, \quad f_2(v) = x_1 + 2x_2 + 3x_3 + 4x_4.$$

The kernel of f_1 is given by the solutions of the equation $x_1+x_2+x_3+x_4 = 0$, therefore it has dimension 3 and a basis is given by the vectors $(-1,1,0,0)$, $(-1,0,1,0)$ and $(-1,0,0,1)$. Similarly $\operatorname{Ker}(f_2)$ is given by the solutions of the equation $x_1+2x_2+3x_3+4x_4 = 0$, it has dimension 3 and a basis is given by the vectors $(-2,1,0,0)$, $(-3,0,1,0)$ and $(-4,0,0,1)$.

The intersection $\operatorname{Ker}(f_1) \cap \operatorname{Ker}(f_2)$ is given by the solutions of the linear system

$$\begin{cases} x_1 + x_2 + x_3 + x_4 = 0 \\ x_1 + 2x_2 + 3x_3 + 4x_4 = 0 \end{cases} \tag{2.6.1}$$

This system is equivalent to

$$\begin{cases} x_1 = x_3 + 2x_4 \\ x_2 = -2x_3 - 3x_4 \end{cases}$$

from which we deduce that $\operatorname{Ker}(f_1) \cap \operatorname{Ker}(f_2)$ has dimension 2 and a basis consists of the vectors $(1,-2,1,0)$ and $(2,-3,0,1)$.

From Grassmann's formula we get $\dim(\mathrm{Ker}(f_1) + \mathrm{Ker}(f_2)) = 4$, hence $\mathrm{Ker}(f_1) + \mathrm{Ker}(f_2) = \mathbb{R}^4$.

If we want $U_1 \oplus U_2 = \mathbb{R}^4$, we can take two subspaces U_1 and U_2 of dimension 2 (or one of dimension 3 and the other of dimension 1); of course we have to make sure that $U_1 \cap U_2 = \{\mathbf{0}\}$.

We can take, for example,

$$U_1 = \langle(-1,1,0,0),(-1,0,1,0)\rangle, \quad U_2 = \langle(-3,0,1,0),(-4,0,0,1)\rangle$$

It is easy to verify that the matrix

$$\begin{pmatrix} -1 & 1 & 0 & 0 \\ -1 & 0 & 1 & 0 \\ -3 & 0 & 1 & 0 \\ -4 & 0 & 0 & 1 \end{pmatrix}$$

has rank 4, hence $(-1,1,0,0)$, $(-1,0,1,0)$, $(-3,0,1,0)$, $(-4,0,0,1)$ are linearly independent vectors. From this we deduce that $U_1 + U_2$ has dimension 4 and, consequently, $\dim(U_1 \cap U_2) = 0$.

The set $f_1^{-1}(1) \cap f_2^{-1}(1)$ consists of the solutions of the system

$$\begin{cases} x_1 + x_2 + x_3 + x_4 = 1 \\ x_1 + 2x_2 + 3x_3 + 4x_4 = 1 \end{cases} \tag{2.6.2}$$

This is an inhomogeneous linear system, whose associated homogeneous system is (2.6.1), which we have already solved. From the theory of linear systems, we know that the solutions of (2.6.2) are obtained by adding a particular solution of (2.6.2) to the solutions of the associated homogeneous system. One particular solution of the system (2.6.2) is found very easily: we set $x_1 = 1$ and $x_2 = x_3 = x_4 = 0$. We can thus conclude that

$$f_1^{-1}(1) \cap f_2^{-1}(1) = (1,0,0,0) + \langle(1,-2,1,0),(2,-3,0,1)\rangle.$$

Any linear function $g\colon \mathbb{R} \to \mathbb{R}^4$ is given by

$$g(t) = t\,(\alpha,\beta,\gamma,\delta) = (\alpha t, \beta t, \gamma t, \delta t),$$

for some vector $(\alpha,\beta,\gamma,\delta) \in \mathbb{R}^4$. Therefore we have:

$$\begin{aligned} (f_1 \circ g)(t) &= f_1(\alpha t, \beta t, \gamma t, \delta t) = (\alpha + \beta + \gamma + \delta)\,t \\ (f_2 \circ g)(t) &= f_2(\alpha t, \beta t, \gamma t, \delta t) = (\alpha + 2\beta + 3\gamma + 4\delta)\,t. \end{aligned}$$

To make $f_1 \circ g$ and $f_2 \circ g$ be the identity, we have to choose $\alpha, \beta, \gamma, \delta$ so that we have $(\alpha + \beta + \gamma + \delta)\, t = t$ and $(\alpha + 2\beta + 3\gamma + 4\delta)\, t = t$, for each t. Note that it is enough to set $\alpha = 1$, $\beta = \gamma = \delta = 0$. Therefore the linear function $g\colon \mathbb{R} \to \mathbb{R}^4$, given by $g(t) = (t, 0, 0, 0)$, satisfies the requirements.

► **Exercise 2.6.18.** Let $V = M_2(\mathbb{R})$ be the vector space of square matrices of order 2 with real entries. Let $U \subset V$ be the subspace of the matrices A such that the vector $(1, -2)$ belongs to the kernel of A. Let $W \subset V$ be the subspace formed by the matrices B such that the image of B is contained in the line with equation $y = 2x$.

(a) Find the dimension and a basis of U and of W.

(b) Find the dimension and a basis of $U + W$ and of $U \cap W$.

(c) Let $C = \left(\begin{smallmatrix} 4 & 2 \\ t & -1 \end{smallmatrix}\right)$. Find for which value of t the set $\{\, C + A \mid \text{for any } A \in U \,\}$ is a vector subspace of V.

(d) Find (if possible) a basis of a vector subpace $L \subset V$ such that $U \oplus L = W \oplus L = U + W$.

Solution. If we write $A = \left(\begin{smallmatrix} a & b \\ c & d \end{smallmatrix}\right)$, requiring that $A \in U$ is equivalent to require that

$$\begin{pmatrix} a & b \\ c & d \end{pmatrix} \begin{pmatrix} 1 \\ -2 \end{pmatrix} = \begin{pmatrix} 0 \\ 0 \end{pmatrix}$$

from which we obtain $a = 2b$ and $c = 2d$. U is therefore the vector subspace formed by the matrices $\left(\begin{smallmatrix} 2b & b \\ 2d & d \end{smallmatrix}\right)$, for $b, d \in \mathbb{R}$. From this we deduce that U has dimension 2 and a basis is given by the matrices $\left(\begin{smallmatrix} 2 & 1 \\ 0 & 0 \end{smallmatrix}\right)$ and $\left(\begin{smallmatrix} 0 & 0 \\ 2 & 1 \end{smallmatrix}\right)$.

Recall that if $B = \left(\begin{smallmatrix} a' & b' \\ c' & d' \end{smallmatrix}\right)$, the image of B is the vector subspace generated by the columns of B, i.e., in this case, by the vectors (a', c') and (b', d'). Requiring that the image of B is contained in the line of equation $y = 2x$, means that $c' = 2a'$ and $d' = 2b'$. W is therefore the vector subspace formed by the matrices $\left(\begin{smallmatrix} a' & b' \\ 2a' & 2b' \end{smallmatrix}\right)$, for $a', b' \in \mathbb{R}$. From this we deduce that W has dimension 2 and a basis is given by the matrices $\left(\begin{smallmatrix} 1 & 0 \\ 2 & 0 \end{smallmatrix}\right)$ and $\left(\begin{smallmatrix} 0 & 1 \\ 0 & 2 \end{smallmatrix}\right)$.

A matrix belonging to $U \cap W$ can be written both in the form $\left(\begin{smallmatrix} 2b & b \\ 2d & d \end{smallmatrix}\right)$ and in the form $\left(\begin{smallmatrix} a' & b' \\ 2a' & 2b' \end{smallmatrix}\right)$. Therefore we obtain the matrix equation

$$\begin{pmatrix} 2b & b \\ 2d & d \end{pmatrix} = \begin{pmatrix} a' & b' \\ 2a' & 2b' \end{pmatrix}$$

whose solutions are

$$\begin{cases} a' = 2b \\ b' = b \\ d = 2b \end{cases}$$

for $b \in \mathbb{R}$. This means that $U \cap W$ has dimension 1 and a basis is given by the matrix $\left(\begin{smallmatrix} 2 & 1 \\ 4 & 2 \end{smallmatrix}\right)$. From Grassmann's formula it follows that $U + W$ has dimension 3. As a basis of $U + W$ we can take three of the following four matrices: $\left(\begin{smallmatrix} 2 & 1 \\ 0 & 0 \end{smallmatrix}\right)$, $\left(\begin{smallmatrix} 0 & 0 \\ 2 & 1 \end{smallmatrix}\right)$, $\left(\begin{smallmatrix} 1 & 0 \\ 2 & 0 \end{smallmatrix}\right)$, $\left(\begin{smallmatrix} 0 & 1 \\ 0 & 2 \end{smallmatrix}\right)$. It is easy to verify that any three of these four matrices are linearly independent.

For the set $\{ C + A \mid \text{for each } A \in U \}$ to be a vector subspace of V it is necessary that it contains the zero matrix. There must therefore exist a matrix $A \in U$ such that $C + A = \mathbf{0}$. But this means that $C = -A$ and therefore $C \in U$ (since U is a vector subspace). Hence we must have

$$\begin{pmatrix} 4 & 2 \\ t & -1 \end{pmatrix} = \begin{pmatrix} 2b & b \\ 2d & d \end{pmatrix}$$

for some $b, d \in \mathbb{R}$: this is possible if and only if $t = -2$. Therefore, for $t = -2$ we have $C \in U$, but in this case the set $\{ C + A \mid \text{for each } A \in U \}$ coincides with U and therefore is a vector subspace.

Finally, to find a vector subspace L such that $U \oplus L = W \oplus L = U + W$, we observe that L must necessarily have dimension 1 (because $\dim U = 2$ and $\dim(U + W) = 3$). As a basis of L we can take, for example, the matrix $\left(\begin{smallmatrix} 2 & 1 \\ 0 & 0 \end{smallmatrix}\right) + \left(\begin{smallmatrix} 1 & 0 \\ 2 & 0 \end{smallmatrix}\right) = \left(\begin{smallmatrix} 3 & 1 \\ 2 & 0 \end{smallmatrix}\right)$. It is easy to verify that this matrix satisfies the requests.

► **Exercise 2.6.19.** Let $f\colon \mathbb{R}^3 \to \mathbb{R}^2$ be the linear function given by the matrix (with respect to the canonical bases)

$$A = \begin{pmatrix} 1 & 2 & -1 \\ 2 & t & -2 \end{pmatrix}$$

(a) Find for which value of t the image of f has dimension 1. For this value of t find a basis of $\operatorname{Ker}(f)$ and a basis of $\operatorname{Im}(f)$.

(b) For the value of t found in (a) determine a basis $\{v_1, v_2\}$ of $\mathbb{R}^2$ such that the matrix of f with respect to the canonical basis of $\mathbb{R}^3$ and the basis $\{v_1, v_2\}$ of $\mathbb{R}^2$ has the second row equal to zero.

(c) Find for which values of t there exists a linear function $g\colon \mathbb{R}^2 \to \mathbb{R}^3$ such that $f \circ g\colon \mathbb{R}^2 \to \mathbb{R}^2$ is the identity.

(d) Now we set $t = 0$. Let $h\colon \mathbb{R}^2 \to \mathbb{R}^3$ be the linear function defined by setting

$$h(v) \cdot w = v \cdot f(w), \qquad \forall v \in \mathbb{R}^2, w \in \mathbb{R}^3.$$

Write the matrix of h with respect to the canonical bases.

Solution. The dimension of the image of f is the rank of the matrix A. This rank is equal to 1 if and only if $t = 4$ (in which case the second row of A is twice the first). For $t = 4$ the matrix A is

$$A = \begin{pmatrix} 1 & 2 & -1 \\ 2 & 4 & -2 \end{pmatrix}$$

The image of f is generated by the columns of A and therefore a basis of $\mathrm{Im}(f)$ is given by the vector $(1, 2)$. The kernel of f is given by the solutions of the equation $x_1 + 2x_2 - x_3 = 0$, from which we obtain $x_3 = x_1 + 2x_2$. $\mathrm{Ker}(f)$ therefore has dimension 2 and a basis is given by the vectors $(1, 0, 1)$ and $(0, 1, 2)$.

Let $\{v_1, v_2\}$ be another basis of $\mathbb{R}^2$. We can therefore write the vectors $f(e_1)$, $f(e_2)$, $f(e_3)$ as a linear combination of v_1 and v_2:

$$\begin{aligned} f(e_1) &= (1, 2) = b_{11}v_1 + b_{21}v_2 \\ f(e_2) &= (2, 4) = b_{12}v_1 + b_{22}v_2 \\ f(e_3) &= (-1, -2) = b_{13}v_1 + b_{23}v_2. \end{aligned}$$

The matrix of f with respect to the canonical basis of the domain and the basis $\{v_1, v_2\}$ of the codomain is

$$B = \begin{pmatrix} b_{11} & b_{12} & b_{13} \\ b_{21} & b_{22} & b_{23} \end{pmatrix}$$

If we want the matrix B to have the second row equal to zero, we must choose the vectors v_1 and v_2 so that we have

$$\begin{aligned} f(e_1) &= (1, 2) = b_{11}v_1 \\ f(e_2) &= (2, 4) = b_{12}v_1 \\ f(e_3) &= (-1, -2) = b_{13}v_1. \end{aligned}$$

A possible choice of v_1 is $v_1 = (1, 2)$, in which case we have $b_{11} = 1$, $b_{12} = 2$ and $b_{13} = -1$. For the choice of v_2 there is no restriction, except that it must be linearly independent of v_1: we can therefore take, for example, $v_2 = (0, 1)$. With this choice of v_1 and v_2 the matrix B becomes

$$B = \begin{pmatrix} 1 & 2 & -1 \\ 0 & 0 & 0 \end{pmatrix}$$

A linear function $g\colon \mathbb{R}^2 \to \mathbb{R}^3$ such that $f \circ g\colon \mathbb{R}^2 \to \mathbb{R}^2$ is the identity exists if and only if f is surjective, i.e., if and only if the matrix A has rank 2, which occurs for $t \neq 4$.

For $t = 0$, the matrix of f becomes

$$A = \begin{pmatrix} 1 & 2 & -1 \\ 2 & 0 & -2 \end{pmatrix}$$

To determine the matrix H of the function $h\colon \mathbb{R}^2 \to \mathbb{R}^3$ with respect to the canonical bases, we need to compute $h(1,0)$ and $h(0,1)$. If we set $h(1,0) = (a_1, a_2, a_3)$, applying the equality $h(v) \cdot w = v \cdot f(w)$ with $v = (1,0)$ and $w = (1,0,0)$, we find $a_1 = 1$. Repeating the same computation with $v = (1,0)$ and $w = (0,1,0)$ we find $a_2 = 2$, while with $v = (1,0)$ and $w = (0,0,1)$ we find $a_3 = -1$. Similarly, if we set $h(0,1) = (b_1, b_2, b_3)$, applying the equality $h(v) \cdot w = v \cdot f(w)$ with $v = (0,1)$ and $w = (1,0,0)$ we get $b_1 = 2$, with $v = (0,1)$ and $w = (0,1,0)$ we get $b_2 = 0$, while with $v = (0,1)$ and $w = (0,0,1)$ we get $b_3 = -2$. It follows that the matrix is

$$H = \begin{pmatrix} 1 & 2 \\ 2 & 0 \\ -1 & -2 \end{pmatrix}$$

which is precisely the transpose of A.

► **Exercise 2.6.20.** In the vector space of the linear functions $f\colon \mathbb{R}^4 \to \mathbb{R}$ let V be the subspace generated by the functions f_1, f_2, f_3, where:

$$\begin{aligned} f_1(x_1, \ldots, x_4) &= 2x_1 - x_2 + x_4, \\ f_2(x_1, \ldots, x_4) &= x_1 + 2x_2 - x_3 + 2x_4, \\ f_3(x_1, \ldots, x_4) &= 3x_1 - 4x_2 + x_3. \end{aligned}$$

(a) Find the dimension and a basis of V.

(b) Determine a linear function $f \in V$ such that $f(1,0,0,0) = 5$ and $f(0,0,0,1) = 4$.

(c) Find the intersection of the kernels of all $f \in V$.

(d) Let $W = \{ f\colon \mathbb{R}^4 \to \mathbb{R} \mid f \text{ is linear and } f(1,1,-2,-2) = 0 \}$. Find a basis of $V \cap W$.

Solution. We denote by A_1, A_2, A_3 the matrices of f_1, f_2, f_3 with respect to the canonical bases (these matrices are actually row vectors). We have:

$$A_1 = (2,-1,0,1), \qquad A_2 = (1,2,-1,2), \qquad A_3 = (3,-4,1,0).$$

These three vectors are linearly dependent, since $A_3 = 2A_1 - A_2$ (and thus also $f_3 = 2f_1 - f_2$), while A_1 and A_2 are linearly independent. From this we deduce that V has dimension 2 and a basis is given by f_1 and f_2.

Since V is generated by f_1 and f_2, each $f \in V$ can be written in the form $f = a_1 f_1 + a_2 f_2$ and therefore the matrix A of such f is

$$A = a_1 A_1 + a_2 A_2 = (2a_1 + a_2, -a_1 + 2a_2, -a_2, a_1 + 2a_2).$$

Explicitly, the function f is given by

$$f(x_1, \ldots, x_4) = (2a_1 + a_2)x_1 + (-a_1 + 2a_2)x_2 - a_2 x_3 + (a_1 + 2a_2)x_4.$$

The equalities $f(1,0,0,0) = 5$ and $f(0,0,0,1) = 4$ translate into the equations

$$\begin{cases} 2a_1 + a_2 = 5 \\ a_1 + 2a_2 = 4 \end{cases}$$

from which we obtain $a_1 = 2$ and $a_2 = 1$. The function f is therefore

$$f(x_1, \ldots, x_4) = 5x_1 - x_3 + 4x_4.$$

The kernel of f_1 is the subspace of $\mathbb{R}^4$ of equation $2x_1 - x_2 + x_4 = 0$, while the kernel of f_2 is the subspace of equation $x_1 + 2x_2 - x_3 + 2x_4 = 0$. The intersection $\operatorname{Ker}(f_1) \cap \operatorname{Ker}(f_2)$ is given by the solutions of the system of linear equations

$$\begin{cases} 2x_1 - x_2 + x_4 = 0 \\ x_1 + 2x_2 - x_3 + 2x_4 = 0 \end{cases}$$

and therefore it is the subspace of $\mathbb{R}^4$ generated by the vectors $u_1 = (1,2,5,0)$ and $u_2 = (0,1,4,1)$. Recalling that every $f \in V$ can be written in the form $f = a_1 f_1 + a_2 f_2$, we have $f(u_1) = 0$ and $f(u_2) = 0$. This means that the vectors u_1 and u_2 belong to the kernel of each function $f \in V$ and, therefore, the intersection of the kernels of all $f \in V$ is the subspace of $\mathbb{R}^4$ generated by $u_1 = (1, 2, 5, 0)$ and $u_2 = (0, 1, 4, 1)$.

The elements of $V \cap W$ are the linear functions f which can be written in the form $f = a_1 f_1 + a_2 f_2$ and which satisfy the condition $f(1, 1, -2, -2) = 0$. Recalling that a function f of the form $f = a_1 f_1 + a_2 f_2$ is given by

$$f(x_1, \dots, x_4) = (2a_1 + a_2)x_1 + (-a_1 + 2a_2)x_2 - a_2 x_3 + (a_1 + 2a_2)x_4,$$

the request $f(1, 1, -2, -2) = 0$ translates into the equation $a_2 - a_1 = 0$, i.e., $a_1 = a_2$. We thus conclude that the subspace $V \cap W$ has dimension 1 and a basis is given by the function f which is obtained by setting, for example, $a_1 = a_2 = 1$, i.e., $f(x_1, \ dots, x_4) = 3x_1 + x_2 - x_3 + 3x_4$.

3

The Determinant

3.1 The determinant: Gaussian elimination

► **Exercise 3.1.1.** Using the Gaussian elimination method, transform the following matrix into an upper (or lower) triangular matrix.

$$A = \begin{pmatrix} 2 & -4 & 6 & -2 \\ 1 & -1 & b+3 & -2 \\ -3 & 3a+6 & 3a-6 & 3 \\ 1 & a-b-1 & a-b+4 & 0 \end{pmatrix}$$

Using the result thus obtained, compute the determinant of A and find for which values of the real parameters a and b the matrix A has rank 3.

Solution. We recall that the elementary operations on the rows, or on the columns, do not alter the rank of a matrix, however some of these operations modify the determinant. Since this problem also requires the computation of the determinant of the matrix A, it will be necessary to take into account any changes to the determinant caused by the elementary operations (on the rows or on the columns) that we are going to perform.

We start by observing that all the elements of the first row of A are divisible by 2, while those of the third row are divisible by 3. If we "collect" the 2 and the 3, we get

$$\det A = 2 \cdot 3 \cdot \begin{vmatrix} 1 & -2 & 3 & -1 \\ 1 & -1 & b+3 & -2 \\ -1 & a+2 & a-2 & 1 \\ 1 & a-b-1 & a-b+4 & 0 \end{vmatrix}$$

Now we carry out the following elementary operations: we subtract the first from the second row, we add the first to the third row and we subtract the first from the fourth row. Recall that these operations do not change the determinant. We therefore have:

$$\det A = 6 \cdot \begin{vmatrix} 1 & -2 & 3 & -1 \\ 0 & 1 & b & -1 \\ 0 & a & a+1 & 0 \\ 0 & a-b+1 & a-b+1 & 1 \end{vmatrix}$$

Now let's swap the second column with the fourth; we observe that in doing so the determinant changes sign:

$$\det A = -6 \cdot \begin{vmatrix} 1 & -1 & 3 & -2 \\ 0 & -1 & b & 1 \\ 0 & 0 & a+1 & a \\ 0 & 1 & a-b+1 & a-b+1 \end{vmatrix}$$

The next step is to add the second row to the fourth (the determinant does not change):

$$\det A = -6 \cdot \begin{vmatrix} 1 & -1 & 3 & -2 \\ 0 & -1 & b & 1 \\ 0 & 0 & a+1 & a \\ 0 & 0 & a+1 & a-b+2 \end{vmatrix}$$

Finally, we subtract the third from the fourth row (the determinant does not change):

$$\det A = -6 \cdot \begin{vmatrix} 1 & -1 & 3 & -2 \\ 0 & -1 & b & 1 \\ 0 & 0 & a+1 & a \\ 0 & 0 & 0 & -b+2 \end{vmatrix}$$

We have thus transformed the matrix A into an upper triangular matrix, as required. The computation of the determinant is now immediate:

$$\det A = 6(a+1)(2-b).$$

The determinant of A therefore vanishes for $a = -1$, or for $b = 2$. We analyze the two cases separately.

If $a = -1$, the previous matrix becomes

$$\begin{pmatrix} 1 & -1 & 3 & -2 \\ 0 & -1 & b & 1 \\ 0 & 0 & 0 & -1 \\ 0 & 0 & 0 & -b+2 \end{pmatrix}$$

It has rank 3 for any value of b.

Instead if $b = 2$, we get the matrix

$$\begin{pmatrix} 1 & -1 & 3 & -2 \\ 0 & -1 & 2 & 1 \\ 0 & 0 & a+1 & a \\ 0 & 0 & 0 & 0 \end{pmatrix}$$

Since the elements $a + 1$ and a cannot both vanish, this matrix has rank 3 for each value of a.

Therefore we conclude that the matrix A has rank 3 if $a = -1$ or if $b = 2$ (in all other cases we have $\det A \neq 0$, therefore A has rank 4).

► **Exercise 3.1.2.** Using the Gaussian elimination method, transform the following matrix into an upper (or lower) triangular matrix.

$$A = \begin{pmatrix} 2 & 6-2a & 2a+4 & -4 \\ -1 & 2a+b-4 & -2a-4 & 2 \\ 0 & 4-a-2b & a-4 & 2 \\ 1 & a+6 & -a-11 & 1 \end{pmatrix}$$

Using the result thus obtained, compute the determinant of A and find for which values of the real parameters a and b the matrix A has rank 3.

Solution. We recall that the elementary operations on the rows, or on the columns, do not alter the rank of a matrix, however some of these operations modify the determinant. Since this problem also requires the computation of the determinant of the matrix A, it will be necessary to take into account any changes to the determinant caused by the elementary operations (on the rows or on the columns) that we are going to perform.

We start by noting that all the elements of the first row of A are divisible by 2, so we have

$$\det A = 2 \begin{vmatrix} 1 & 3-a & a+2 & -2 \\ -1 & 2a+b-4 & -2a-4 & 2 \\ 0 & 4-a-2b & a-4 & 2 \\ 1 & a+6 & -a-11 & 1 \end{vmatrix}$$

Now we carry out the following elementary operations: we add the first to the second row and we subtract the first from the fourth row (remember that these operations do not modify the determinant). We therefore have:

$$\det A = 2\begin{vmatrix} 1 & 3-a & a+2 & -2 \\ 0 & a+b-1 & -a-2 & 0 \\ 0 & 4-a-2b & a-4 & 2 \\ 0 & 2a+3 & -2a-13 & 3 \end{vmatrix}$$

Now let's swap the second column with the fourth; we observe that in doing so the determinant changes sign:

$$\det A = -2\begin{vmatrix} 1 & -2 & a+2 & 3-a \\ 0 & 0 & -a-2 & a+b-1 \\ 0 & 2 & a-4 & 4-a-2b \\ 0 & 3 & -2a-13 & 2a+3 \end{vmatrix}$$

To simplify the third column we decide to add the fourth column to it (the determinant does not change):

$$\det A = -2\begin{vmatrix} 1 & -2 & 5 & 3-a \\ 0 & 0 & b-3 & a+b-1 \\ 0 & 2 & -2b & 4-a-2b \\ 0 & 3 & -10 & 2a+3 \end{vmatrix}$$

We now swap the second row with the third row (so the determinant changes sign) and then "collect" the 2 from what is now the second row:

$$\det A = 4\begin{vmatrix} 1 & -2 & 5 & 3-a \\ 0 & 1 & -b & 2-\frac{a}{2}-b \\ 0 & 0 & b-3 & a+b-1 \\ 0 & 3 & -10 & 2a+3 \end{vmatrix}$$

The next step is to add the second row multiplied by -3 to the fourth row (the determinant does not change):

$$\det A = 4\begin{vmatrix} 1 & -2 & 5 & 3-a \\ 0 & 1 & -b & 2-\frac{a}{2}-b \\ 0 & 0 & b-3 & a+b-1 \\ 0 & 0 & 3b-10 & \frac{7}{2}a+3b-3 \end{vmatrix}$$

Now let's try to get a pivot equal to 1 in the third row. To do this we multiply the third row by 3 (this modifies the determinant):

$$\det A = \frac{4}{3}\begin{vmatrix} 1 & -2 & 5 & 3-a \\ 0 & 1 & -b & 2-\frac{a}{2}-b \\ 0 & 0 & 3b-9 & 3a+3b-3 \\ 0 & 0 & 3b-10 & \frac{7}{2}a+3b-3 \end{vmatrix}$$

and now we subtract the fourth row from the third (the determinant does not change):

$$\det A = \frac{4}{3}\begin{vmatrix} 1 & -2 & 5 & 3-a \\ 0 & 1 & -b & 2-\frac{a}{2}-b \\ 0 & 0 & 1 & -\frac{1}{2}\,a \\ 0 & 0 & 3b-10 & \frac{7}{2}\,a+3b-3 \end{vmatrix}$$

Finally, we subtract the third row multiplied by $3b-10$ from the fourth row (the determinant does not change):

$$\det A = \frac{4}{3}\begin{vmatrix} 1 & -2 & 5 & 3-a \\ 0 & 1 & -b & 2-\frac{a}{2}-b \\ 0 & 0 & 1 & -\frac{1}{2}\,a \\ 0 & 0 & 0 & \frac{3}{2}\,(a+2)(b-1) \end{vmatrix}$$

We have thus transformed the matrix A into an upper triangular matrix, as required. The computation of the determinant is now immediate:

$$\det A = \frac{4}{3}\cdot\frac{3}{2}\,(a+2)(b-1) = 2(a+2)(b-1).$$

The determinant of A therefore vanishes for $a=-2$, or for $b=1$. Since the pivots of the first three rows are non-zero (they are all equal to 1), we conclude that, if $a=-2$ or $b=1$, the rank of the matrix A is 3, otherwise this rank is equal to 4.

3.2 Laplace formula

► **Exercise 3.2.1.** Compute

$$\det\begin{pmatrix} 2 & 0 & 3 & -1 \\ 4 & 1 & 2 & 0 \\ 0 & 3 & 0 & 1 \\ 2 & 0 & 2 & -1 \end{pmatrix}$$

Solution. Using Laplace rule along the third row, we have:

$$\det\begin{pmatrix}2&0&3&-1\\4&1&2&0\\0&3&0&1\\2&0&2&-1\end{pmatrix}=-3\det\begin{pmatrix}2&3&-1\\4&2&0\\2&2&-1\end{pmatrix}-\det\begin{pmatrix}2&0&3\\4&1&2\\2&0&2\end{pmatrix}$$
$$=-3\left[-\det\begin{pmatrix}4&2\\2&2\end{pmatrix}-\det\begin{pmatrix}2&3\\4&2\end{pmatrix}\right]-\det\begin{pmatrix}2&3\\2&2\end{pmatrix}$$
$$=-3(-4+8)+2=-10.$$

▶ **Exercise 3.2.2.** Compute

$$\det\begin{pmatrix}2&3&4&-1&2\\2&3&4&-1&1\\2&3&5&1&1\\2&1&1&1&1\\1&2&-1&2&-1\end{pmatrix}$$

Solution. In this case it is not advisable to apply Laplace rule immediately, given the amount of calculations required. Recalling instead the properties of the determinant we can try to reduce the matrix to a simpler form: this means, in practice, trying to make several zeros appear among the elements of the matrix. This can be done, for example, by subtracting the second row from the first, and remembering that this operation does not change the determinant. We therefore have:

$$\det\begin{pmatrix}2&3&4&-1&2\\2&3&4&-1&1\\2&3&5&1&1\\2&1&1&1&1\\1&2&-1&2&-1\end{pmatrix}=\det\begin{pmatrix}0&0&0&0&1\\2&3&4&-1&1\\2&3&5&1&1\\2&1&1&1&1\\1&2&-1&2&-1\end{pmatrix}$$

Now we can use Laplace formula along the first row:

$$\det\begin{pmatrix}2&3&4&-1&2\\2&3&4&-1&1\\2&3&5&1&1\\2&1&1&1&1\\1&2&-1&2&-1\end{pmatrix}=\det\begin{pmatrix}2&3&4&-1\\2&3&5&1\\2&1&1&1\\1&2&-1&2\end{pmatrix}$$

Once again it is convenient to transform this matrix, for example by subtracting the second row from the first. Therefore, we get:

$$\det\begin{pmatrix}2 & 3 & 4 & -1\\ 2 & 3 & 5 & 1\\ 2 & 1 & 1 & 1\\ 1 & 2 & -1 & 2\end{pmatrix} = \det\begin{pmatrix}0 & 0 & -1 & -2\\ 2 & 3 & 5 & 1\\ 2 & 1 & 1 & 1\\ 1 & 2 & -1 & 2\end{pmatrix}$$

Now we can use Laplace formula along the first row:

$$\det\begin{pmatrix}2 & 3 & 4 & -1\\ 2 & 3 & 5 & 1\\ 2 & 1 & 1 & 1\\ 1 & 2 & -1 & 2\end{pmatrix} = -\det\begin{pmatrix}2 & 3 & 1\\ 2 & 1 & 1\\ 1 & 2 & 2\end{pmatrix} + 2\det\begin{pmatrix}2 & 3 & 5\\ 2 & 1 & 1\\ 1 & 2 & -1\end{pmatrix}$$

Once again we can make zeros appear in these last two matrices by subtracting, for example, the second row from the first. Therefore, we get:

$$\begin{aligned}\det\begin{pmatrix}2 & 3 & 4 & -1\\ 2 & 3 & 5 & 1\\ 2 & 1 & 1 & 1\\ 1 & 2 & -1 & 2\end{pmatrix} &= -\det\begin{pmatrix}0 & 2 & 0\\ 2 & 1 & 1\\ 1 & 2 & 2\end{pmatrix} + 2\det\begin{pmatrix}0 & 2 & 4\\ 2 & 1 & 1\\ 1 & 2 & -1\end{pmatrix}\\ &= 2\det\begin{pmatrix}2 & 1\\ 1 & 2\end{pmatrix} + 2\left[-2\det\begin{pmatrix}2 & 1\\ 1 & -1\end{pmatrix}\right.\\ &\qquad \left. +4\det\begin{pmatrix}2 & 1\\ 1 & 2\end{pmatrix}\right]\\ &= 6 + 2(6+12) = 42.\end{aligned}$$

3.3 The inverse of a matrix

► **Exercise 3.3.1.** Compute the inverse of the matrix

$$A = \begin{pmatrix}2 & -3 & -1\\ 2 & -1 & -3\\ 1 & -3 & -1\end{pmatrix}$$

Solution. We recall that, given a square matrix $A = (a_{ij})$, the *algebraic complement* of the element a_{ij} is the element A_{ij} which is equal to the product of $(-1)^{i+j}$ times the determinant of the matrix obtained from A by deleting its i-th row and j-th column. The inverse matrix of A is equal to the transpose of the matrix of algebraic complements of the elements of A, divided by the determinant of A.

We compute the determinant of A using Laplace formula along the first row:

$$\det A = 2\begin{vmatrix}-1 & -3\\ -3 & -1\end{vmatrix} + 3\begin{vmatrix}2 & -3\\ 1 & -1\end{vmatrix} - \begin{vmatrix}2 & -1\\ 1 & -3\end{vmatrix}$$
$$= 2(1-9) + 3(-2+3) - (-6+1) = -8.$$

We now write the transpose of the matrix of algebraic complements:

$$\begin{pmatrix} +\begin{vmatrix}-1 & -3\\ -3 & -1\end{vmatrix} & -\begin{vmatrix}-3 & -1\\ -3 & -1\end{vmatrix} & +\begin{vmatrix}-3 & -1\\ -1 & -3\end{vmatrix} \\ -\begin{vmatrix}2 & -3\\ 1 & -1\end{vmatrix} & +\begin{vmatrix}2 & -1\\ 1 & -1\end{vmatrix} & -\begin{vmatrix}2 & -1\\ 2 & -3\end{vmatrix} \\ +\begin{vmatrix}2 & -1\\ 1 & -3\end{vmatrix} & -\begin{vmatrix}2 & -3\\ 1 & -3\end{vmatrix} & +\begin{vmatrix}2 & -3\\ 2 & -1\end{vmatrix} \end{pmatrix} = \begin{pmatrix}-8 & 0 & 8\\ -1 & -1 & 4\\ -5 & 3 & 4\end{pmatrix}$$

The inverse matrix of A is obtained by dividing the matrix just found by the determinant of A. So it is

$$A^{-1} = \begin{pmatrix}1 & 0 & -1\\ 1/8 & 1/8 & -1/2\\ 5/8 & -3/8 & -1/2\end{pmatrix}$$

▶ **Exercise 3.3.2.** Compute the inverse of the matrix

$$A = \begin{pmatrix}1 & 2 & 1\\ 1 & 3 & 0\\ 1 & 2 & 2\end{pmatrix}$$

Solution. In the previous exercise we saw how the inverse of a matrix can be calculated. However, the method used is not very effective, especially for large matrices, because it requires the computation of a large number of determinants.

Another method, much simpler than the previous one even if less "mechanical," for computing the inverse of a matrix (when it exists) consists in writing the identical matrix to the right of the matrix A, obtaining

$$\left(\begin{array}{ccc|ccc}1 & 2 & 1 & 1 & 0 & 0\\ 1 & 3 & 0 & 0 & 1 & 0\\ 1 & 2 & 2 & 0 & 0 & 1\end{array}\right)$$

Now, using only elementary row operations, we try to make the identical matrix appear on the left: in the end, the matrix that we will find on the right will be the inverse of A. In this case, the operations to be performed are, for example, the following: we subtract the first from the second row, and the first from the third row, obtaining

$$\left(\begin{array}{ccc|ccc} 1 & 2 & 1 & 1 & 0 & 0 \\ 0 & 1 & -1 & -1 & 1 & 0 \\ 0 & 0 & 1 & -1 & 0 & 1 \end{array}\right)$$

Now we add the third to the second row, while we subtract the third row from the first, obtaining

$$\left(\begin{array}{ccc|ccc} 1 & 2 & 0 & 2 & 0 & -1 \\ 0 & 1 & 0 & -2 & 1 & 1 \\ 0 & 0 & 1 & -1 & 0 & 1 \end{array}\right)$$

Now we add the second row multiplied by -2 to the first row, obtaining

$$\left(\begin{array}{ccc|ccc} 1 & 0 & 0 & 6 & -2 & -3 \\ 0 & 1 & 0 & -2 & 1 & 1 \\ 0 & 0 & 1 & -1 & 0 & 1 \end{array}\right)$$

Therefore we can conclude that the inverse of A is the matrix

$$A^{-1} = \begin{pmatrix} 6 & -2 & -3 \\ -2 & 1 & 1 \\ -1 & 0 & 1 \end{pmatrix}$$

3.4 Other exercises

► **Exercise 3.4.1.** We denote by A_n the following $n \times n$ matrix:

$$A_n = \begin{pmatrix} 2 & 1 & 0 & 0 & \cdots & 0 \\ 3 & 2 & 1 & 0 & \cdots & 0 \\ 0 & 3 & 2 & 1 & \cdots & 0 \\ \vdots & \ddots & \ddots & \ddots & \ddots & \vdots \\ 0 & \cdots & 0 & 3 & 2 & 1 \\ 0 & \cdots & 0 & 0 & 3 & 2 \end{pmatrix}$$

Find a recursive formula for $\det A_n$, for every integer $n \geq 1$.

Solution. Using Laplace formula along the first row, we get:

$$\begin{aligned} \det A_n &= 2\det A_{n-1} - \det \begin{pmatrix} 3 & 1 & 0 & 0 & \cdots & 0 \\ 0 & 2 & 1 & 0 & \cdots & 0 \\ 0 & 3 & 2 & 1 & \cdots & 0 \\ \vdots & \ddots & \ddots & \ddots & \ddots & \vdots \\ 0 & \cdots & 0 & 3 & 2 & 1 \\ 0 & \cdots & 0 & 0 & 3 & 2 \end{pmatrix} \\ &= 2\det A_{n-1} - 3\det A_{n-2}. \end{aligned}$$

(The second matrix to the right is a $(n-1) \times (n-1)$ matrix, and to compute its determinant we used Laplace formula along the first column. We also assumed, in the previous calculation, that $n \geq 3$).

Now we must compute the determinants for $n = 1$ and $n = 2$:

$$\det A_1 = \det(2) = 2,$$
$$\det A_2 = \det \begin{pmatrix} 2 & 1 \\ 3 & 2 \end{pmatrix} = 1.$$

In this way, all the other determinants can be calculated successively; for example, we have:

$$\det A_3 = 2\det A_2 - 3\det A_1 = -4,$$
$$\det A_4 = 2\det A_3 - 3\det A_2 = -11.$$

Remark: The previous recursive formula allows the calculation of all $\det A_n$, however in order to calculate $\det A_n$ for a particular value of n it is necessary to first calculate all the various $\det A_i$, for each integer $i < n$. To find a formula that allows the direct calculation of $\det A_n$ one can proceed as follows. Let $d_n = \det A_n$; we have $d_1 = 2$, $d_2 = 1$ and $d_n = 2d_{n-1} - 3d_{n-2}$, for every $n \geq 3$. Then we also have $d_{n+1} = 2d_n - 3d_{n-1} = 2(2d_{n-1} - 3d_{n-2}) - 3d_{n-1} = d_{n-1} - 6d_{n-2}$. We therefore obtain the following equation:

$$\begin{pmatrix} d_{n+1} \\ d_n \end{pmatrix} = \begin{pmatrix} 1 & -6 \\ 2 & -3 \end{pmatrix} \begin{pmatrix} d_{n-1} \\ d_{n-2} \end{pmatrix}$$

or, by setting $v_n = \begin{pmatrix} d_{n+1} \\ d_n \end{pmatrix}$ and denoting by B the matrix $\begin{pmatrix} 1 & -6 \\ 2 & -3 \end{pmatrix}$,

$$v_n = Bv_{n-2}$$

Therefore we have $v_3 = Bv_1$, $v_5 = Bv_3 = B^2v_1$, $v_7 = Bv_5 = B^3v_1$, and in general,

$$v_{2n+1} = B^n v_1$$

Recalling that $v_1 = \begin{pmatrix} 1 \\ 2 \end{pmatrix}$, we then discover that the problem of directly calculating the various d_n is reduced to the problem of calculating the powers of a matrix. Obviously the calculation of B^n can be done by multiplying the matrix B by itself n times, but this is equivalent to calculating all the various d_i, for each $i < n$. However, as we will see later, there is a way to compute B^n directly without having to compute all the various powers B^i, $i = 2, \ldots, n-1$.

► **Exercise 3.4.2.** Let A_n be the following $n \times n$ matrix:

$$A_n = \begin{pmatrix} 1+a_1 & 1 & 1 & 1 & \cdots & 1 \\ 1 & 1+a_2 & 1 & 1 & \cdots & 1 \\ 1 & 1 & 1+a_3 & 1 & \cdots & 1 \\ \vdots & \ddots & \ddots & \ddots & \ddots & \vdots \\ 1 & \cdots & 1 & 1 & 1+a_{n-1} & 1 \\ 1 & \cdots & 1 & 1 & 1 & 1+a_n \end{pmatrix}$$

where $a_1, \ldots, a_n \in \mathbb{R}$.

Prove that, for every integer $n \geq 2$, we have

$$\det A_n = a_1 a_2 \cdots a_n + \sum_{i=1}^{n} a_1 a_2 \cdots \hat{a}_i \cdots a_n,$$

where $\hat{a}_i$ means that the element a_i does not appear in the product.

Solution. We set $d_n(a_1, \ldots, a_n) = \det A_n$. Let's check the given formula for $n = 2$:

$$d_2(a_1, a_2) = \det \begin{pmatrix} 1 + a_1 & 1 \\ 1 & 1 + a_2 \end{pmatrix} = a_1 a_2 + a_1 + a_2.$$

So, for $n = 2$, the formula is correct. We then proceed by induction, assuming that the formula is correct for A_{n-1} and trying to prove that then the given formula is also correct for A_n. To calculate the determinant $d_n(a_1, \ldots, a_n)$, it is convenient to subtract the second from the first column (in this way the determinant does not change), obtaining

$$d_n(a_1, \ldots, a_n) = \det \begin{pmatrix} a_1 & 1 & 1 & 1 & \cdots & 1 \\ -a_2 & 1 + a_2 & 1 & 1 & \cdots & 1 \\ 0 & 1 & 1 + a_3 & 1 & \cdots & 1 \\ \vdots & \ddots & \ddots & \ddots & \ddots & \vdots \\ 0 & \cdots & 1 & 1 & 1 + a_{n-1} & 1 \\ 0 & \cdots & 1 & 1 & 1 & 1 + a_n \end{pmatrix}$$

Now we can use Laplace formula along the first column, obtaining

$$d_n(a_1, \ldots, a_n) = a_1 \det \begin{pmatrix} 1 + a_2 & 1 & 1 & 1 & \cdots & 1 \\ 1 & 1 + a_3 & 1 & 1 & \cdots & 1 \\ 1 & 1 & 1 + a_4 & 1 & \cdots & 1 \\ \vdots & \ddots & \ddots & \ddots & \ddots & \vdots \\ 1 & \cdots & 1 & 1 & 1 + a_{n-1} & 1 \\ 1 & \cdots & 1 & 1 & 1 & 1 + a_n \end{pmatrix}$$

$$+ a_2 \det \begin{pmatrix} 1 & 1 & 1 & 1 & \cdots & 1 \\ 1 & 1 + a_3 & 1 & 1 & \cdots & 1 \\ 1 & 1 & 1 + a_4 & 1 & \cdots & 1 \\ \vdots & \ddots & \ddots & \ddots & \ddots & \vdots \\ 1 & \cdots & 1 & 1 & 1 + a_{n-1} & 1 \\ 1 & \cdots & 1 & 1 & 1 & 1 + a_n \end{pmatrix}$$

$$= a_1 d_{n-1}(a_2, a_3, \ldots, a_n) + a_2 d_{n-1}(0, a_3, \ldots, a_n).$$

Using the inductive hypothesis, we have

$$d_{n-1}(a_2, a_3, \ldots, a_n) = a_2 a_3 \cdots a_n + \sum_{i=2}^{n} a_2 a_3 \cdots \hat{a}_i \cdots a_n,$$

hence

$$d_{n-1}(0, a_3, \ldots, a_n) = 0 + \sum_{i=2}^{n} 0 \cdot a_3 \cdots \hat{a}_i \cdots a_n = a_3 \cdots a_n.$$

Substituting into the expression previously found, we have:

$$\begin{aligned} d_n(a_1, \ldots, a_n) &= a_1(a_2 a_3 \cdots a_n + \sum_{i=2}^{n} a_2 a_3 \cdots \hat{a}_i \cdots a_n) + a_2(a_3 \cdots a_n) \\ &= a_1 a_2 \cdots a_n + \sum_{i=2}^{n} a_1 a_2 \cdots \hat{a}_i \cdots a_n + a_2 a_3 \cdots a_n \\ &= a_1 a_2 \cdots a_n + \sum_{i=1}^{n} a_1 a_2 \cdots \hat{a}_i \cdots a_n, \end{aligned}$$

which is exactly what we wanted to prove.

► **Exercise 3.4.3.** Let X be the block matrix

$$X = \begin{pmatrix} A & B \\ 0 & C \end{pmatrix}$$

where A is a $n \times n$ matrix and C is a $m \times m$ matrix.

Prove that $\det X = \det A \cdot \det C$.

Solution. We use the definition of the determinant:

$$\det X = \sum_{\sigma} (\operatorname{sgn} \sigma) x_{1,\sigma(1)} x_{2,\sigma(2)} \cdots x_{n,\sigma(n)} x_{n+1,\sigma(n+1)} \cdots x_{n+m,\sigma(n+m)}$$

where we set $X = (x_{ij})$ and where the sum is done as σ varies among all the permutations of $n + m$ objects.

Based on the block structure of the matrix X, we deduce that $x_{ij} = 0$ if $i > n$ and $j \leq n$, therefore in the product

$$x_{n+1,\sigma(n+1)} \cdots x_{n+m,\sigma(n+m)}$$

there is a zero whenever $\sigma(n+1) \leq n$, or $\sigma(n+2) \leq n$, ..., or $\sigma(n+m) \leq n$. This means that the addends that actually appear in the expression of the determinant of X correspond to the permutations σ such that

$$\sigma(n+1), \sigma(n+2), \ldots, \sigma(n+m) \in \{n+1, \ldots, n+m\}.$$

Consequently we must have also

$$\sigma(1), \sigma(2), \ldots, \sigma(n) \in \{1, \ldots, n\}.$$

This means that the only addends that actually appear in the expression of the determinant of X correspond to the permutations σ that split into the product of two permutations $\sigma = \sigma_1\sigma_2$, where $\sigma_1 \colon \{1, \ldots, n\} \to \{1, \ldots, n\}$ and $\sigma_2 \colon \{n+1, \ldots, n+m\} \to \{n+1, \ldots, n+m\}$. Remember that one also has $\operatorname{sgn}(\sigma) = \operatorname{sgn}(\sigma_1)\operatorname{sgn}(\sigma_2)$. Therefore we get

$$\begin{aligned}
\det X &= \sum_{\sigma_1,\sigma_2} (\operatorname{sgn}\sigma_1)(\operatorname{sgn}\sigma_2) x_{1,\sigma_1(1)} x_{2,\sigma_1(2)} \cdots x_{n,\sigma_1(n)} x_{n+1,\sigma_2(n+1)} \cdots \\
&\qquad \cdots x_{n+m,\sigma_2(n+m)} \\
&= \Big(\sum_{\sigma_1} (\operatorname{sgn}\sigma_1) x_{1,\sigma_1(1)} x_{2,\sigma_1(2)} \cdots x_{n,\sigma_1(n)}\Big) \times \\
&\quad \Big(\sum_{\sigma_2} (\operatorname{sgn}\sigma_2) x_{n+1,\sigma_2(n+1)} x_{n+2,\sigma_2(n+2)} \cdots x_{n+m,\sigma_2(n+m)}\Big) \\
&= \det A \cdot \det C.
\end{aligned}$$

4

Diagonalization of Endomorphisms

4.1 Eigenvalues and eigenvectors

► **Exercise 4.1.1.** Determine if the following matrix is diagonalizable:

$$A = \begin{pmatrix} 1 & 4 & 1 \\ -4 & -7 & 2 \\ 6 & 6 & 0 \end{pmatrix}$$

If so, find a diagonal matrix D and an invertible matrix P such that $A = PDP^{-1}$.

Solution. Let's find the eigenvalues of A. We have:

$$\begin{aligned}
\det(A - \lambda \mathbf{1}) &= \det \begin{pmatrix} 1-\lambda & 4 & 1 \\ -4 & -7-\lambda & 2 \\ 6 & 6 & -\lambda \end{pmatrix} = \det \begin{pmatrix} 1-\lambda & 3+\lambda & 1 \\ -4 & -3-\lambda & 2 \\ 6 & 0 & -\lambda \end{pmatrix} \\
&= \det \begin{pmatrix} 1-\lambda & 3+\lambda & 1 \\ -3-\lambda & 0 & 3 \\ 6 & 0 & -\lambda \end{pmatrix} = -(3+\lambda)\det \begin{pmatrix} -3-\lambda & 3 \\ 6 & -\lambda \end{pmatrix} \\
&= -(3+\lambda)(\lambda^2 + 3\lambda - 18),
\end{aligned}$$

where, in the first step we subtracted the first from the second column, while in the second we added the first to the second row.

Therefore we have $\det(A - \lambda \mathbf{1}) = 0$ for $\lambda = -3$, $\lambda = 3$ and $\lambda = -6$. These are the three eigenvalues of A, which is therefore diagonalizable having three distinct (real) eigenvalues. So we can take as diagonal matrix D, for example, the matrix

$$D = \begin{pmatrix} -6 & 0 & 0 \\ 0 & -3 & 0 \\ 0 & 0 & 3 \end{pmatrix}$$

To find an invertible matrix P such that $A = PDP^{-1}$ it is necessary to determine the eigenvectors of the matrix A.

We start from the eigenvectors corresponding to the eigenvalue $\lambda_1 = -6$; they are the solutions of the system

$$(A - \lambda_1 \mathbf{1}) \begin{pmatrix} x_1 \\ x_2 \\ x_3 \end{pmatrix} = \mathbf{0}.$$

We obtain the system

$$\begin{cases} 7x_1 + 4x_2 + x_3 = 0 \\ -4x_1 - x_2 + 2x_3 = 0 \\ 6x_1 + 6x_2 + 6x_3 = 0 \end{cases}$$

whose solutions are given by $x_2 = -2x_1$ and $x_3 = x_1$. An eigenvector associated with the eigenvalue $\lambda_1 = -6$ is, for example, the vector $v_1 = (1, -2, 1)$. This will form the first column of the matrix P.

For the eigenvalue $\lambda_2 = -3$ we obtain the system

$$(A - \lambda_2 \mathbf{1}) \begin{pmatrix} x_1 \\ x_2 \\ x_3 \end{pmatrix} = \mathbf{0},$$

which is

$$\begin{cases} 4x_1 + 4x_2 + x_3 = 0 \\ -4x_1 - 4x_2 + 2x_3 = 0 \\ 6x_1 + 6x_2 + 3x_3 = 0 \end{cases}$$

whose solutions are $x_1 = -x_2$ and $x_3 = 0$. An eigenvector associated with the eigenvalue $\lambda_2 = -3$ is, for example, the vector $v_2 = (-1, 1, 0)$. This will form the second column of the matrix P.

Finally, for the eigenvalue $\lambda_3 = 3$ we obtain the system

$$(A - \lambda_3 \mathbf{1}) \begin{pmatrix} x_1 \\ x_2 \\ x_3 \end{pmatrix} = \mathbf{0},$$

which is

$$\begin{cases} -2x_1 + 4x_2 + x_3 = 0 \\ -4x_1 - 10x_2 + 2x_3 = 0 \\ 6x_1 + 6x_2 - 3x_3 = 0 \end{cases}$$

whose solutions are $x_2 = 0$ and $x_3 = 2x_1$. An eigenvector associated with the eigenvalue $\lambda_3 = 3$ is, for example, the vector $v_3 = (1, 0, 2)$. This will form the third column of the matrix P. Therefore we have

$$P = \begin{pmatrix} 1 & -1 & 1 \\ -2 & 1 & 0 \\ 1 & 0 & 2 \end{pmatrix}$$

We can now verify, as an exercise, that $AP = PD$, i.e., that $A = PDP^{-1}$, as required.

► **Exercise 4.1.2.** Determine if the following matrix is diagonalizable:

$$A = \begin{pmatrix} 1 & 4 & 0 & -2 \\ 0 & 1 & 0 & 0 \\ 0 & -2 & 1 & 1 \\ 0 & 4 & 0 & -1 \end{pmatrix}$$

If so, find a diagonal matrix D and an invertible matrix P such that $D = P^{-1}AP$.

Solution. Let's find the eigenvalues of A:

$$\begin{aligned} \det(A - \lambda \mathbf{1}) &= \det \begin{pmatrix} 1-\lambda & 4 & 0 & -2 \\ 0 & 1-\lambda & 0 & 0 \\ 0 & -2 & 1-\lambda & 1 \\ 0 & 4 & 0 & -1-\lambda \end{pmatrix} \\ &= (1-\lambda) \det \begin{pmatrix} 1-\lambda & 0 & 0 \\ -2 & 1-\lambda & 1 \\ 4 & 0 & -1-\lambda \end{pmatrix} \\ &= (1-\lambda)^2 \det \begin{pmatrix} 1-\lambda & 1 \\ 0 & -1-\lambda \end{pmatrix} \\ &= (1-\lambda)^3(-1-\lambda) = 0, \end{aligned}$$

whose solutions are $\lambda_1 = 1$, with multiplicity 3, and $\lambda_2 = -1$, with multiplicity 1. Note that the eigenvalues of A are not all distinct, consequently the information we have so far is not sufficient to decide whether A is diagonalizable or not. We then determine the eigenvectors of A.

We start from the eigenvectors corresponding to the eigenvalue $\lambda_1 = 1$; they are the solutions of the system

$$(A - \mathbf{1})\begin{pmatrix} x_1 \\ x_2 \\ x_3 \\ x_4 \end{pmatrix} = \mathbf{0}.$$

This system is

$$\begin{cases} 4x_2 - 2x_4 = 0 \\ -2x_2 + x_4 = 0 \\ 4x_2 - 2x_4 = 0 \end{cases}$$

whose solutions are given by $x_4 = 2x_2$ (x_1, x_2 and x_3 are indeterminate). It thus turns out that the eigenspace corresponding to the eigenvalue $\lambda_1 = 1$ has dimension 3, equal to the multiplicity of the eigenvalue. A basis of this eigenspace is given, for example, by the vectors $v_1 = (1, 0, 0, 0)$, $v_2 = (0, 1, 0, 2)$ and $v_3 = (0, 0, 1, 0)$.

For the eigenvalue $\lambda_2 = -1$ we obtain the system

$$(A + \mathbf{1})\begin{pmatrix} x_1 \\ x_2 \\ x_3 \\ x_4 \end{pmatrix} = \mathbf{0},$$

that is

$$\begin{cases} 2x_1 + 4x_2 - 2x_4 = 0 \\ 2x_2 = 0 \\ -2x_2 + 2x_3 + x_4 = 0 \\ 4x_2 = 0 \end{cases}$$

whose solutions are $x_1 = -2x_3$, $x_2 = 0$ and $x_4 = -2x_3$. The eigenspace corresponding to the eigenvalue $\lambda_2 = -1$ therefore has dimension 1, which is equal to the multiplicity of the eigenvalue. A basis of this eigenspace is given, for example, by the vector $v_4 = (-2, 0, 1, -2)$.

In conclusion, the matrix A is diagonalizable and the two required matrices D and P are, for example,

$$D = \begin{pmatrix} 1 & 0 & 0 & 0 \\ 0 & 1 & 0 & 0 \\ 0 & 0 & 1 & 0 \\ 0 & 0 & 0 & -1 \end{pmatrix} \qquad P = \begin{pmatrix} 1 & 0 & 0 & -2 \\ 0 & 1 & 0 & 0 \\ 0 & 0 & 1 & 1 \\ 0 & 2 & 0 & -2 \end{pmatrix}$$

We can now verify, as an exercise, that $AP = PD$, i.e., that $D = P^{-1}AP$, as required.

► **Exercise 4.1.3.** Determine for which values of the real parameter t the matrix

$$A = \begin{pmatrix} 3 & -1 & -1 \\ t-1 & t+3 & t+1 \\ -t & -t & 2-t \end{pmatrix}$$

is diagonalizable. For each of these values find a diagonal matrix D and an invertible matrix P such that $A = PDP^{-1}$.

Solution. Let's find the eigenvalues of A. We have:

$$\begin{aligned} \det(A - \lambda\mathbf{1}) &= \det \begin{pmatrix} 3-\lambda & -1 & -1 \\ t-1 & t+3-\lambda & t+1 \\ -t & -t & 2-t-\lambda \end{pmatrix} \\ &= \det \begin{pmatrix} 4-\lambda & -1 & -1 \\ -4+\lambda & t+3-\lambda & t+1 \\ 0 & -t & 2-t-\lambda \end{pmatrix} \\ &= \det \begin{pmatrix} 4-\lambda & -1 & -1 \\ 0 & t+2-\lambda & t \\ 0 & -t & 2-t-\lambda \end{pmatrix} \\ &= (4-\lambda)\det \begin{pmatrix} t+2-\lambda & t \\ -t & 2-t-\lambda \end{pmatrix} \\ &= (4-\lambda)(\lambda-2)^2, \end{aligned}$$

where, in the first step we subtracted the second from the first column, while in the second we added the first to the second row.

The solutions of $\det(A - \lambda\mathbf{1}) = 0$ are $\lambda_1 = 2$, with multiplicity 2, and $\lambda_2 = 4$, with multiplicity 1. Note that the eigenvalues of A are not all distinct, consequently the information we have so far is not sufficient to decide whether A is diagonalizable or not. We then determine the eigenvectors of A.

We start from the eigenvectors corresponding to the eigenvalue $\lambda_1 = 2$; they are the solutions of the system

$$(A - 2\mathbf{1}) \begin{pmatrix} x_1 \\ x_2 \\ x_3 \end{pmatrix} = \mathbf{0}.$$

We obtain the system

$$\begin{cases} x_1 - x_2 - x_3 = 0 \\ (t-1)x_1 + (t+1)x_2 + (t+1)x_3 = 0 \\ -tx_1 - tx_2 - tx_3 = 0 \end{cases}$$

If $t \neq 0$, the solutions are $x_1 = 0$ and $x_2 = -x_3$, from which we deduce that the eigenspace corresponding to the eigenvalue $\lambda_1 = 2$ has dimension 1. Since the dimension of the eigenspace is smaller than the multiplicity of the corresponding eigenvalue, we can conclude that, for $t \neq 0$, the matrix A is not diagonalizable.

If instead $t = 0$, the solutions of the linear system are given by $x_1 = x_2 + x_3$. In this case the eigenspace corresponding to the eigenvalue $\lambda_1 = 2$ has dimension 2, which is equal to the multiplicity of the eigenvalue. Since the eigenvalue $\lambda_2 = 4$ has multiplicity 1, we can conclude that the matrix A is diagonalizable only for $t = 0$. A basis of the eigenspace corresponding to the eigenvalue $\lambda_1 = 2$ is given, for example, by the vectors $v_1 = (1, 1, 0)$ and $v_2 = (1, 0, 1)$, which will form the first two columns of the matrix P.

For the eigenvalue $\lambda_2 = 4$ we obtain the system

$$(A - 4\mathbf{1}) \begin{pmatrix} x_1 \\ x_2 \\ x_3 \end{pmatrix} = \mathbf{0},$$

that is

$$\begin{cases} -x_1 - x_2 - x_3 = 0 \\ (t-1)x_1 + (t-1)x_2 + (t+1)x_3 = 0 \\ -tx_1 - tx_2 - (t+2)x_3 = 0 \end{cases}$$

Let's solve it just for the value of interest, $t = 0$. We find $x_1 = -x_2$ and $x_3 = 0$. An eigenvector corresponding to the eigenvalue $\lambda_2 = 4$ is, for example, the vector $v_3 = (-1, 1, 0)$. This will form the third column of the matrix P.

In conclusion, the matrix A is diagonalizable only for $t = 0$, in which case the two required matrices D and P are, for example,

$$D = \begin{pmatrix} 2 & 0 & 0 \\ 0 & 2 & 0 \\ 0 & 0 & 4 \end{pmatrix} \qquad P = \begin{pmatrix} 1 & 1 & -1 \\ 1 & 0 & 1 \\ 0 & 1 & 0 \end{pmatrix}$$

► **Exercise 4.1.4.** Compute A^n, for every integer $n \geq 1$, where A is the matrix

$$A = \begin{pmatrix} 5/2 & -1 \\ 3 & -1 \end{pmatrix}$$

Solution. Let's find the eigenvalues of A:

$$\begin{aligned} \det(A - \lambda \mathbf{1}) &= \det \begin{pmatrix} 5/2 - \lambda & -1 \\ 3 & -1 - \lambda \end{pmatrix} \\ &= \lambda^2 - \tfrac{3}{2}\lambda + \tfrac{1}{2} = 0, \end{aligned}$$

whose solutions are $\lambda_1 = 1/2$ and $\lambda_2 = 1$. Since the eigenvalues of A are all distinct, A is diagonalizable.

Now let us determine the eigenvectors of A. We start from the eigenvectors corresponding to the eigenvalue $\lambda_1 = 1/2$; they are the solutions of the system

$$(A - \tfrac{1}{2}\mathbf{1}) \begin{pmatrix} x_1 \\ x_2 \end{pmatrix} = \mathbf{0}.$$

This gives us the system

$$\begin{cases} 2x_1 - x_2 = 0 \\ 3x_1 - \tfrac{3}{2}x_2 = 0 \end{cases}$$

whose solutions are $x_2 = 2x_1$. An eigenvector for the eigenvalue $\lambda_1 = 1/2$ is, for example, $v_1 = (1, 2)$.

For the eigenvalue $\lambda_2 = 1$ we obtain the system

$$(A - \mathbf{1}) \begin{pmatrix} x_1 \\ x_2 \end{pmatrix} = \mathbf{0},$$

that is

$$\begin{cases} \tfrac{3}{2}x_1 - x_2 = 0 \\ 3x_1 - 2x_2 = 0 \end{cases}$$

whose solutions are $x_1 = \frac{2}{3}x_2$. An eigenvector for the eigenvalue $\lambda_2 = 1$ is, for example, $v_2 = (2, 3)$.

This means that A is diagonalizable, and if we set

$$D = \begin{pmatrix} 1/2 & 0 \\ 0 & 1 \end{pmatrix} \qquad P = \begin{pmatrix} 1 & 2 \\ 2 & 3 \end{pmatrix}$$

we have $A = PDP^{-1}$. Let's compute P^{-1}. We have $\det P = -1$, hence

$$P^{-1} = \begin{pmatrix} -3 & 2 \\ 2 & -1 \end{pmatrix}$$

Now it is easy to compute A^n. In fact, we have:

$$\begin{aligned} A^n &= (PDP^{-1})^n = (PDP^{-1})(PDP^{-1})\cdots(PDP^{-1}) = PD^nP^{-1} \\ &= \begin{pmatrix} 1 & 2 \\ 2 & 3 \end{pmatrix} \begin{pmatrix} \frac{1}{2^n} & 0 \\ 0 & 1^n \end{pmatrix} \begin{pmatrix} -3 & 2 \\ 2 & -1 \end{pmatrix} \\ &= \begin{pmatrix} 4 - \frac{3}{2^n} & \frac{1}{2^{n-1}} - 2 \\ 6 - \frac{3}{2^{n-1}} & \frac{1}{2^{n-2}} - 3 \end{pmatrix}. \end{aligned}$$

► **Exercise 4.1.5.** Compute $\exp(A)$, where A is the matrix

$$A = \begin{pmatrix} 3 & -2 & 0 \\ 0 & 0 & -2 \\ -3 & 2 & -2 \end{pmatrix}$$

Solution. Recall that the exponential of a square matrix, $\exp(A)$, is defined as the sum of the following series (which converges, for each matrix A):

$$\exp(A) = \sum_{n=0}^{+\infty} \frac{1}{n!} A^n.$$

From this formula it follows that, if D is a diagonal matrix whose elements on the main diagonal are $d_1, \dots, d_r$, then $\exp(D)$ is a diagonal matrix, having on the main diagonal the elements $e^{d_1}, \dots, e^{d_r}$.

If A is similar to a diagonal matrix, $A = PDP^{-1}$, where D is a diagonal matrix and P is an invertible matrix. In this case it is easy to compute $\exp(A)$. In fact, we have:

$$\begin{aligned} \exp(A) &= \sum_{n=0}^{+\infty} \frac{1}{n!} A^n = \sum_{n=0}^{+\infty} \frac{1}{n!} (PDP^{-1})^n \\ &= \sum_{n=0}^{+\infty} \frac{1}{n!} PD^nP^{-1} = P\left(\sum_{n=0}^{+\infty} \frac{1}{n!} D^n\right) P^{-1} \\ &= P \exp(D) P^{-1}, \end{aligned}$$

therefore the computation of $\exp(A)$ reduces to the computation of P, $\exp(D)$ and P^{-1}.

Let's find the eigenvalues of A:

$$\begin{aligned}\det(A-\lambda\mathbf{1}) &= \det\begin{pmatrix}3-\lambda & -2 & 0\\ 0 & -\lambda & -2\\ -3 & 2 & -2-\lambda\end{pmatrix}\\ &= \lambda(-\lambda^2+\lambda+2)=0,\end{aligned}$$

whose solutions are $\lambda_1 = 0$, $\lambda_2 = -1$ and $\lambda_3 = 2$. Since the eigenvalues of A are all distinct, A is diagonalizable.

We now determine the eigenvectors of A. We start from the eigenvectors corresponding to the eigenvalue $\lambda_1 = 0$; they are the solutions of the system

$$A\begin{pmatrix}x_1\\ x_2\\ x_3\end{pmatrix} = \mathbf{0}.$$

This is the system

$$\begin{cases}3x_1 - 2x_2 = 0\\ -2x_3 = 0\\ -3x_1 + 2x_2 - 2x_3 = 0\end{cases}$$

whose solutions are given by $x_1 = \frac{2}{3}x_2$ and $x_3 = 0$. An eigenvector for the eigenvalue $\lambda_1 = 0$ is, for example, $v_1 = (2, 3, 0)$.

For the eigenvalue $\lambda_2 = -1$ we obtain the system

$$(A+\mathbf{1})\begin{pmatrix}x_1\\ x_2\\ x_3\end{pmatrix} = \mathbf{0},$$

that is

$$\begin{cases}4x_1 - 2x_2 = 0\\ x_2 - 2x_3 = 0\\ -3x_1 + 2x_2 - x_3 = 0\end{cases}$$

whose solutions are $x_1 = x_3$ and $x_2 = 2x_3$. An eigenvector for the eigenvalue $\lambda_2 = -1$ is, for example, $v_2 = (1, 2, 1)$.

For the eigenvalue $\lambda_3 = 2$ we obtain the system

$$(A-2\mathbf{1})\begin{pmatrix}x_1\\ x_2\\ x_3\end{pmatrix} = \mathbf{0},$$

that is

$$\begin{cases} x_1 - 2x_2 = 0 \\ -2x_2 - 2x_3 = 0 \\ -3x_1 + 2x_2 - 4x_3 = 0 \end{cases}$$

whose solutions are $x_1 = 2x_2$ and $x_3 = -x_2$. An eigenvector for the eigenvalue $\lambda_3 = 2$ is, for example, $v_3 = (2, 1, -1)$.

Therefore A is diagonalizable, and if we set

$$D = \begin{pmatrix} 0 & 0 & 0 \\ 0 & -1 & 0 \\ 0 & 0 & 2 \end{pmatrix} \qquad P = \begin{pmatrix} 2 & 1 & 2 \\ 3 & 2 & 1 \\ 0 & 1 & -1 \end{pmatrix}$$

we have $A = PDP^{-1}$. Now we compute P^{-1}. We have $\det P = 3$ and

$$P^{-1} = \begin{pmatrix} -1 & 1 & -1 \\ 1 & -2/3 & 4/3 \\ 1 & -2/3 & 1/3 \end{pmatrix}$$

From what we have seen above, we have therefore:

$$\begin{aligned} \exp(A) &= P\exp(D)P^{-1} \\ &= \begin{pmatrix} 2 & 1 & 2 \\ 3 & 2 & 1 \\ 0 & 1 & -1 \end{pmatrix} \begin{pmatrix} e^0 & 0 & 0 \\ 0 & e^{-1} & 0 \\ 0 & 0 & e^2 \end{pmatrix} \begin{pmatrix} -1 & 1 & -1 \\ 1 & -2/3 & 4/3 \\ 1 & -2/3 & 1/3 \end{pmatrix} \\ &= \begin{pmatrix} -2 + \frac{1}{e} + 2e^2 & 2 - \frac{2}{3e} - \frac{4}{3}e^2 & -2 + \frac{4}{3e} + \frac{2}{3}e^2 \\ -3 + \frac{2}{e} + e^2 & 3 - \frac{4}{3e} - \frac{2}{3}e^2 & -3 + \frac{8}{3e} + \frac{1}{3}e^2 \\ \frac{1}{e} - e^2 & -\frac{2}{3e} + \frac{2}{3}e^2 & \frac{4}{3e} - \frac{1}{3}e^2 \end{pmatrix}. \end{aligned}$$

► **Exercise 4.1.6.** Let A be a nonzero square matrix such that $A^n = \mathbf{0}$, for some integer $n \geq 2$. Prove that all eigenvalues of A are zero.

Solution. Let $\lambda \in \mathbb{C}$ be an eigenvalue of A (recall that the complex field $\mathbb{C}$ is algebraically closed, therefore every square matrix A has all its eigenvalues in $\mathbb{C}$) and let v be a corresponding eigenvector. We therefore have $Av = \lambda v$ and, consequently, $A^2v = A(Av) = A(\lambda v) = \lambda Av = \lambda^2 v$. This means that if λ is an eigenvalue of A, then λ^2 is an eigenvalue of A^2. Analogously it can be proved that $A^3v = \lambda^3 v$ and, in general, that $A^n v = \lambda^n v$, for any integer $n \geq 2$. That is, if λ is an eigenvalue of A, then λ^n is an eigenvalue of A^n. Assuming now that $A^n = \mathbf{0}$, we deduce that $\lambda^n = 0$, and therefore $\lambda = 0$, for each eigenvalue λ of A, i.e., that all eigenvalues of A are zero.

4.2 Jordan canonical form

▶ **Exercise 4.2.1.** Determine the Jordan canonical form of the endomorphism $\phi\colon \mathbb{R}^3 \to \mathbb{R}^3$ whose matrix, with respect to the canonical basis, is

$$A = \begin{pmatrix} 3 & 4 & 4 \\ -3 & -6 & -8 \\ 3 & 7 & 9 \end{pmatrix}$$

Moreover, determine a basis of $\mathbb{R}^3$ with respect to which the matrix of ϕ is the Jordan canonical form.

Solution. Let's find the eigenvalues of A:

$$\begin{aligned}\det(\lambda \mathbf{1} - A) &= \det \begin{pmatrix} \lambda - 3 & -4 & -4 \\ 3 & \lambda + 6 & 8 \\ -3 & -7 & \lambda - 9 \end{pmatrix} \\ &= (\lambda - 3) \begin{vmatrix} \lambda + 6 & 8 \\ -7 & \lambda - 9 \end{vmatrix} + 4 \begin{vmatrix} 3 & 8 \\ -3 & \lambda - 9 \end{vmatrix} - 4 \begin{vmatrix} 3 & \lambda + 6 \\ -3 & -7 \end{vmatrix} \\ &= \lambda^3 - 6\lambda^2 + 11\lambda - 6 = 0,\end{aligned}$$

whose solutions are $\lambda_1 = 1$, $\lambda_2 = 2$ and $\lambda_3 = 3$. The matrix A therefore has 3 distinct eigenvalues and, consequently, its Jordan canonical form is

$$J = \begin{pmatrix} 1 & 0 & 0 \\ 0 & 2 & 0 \\ 0 & 0 & 3 \end{pmatrix}$$

A basis of $\mathbb{R}^3$ with respect to which the matrix of ϕ is J consists of the eigenvectors associated with the eigenvalues 1, 2 and 3 of A.

Let us now determine the eigenvectors corresponding to the eigenvalue $\lambda_1 = 1$; they are the solutions of the system

$$(A - \lambda_1 \mathbf{1}) \begin{pmatrix} x_1 \\ x_2 \\ x_3 \end{pmatrix} = \mathbf{0}.$$

This is the system

$$\begin{cases} 2x_1 + 4x_2 + 4x_3 = 0 \\ -3x_1 - 7x_2 - 8x_3 = 0 \\ 3x_1 + 7x_2 + 8x_3 = 0 \end{cases}$$

whose solutions are given by $x_1 = 2x_3$ and $x_2 = -2x_3$. An eigenvector for the eigenvalue $\lambda_1 = 1$ is, for example, $v_1 = (2, -2, 1)$. This will be the first vector of a Jordan basis.

For the eigenvalue $\lambda_2 = 2$ we obtain the system

$$(A - \lambda_2 \mathbf{1}) \begin{pmatrix} x_1 \\ x_2 \\ x_3 \end{pmatrix} = \mathbf{0},$$

that is

$$\begin{cases} x_1 + 4x_2 + 4x_3 = 0 \\ -3x_1 - 8x_2 - 8x_3 = 0 \\ 3x_1 + 7x_2 + 7x_3 = 0 \end{cases}$$

whose solutions are $x_1 = 0$ and $x_2 = -x_3$. An eigenvector for the eigenvalue $\lambda_2 = 2$ is, for example, $v_2 = (0, -1, 1)$. This will be the second vector of a Jordan basis.

Finally, for the eigenvalue $\lambda_3 = 3$ we obtain the system

$$(A - \lambda_3 \mathbf{1}) \begin{pmatrix} x_1 \\ x_2 \\ x_3 \end{pmatrix} = \mathbf{0},$$

that is

$$\begin{cases} 4x_2 + 4x_3 = 0 \\ -3x_1 - 9x_2 - 8x_3 = 0 \\ 3x_1 + 7x_2 + 6x_3 = 0 \end{cases}$$

whose solutions are $x_1 = \frac{1}{3}x_3$ and $x_2 = -x_3$. An eigenvector for the eigenvalue $\lambda_3 = 3$ is, for example, $v_3 = (1, -3, 3)$. This will be the third vector of a Jordan basis.

If we denote by S the matrix whose columns contain the components of the vectors of the Jordan basis,

$$S = \begin{pmatrix} 2 & 0 & 1 \\ -2 & -1 & -3 \\ 1 & 1 & 3 \end{pmatrix}$$

we have $AS = SJ$, from which it follows that $A = SJS^{-1}$.

► **Exercise 4.2.2.** Determine the Jordan canonical form of the endomorphism $\phi\colon \mathbb{R}^4 \to \mathbb{R}^4$ whose matrix, with respect to the canonical bases, is

$$A = \begin{pmatrix} -1 & 0 & 0 & 0 \\ 0 & 1 & 0 & 0 \\ 1 & -3 & -1 & -1 \\ 0 & -2 & 0 & 1 \end{pmatrix}$$

Moreover, determine a basis of $\mathbb{R}^4$ with respect to which the matrix of ϕ is the Jordan canonical form.

Solution. Let's find the eigenvalues of A:

$$\begin{aligned} \det(\lambda\mathbf{1} - A) &= \det \begin{pmatrix} \lambda+1 & 0 & 0 & 0 \\ 0 & \lambda-1 & 0 & 0 \\ -1 & 3 & \lambda+1 & 1 \\ 0 & 2 & 0 & \lambda-1 \end{pmatrix} \\ &= (\lambda+1) \begin{vmatrix} \lambda+1 & 0 & 0 \\ 0 & \lambda-1 & 0 \\ 0 & 2 & \lambda-1 \end{vmatrix} \\ &= (\lambda+1)^2(\lambda-1)^2. \end{aligned}$$

The eigenvalues of A are therefore $\lambda_1 = -1$, with multiplicity 2, and $\lambda_2 = 1$, also with multiplicity 2.

Let us now determine the eigenvectors corresponding to the eigenvalue $\lambda_1 = -1$; they are the solutions of the system

$$(A - \lambda_1\mathbf{1}) \begin{pmatrix} x_1 \\ x_2 \\ x_3 \\ x_4 \end{pmatrix} = \mathbf{0}.$$

This is the system

$$\begin{cases} 2x_2 = 0 \\ x_1 - 3x_2 - x_4 = 0 \\ -2x_2 + 2x_4 = 0 \end{cases}$$

whose solutions are given by $x_1 = x_2 = x_4 = 0$ and any x_3. The set of solutions is therefore a vector subspace of dimension 1, which is precisely the eigenspace associated with the eigenvalue $\lambda_1 = -1$. A basis is given, for example, by the vector $v_1 = e_3 = (0, 0, 1, 0)$. This will be the first vector of a Jordan basis.

It can now be deduced that in the Jordan canonical form of A a Jordan block of the following form will appear:

$$\begin{pmatrix} -1 & 1 \\ 0 & -1 \end{pmatrix}$$

because the eigenvalue -1 has multiplicity 2, but there is only one linearly independent eigenvector corresponding to this eigenvalue.

Consequently, as the second vector v_2 of a Jordan basis, we will have to take a vector that satisfies the following condition:

$$Av_2 = -v_2 + v_1,$$

that is

$$(A + \mathbf{1})v_2 = v_1.$$

We thus obtain a system of linear equations which is equal to the previous system in which we replaced the column of zeros to the right of the equal sign with the column of the components of the vector v_1.

This gives the system

$$\begin{cases} 2x_2 = 0 \\ x_1 - 3x_2 - x_4 = 1 \\ -2x_2 + 2x_4 = 0 \end{cases}$$

whose solutions are given by $x_1 = 1$, $x_2 = x_4 = 0$ and any x_3. We can therefore take as the second vector of a Jordan basis the vector $v_2 = (1, 0, 0, 0)$.

Let us now consider the eigenvalue $\lambda_2 = 1$. The corresponding eigenvectors are found by solving the system

$$(A - \lambda_2 \mathbf{1}) \begin{pmatrix} x_1 \\ x_2 \\ x_3 \\ x_4 \end{pmatrix} = \mathbf{0}.$$

This gives the system

$$\begin{cases} -2x_1 = 0 \\ x_1 - 3x_2 - 2x_3 - x_4 = 0 \\ -2x_2 = 0 \end{cases}$$

whose solutions are $x_1 = x_2 = 0$ and $x_4 = -2x_3$. The set of solutions is therefore a vector subspace of dimension 1, which is precisely the eigenspace associated with the eigenvalue $\lambda_2 = 1$. A basis is given, for example, by the vector $v_3 = (0, 0, 1, -2)$. This will be the third vector of a Jordan basis.

It can now be deduced that in the Jordan canonical form of A a Jordan block of the following form will appear:

$$\begin{pmatrix} 1 & 1 \\ 0 & 1 \end{pmatrix}$$

because the eigenvalue 1 has multiplicity 2, but there is only one linearly independent eigenvector corresponding to this eigenvalue.

Consequently, as the fourth vector v_4 of a Jordan basis, we will have to take a vector that satisfies the following equation:

$$Av_4 = v_4 + v_3,$$

that is

$$(A - \mathbf{1})v_4 = v_3.$$

We thus obtain a system of linear equations which is equal to the previous system in which we replaced the column of zeros to the right of the equal sign with the column of the components of the vector v_3.

This gives the system

$$\begin{cases} -2x_1 = 0 \\ x_1 - 3x_2 - 2x_3 - x_4 = 1 \\ -2x_2 = -2 \end{cases}$$

whose solutions are $x_1 = 0$, $x_2 = 1$ and $x_4 = -2x_3 - 4$. We can therefore take as the fourth vector of a Jordan basis the vector $v_4 = (0, 1, 0, -4)$.

We conclude that the vectors v_1, v_2, v_3 and v_4 constitute a Jordan basis, with respect to which the endomorphism ϕ has the Jordan matrix

$$J = \begin{pmatrix} -1 & 1 & 0 & 0 \\ 0 & -1 & 0 & 0 \\ 0 & 0 & 1 & 1 \\ 0 & 0 & 0 & 1 \end{pmatrix}$$

If we denote by S the matrix whose columns contain the components of the vectors of the Jordan basis,

$$S = \begin{pmatrix} 0 & 1 & 0 & 0 \\ 0 & 0 & 0 & 1 \\ 1 & 0 & 1 & 0 \\ 0 & 0 & -2 & -4 \end{pmatrix}$$

we have $AS = SJ$, from which it follows that $A = SJS^{-1}$.

► **Exercise 4.2.3.** Determine the Jordan canonical form of the endomorphism $\phi\colon \mathbb{R}^4 \to \mathbb{R}^4$ whose matrix, with respect to the canonical bases, is

$$A = \begin{pmatrix} 2 & 0 & 0 & 0 \\ -4 & 2 & -4 & -3 \\ 4 & 0 & 5 & 3 \\ -3 & 0 & -2 & 0 \end{pmatrix}$$

Moreover, determine a basis of $\mathbb{R}^4$ with respect to which the matrix of ϕ is the Jordan canonical form.

Solution. Let's find the eigenvalues of A:

$$\begin{aligned} \det(\lambda\mathbf{1} - A) &= \det \begin{pmatrix} \lambda-2 & 0 & 0 & 0 \\ 4 & \lambda-2 & 4 & 3 \\ -4 & 0 & \lambda-5 & -3 \\ 3 & 0 & 2 & \lambda \end{pmatrix} \\ &= (\lambda-2)^2 \begin{vmatrix} \lambda-5 & -3 \\ 2 & \lambda \end{vmatrix} \\ &= (\lambda-2)^2(\lambda^2 - 5\lambda + 6) \\ &= (\lambda-2)^3(\lambda-3). \end{aligned}$$

The eigenvalues of A are therefore $\lambda_1 = 2$, with multiplicity 3, and $\lambda_2 = 3$, with multiplicity 1.

Let us now determine the eigenvectors corresponding to the eigenvalue $\lambda_1 = 2$; they are the solutions of the system

$$(A - \lambda_1\mathbf{1}) \begin{pmatrix} x_1 \\ x_2 \\ x_3 \\ x_4 \end{pmatrix} = \mathbf{0}.$$

This is the system

$$\begin{cases} -4x_1 - 4x_3 - 3x_4 = 0 \\ 4x_1 + 3x_3 + 3x_4 = 0 \\ -3x_1 - 2x_3 - 2x_4 = 0 \end{cases}$$

whose solutions are given by $x_1 = x_3 = x_4 = 0$ and any x_2. The set of solutions is therefore a vector subspace of dimension 1, which is precisely the eigenspace associated with the eigenvalue $\lambda_1 = 2$. A basis is given, for example, by the vector $v_1 = e_2 = (0, 1, 0, 0)$. This will be the first vector of a Jordan basis.

It can now be deduced that in the Jordan canonical form of A a Jordan block of the following form will appear:

$$\begin{pmatrix} 2 & 1 & 0 \\ 0 & 2 & 1 \\ 0 & 0 & 2 \end{pmatrix}$$

because the eigenvalue 2 has multiplicity 3, but there is only one linearly independent eigenvector corresponding to this eigenvalue.

Consequently, as the second vector v_2 of a Jordan basis, we will have to take a vector that satisfies the following condition:

$$Av_2 = 2v_2 + v_1,$$

that is

$$(A - 2\mathbf{1})v_2 = v_1.$$

We thus obtain a system of linear equations which is equal to the previous system in which we replaced the column of zeros to the right of the equal sign with the column of the components of the vector v_1.

The solutions of this system are given by $x_1 = 0$, $x_3 = -1$, $x_4 = 1$ and any x_2. We can therefore take as the second vector of a Jordan basis the vector $v_2 = (0, 0, -1, 1)$.

Similarly, as the third vector v_3 of a Jordan basis, we will have to take a vector that satisfies the following condition:

$$Av_3 = 2v_3 + v_2,$$

that is

$$(A - 2\mathbf{1})v_3 = v_2.$$

We thus obtain a system of linear equations which is equal to the previous system in which we replaced the column to the right of the equal sign with the column of the components of the vector v_2.

The solutions of this system are given by $x_1 = -1$, $x_3 = 1$, $x_4 = 0$ and any x_2. We can therefore take as the third vector of a Jordan basis the vector $v_3 = (-1, 0, 1, 0)$.

Let us now consider the eigenvalue $\lambda_2 = 3$. The corresponding eigenvectors are found by solving the system

$$(A - \lambda_2 \mathbf{1}) \begin{pmatrix} x_1 \\ x_2 \\ x_3 \\ x_4 \end{pmatrix} = \mathbf{0}.$$

This gives the system

$$\begin{cases} -x_1 = 0 \\ -4x_1 - x_2 - 4x_3 - 3x_4 = 0 \\ 4x_1 + 2x_3 + 3x_4 = 0 \end{cases}$$

whose solutions are $x_1 = 0$, $x_2 = -2x_3$ and $x_4 = -\frac{2}{3}\, x_3$. The set of solutions is therefore a vector subspace of dimension 1, which is precisely the eigenspace associated with the eigenvalue $\lambda_2 = 3$. A basis is given, for example, by the vector $v_4 = (0, -6, 3, -2)$. This will be the fourth vector of a Jordan basis.

We therefore conclude that the vectors v_1, v_2, v_3 and v_4 constitute a Jordan basis, with respect to which the endomorphism ϕ has the Jordan matrix

$$J = \begin{pmatrix} 2 & 1 & 0 & 0 \\ 0 & 2 & 1 & 0 \\ 0 & 0 & 2 & 0 \\ 0 & 0 & 0 & 3 \end{pmatrix}$$

If we denote by S the matrix whose columns contain the components of the vectors of the Jordan basis,

$$S = \begin{pmatrix} 0 & 0 & -1 & 0 \\ 1 & 0 & 0 & -6 \\ 0 & -1 & 1 & 3 \\ 0 & 1 & 0 & -2 \end{pmatrix}$$

we have $AS = SJ$, from which it follows that $A = SJS^{-1}$. Verify now, as an exercise, that

$$S^{-1} = \begin{pmatrix} 6 & 1 & 6 & 6 \\ 2 & 0 & 2 & 3 \\ -1 & 0 & 0 & 0 \\ 1 & 0 & 1 & 1 \end{pmatrix}$$

and check the equality $A = SJS^{-1}$ with a direct computation.

► **Exercise 4.2.4.** Determine the Jordan canonical form of the endomorphism $\phi\colon \mathbb{R}^4 \to \mathbb{R}^4$ whose matrix, with respect to the canonical bases, is

$$A = \begin{pmatrix} 0 & 0 & 2 & -4 \\ 1 & 2 & -1 & 2 \\ 0 & 0 & 2 & 0 \\ 1 & 0 & -1 & 4 \end{pmatrix}$$

Moreover, determine a basis of $\mathbb{R}^4$ with respect to which the matrix of ϕ is the Jordan canonical form.

Solution. Let's find the eigenvalues of A:

$$\begin{aligned} \det(\lambda\mathbf{1} - A) &= \det\begin{pmatrix} \lambda & 0 & -2 & 4 \\ -1 & \lambda-2 & 1 & -2 \\ 0 & 0 & \lambda-2 & 0 \\ -1 & 0 & 1 & \lambda-4 \end{pmatrix} \\ &= (\lambda-2)\begin{vmatrix} \lambda & -2 & 4 \\ 0 & \lambda-2 & 0 \\ -1 & 1 & \lambda-4 \end{vmatrix} \\ &= (\lambda-2)^2\begin{vmatrix} \lambda & 4 \\ -1 & \lambda-4 \end{vmatrix} \\ &= (\lambda-2)^4. \end{aligned}$$

So A has a unique eigenvalue $\lambda = 2$, with multiplicity 4.

Let us now determine the eigenvectors corresponding to the eigenvalue $\lambda = 2$; they are the solutions of the system

$$(A - \lambda\mathbf{1})\begin{pmatrix} x_1 \\ x_2 \\ x_3 \\ x_4 \end{pmatrix} = \mathbf{0}.$$

This is the system

$$\begin{cases} -2x_1 + 2x_3 - 4x_4 = 0 \\ x_1 - x_3 + 2x_4 = 0 \\ x_1 - x_3 + 2x_4 = 0 \end{cases}$$

whose solutions are given by $x_1 = x_3 - 2x_4$ and any x_2, x_3, x_4. The set of solutions is therefore a vector subspace of dimension 3, which is precisely the eigenspace associated with the eigenvalue $\lambda = 2$. A basis is given, for example, by the vectors $v_1 = (0, 1, 0, 0)$, $v_2 = (1, 0, 1, 0)$ and $v_3 = (-2, 0, 0, 1)$.

From what we have seen we deduce that the Jordan canonical form of A is (up to a reordering of the Jordan blocks)

$$J = \begin{pmatrix} 2 & 1 & 0 & 0 \\ 0 & 2 & 0 & 0 \\ 0 & 0 & 2 & 0 \\ 0 & 0 & 0 & 2 \end{pmatrix}$$

In this case it is not possible to apply the method described in the previous two exercises to determine a Jordan basis. In fact we should now determine a generalized eigenvector v_4 for the eigenvalue $\lambda = 2$, i.e., v_4 must satisfy the equation

$$Av_4 = 2v_4 + v,$$

where v is an eigenvector for the eigenvalue $\lambda = 2$. But now there is a subspace of eigenvectors of dimension 3, so we don't know how to choose our v! (Verify, as an exercise, that none of the three systems $Av_4 = 2v_4 + v_1$, $Av_4 = 2v_4 + v_2$ and $Av_4 = 2v_4 + v_3$ has solutions. Therefore it is not possible to find a Jordan basis using the three vectors v_1, v_2 and v_3 determined previously; which is quite natural, since the vectors v_1, v_2 and v_3 have been taken *at random* within a space of dimension 3).

So let's use another method. We denote by w_1 and w_2 the first two vectors of the Jordan basis (that we want to determine). Observing the previously written Jordan matrix J, we deduce that w_1 is an eigenvector for the eigenvalue $\lambda = 2$, i.e., we must have

$$(A - 2\mathbf{1})w_1 = \mathbf{0},$$

while w_2 is a generalized eigenvector for the eigenvalue $\lambda = 2$, i.e., we must have

$$(A - 2\mathbf{1})w_2 = w_1.$$

Multiplying this last equation on the left by $A - 2$, we obtain

$$(A - 2\mathbf{1})^2 w_2 = (A - 2\mathbf{1})w_1 = \mathbf{0},$$

hence w_2 belongs to the kernel of $(\phi - 2\,\mathrm{id})^2$ (i.e., $(A - 2\mathbf{1})^2 w_2 = \mathbf{0}$) but does not belong to the kernel of $\phi - 2\,\mathrm{id}$ (i.e., $(A - 2\mathbf{1})w_2 \neq \mathbf{0}$). To determine w_2 we therefore take any vector such that $(A - 2\mathbf{1})^2 w_2 = \mathbf{0}$ and $(A - 2\mathbf{1})w_2 \neq \mathbf{0}$. Given that we have already determined the kernel of $A - 2\mathbf{1}$, and given that $(A - 2\mathbf{1})^2 = \mathbf{0}$ (as can be seen by observing the Jordan matrix J, or as can be verified with a direct computation), it is easy to find a vector w_2 that satisfies the required conditions: for example the vector $w_2 = (2,1,1,1)$. Knowing w_2 immediately determines w_1; just remember that $(A - 2\mathbb{1})w_2 = w_1$. Therefore we have

$$w_1 = (A - 2\mathbb{1})w_2 = (-6,3,0,3).$$

These are the first two vectors of a Jordan basis. The remaining vectors, w_3 and w_4, must now satisfy the following two conditions: they must be eigenvectors corresponding to the eigenvalue $\lambda = 2$ and they must be such that $\{w_1, w_2, w_3, w_4\}$ is a basis of $\mathbb{R}^4$. It is easy to verify that the vectors $w_3 = (1,0,1,0)$ and $w_4 = (0,1,0,0)$ satisfy the required conditions.

In conclusion: the vectors w_1, w_2, w_3 and w_4 (in this order) form a Jordan basis, with respect to which the endomorphism ϕ has the Jordan matrix

$$J = \begin{pmatrix} 2 & 1 & 0 & 0 \\ 0 & 2 & 0 & 0 \\ 0 & 0 & 2 & 0 \\ 0 & 0 & 0 & 2 \end{pmatrix}$$

If we denote by S the matrix whose columns contain the components of the vectors of the Jordan basis,

$$S = \begin{pmatrix} -6 & 2 & 1 & 0 \\ 3 & 1 & 0 & 1 \\ 0 & 1 & 1 & 0 \\ 3 & 1 & 0 & 0 \end{pmatrix}$$

we have $AS = SJ$, from which it follows that $A = SJS^{-1}$. Verify now, as an exercise, that

$$S^{-1} = \begin{pmatrix} -\frac{1}{9} & 0 & \frac{1}{9} & \frac{1}{9} \\ \frac{1}{3} & 0 & -\frac{1}{3} & \frac{2}{3} \\ -\frac{1}{3} & 0 & \frac{4}{3} & -\frac{2}{3} \\ 0 & 1 & 0 & -1 \end{pmatrix}$$

and verify the equality $A = SJS^{-1}$ with a direct computation.

► **Exercise 4.2.5.** Let A be a $n \times n$ matrix with real entries such that $A^2 = A$. Prove that A is diagonalizable and that its eigenvalues are only 0 or 1. Prove further that if B is also a matrix such that $B^2 = B$, then A and B are similar if and only if they have the same rank.

Solution. Let's work in the complex field $\mathbb{C}$, so that there is a Jordan canonical form J for the matrix A, and we have $A = SJS^{-1}$, for some invertible matrix S (with entries in $\mathbb{C}$). It follows that $A^2 = SJS^{-1}SJS^{-1} = SJ^2S^{-1}$, hence the equality $A^2 = A$ is equivalent to $J^2 = J$. Recalling the form of the Jordan matrix J, we immediately see that $J^2 = J$ is possible if and only if J is diagonal (and therefore A is diagonalizable) and, in this case, for every eigenvalue λ of A we must have $\lambda^2 = \lambda$, from which it follows that $\lambda = 0$ or $\lambda = 1$. It follows that A is similar to the matrix

$$J = \begin{pmatrix} \mathbf{1}_r & \mathbf{0} \\ \mathbf{0} & \mathbf{0}_{n-r} \end{pmatrix}$$

where $\mathbf{1}_r$ denotes the identity matrix of order r and $\mathbf{0}_{n-r}$ is the zero matrix of order $n - r$.

Note that the integer r coincides with the rank of A, therefore, if B is another matrix with the same properties as A, it will be similar to J (and therefore to A) if and only if its rank is r.

► **Exercise 4.2.6.** Let A be a non-degenerate, skew-symmetric $n \times n$ matrix with real entries. Prove that all the eigenvalues of A are purely imaginary.

Solution. A is skew-symmetric, i.e., we have ${}^tA = -A$. It follows that A^2 is symmetric:

$${}^t(A^2) = ({}^tA)^2 = (-A)^2 = A^2.$$

Recall that every symmetric matrix with real entries is diagonalizable and all its eigenvalues are real. Therefore all eigenvalues of A^2 are real. We now prove that they are all negative.

Let $v \in \mathbb{R}^n$ be a nonzero vector; we have

$${}^tv(A^2)v = ({}^tvA)(Av) = (-{}^tv{}^tA)(Av) = -{}^t(Av)(Av) = -w \cdot w < 0,$$

where we set $w = Av$ and we recalled that the usual scalar product in $\mathbb{R}^n$ is positive definite (note that $w \neq \mathbf{0}$ because A is assumed to be non-degenerate).

This means that A^2 is negative definite, hence all its eigenvalues are strictly negative. Consequently, the eigenvalues of A, being the square roots of the eigenvalues of A^2, are purely imaginary.

We also propose another proof. Let $\lambda \in \mathbb{C}$ be an eigenvalue of A, and let $v \in \mathbb{C}^n$ be a corresponding eigenvector; therefore we have $Av = \lambda v$.

Conjugating both sides of this equality, and remembering that A is a matrix with real entries, i.e., $\bar{A} = A$, we obtain $A\bar{v} = \bar{\lambda}\bar{v}$. We now multiply both sides on the left by ${}^t v$, obtaining ${}^t v A\bar{v} = \bar{\lambda}{}^t v\bar{v}$, then we transpose both sides of the equality:

$$ {}^t({}^t v A\bar{v}) = {}^t\bar{v}{}^t A v = {}^t(\bar{\lambda}{}^t v\bar{v}) = \bar{\lambda}{}^t\bar{v}v. $$

Recall that A is skew-symmetric, i.e., ${}^t A = -A$. From the previous equality we therefore obtain $-{}^t\bar{v}Av = \bar{\lambda}{}^t\bar{v}v$, from which (remembering that $Av = \lambda v$) it follows that $-\lambda{}^t\bar{v}v = \bar{\lambda}{}^t\bar{v}v$. In conclusion we have $\bar{\lambda} = -\lambda$ and, given that $\lambda \neq 0$ because A is non-degenerate, this means that the eigenvalue λ is purely imaginary.

4.3 Recap exercises

► **Exercise 4.3.1.** Find $t \in \mathbb{R}$ so that the matrix

$$A = \begin{pmatrix} -1 & 3 & 2 \\ 0 & -5 & -4 \\ 0 & t-1 & -3 \end{pmatrix}$$

has -7 as an eigenvalue. For this value of t, determine the eigenvalues and eigenvectors of A and establish whether A is diagonalizable.

Solution. The matrix A has an eigenvalue equal to -7 if and only if

$$\det\big(A - (-7)\mathbf{1}\big) = 0.$$

We have:

$$\begin{aligned} \det\big(A - (-7)\mathbf{1}\big) &= \begin{vmatrix} 6 & 3 & 2 \\ 0 & 2 & -4 \\ 0 & t-1 & 4 \end{vmatrix} \\ &= 6 \begin{vmatrix} 2 & -4 \\ t-1 & 4 \end{vmatrix} \\ &= 24(t+1). \end{aligned}$$

This determinant is zero if and only if $t = -1$. Thus we conclude that -7 is an eigenvalue of A if and only if $t = -1$.

For this value of t the matrix becomes

$$A = \begin{pmatrix} -1 & 3 & 2 \\ 0 & -5 & -4 \\ 0 & -2 & -3 \end{pmatrix}$$

We now compute the eigenvalues of this matrix:

$$\begin{aligned} \det(A - \lambda\mathbf{1}) &= \begin{vmatrix} -1-\lambda & 3 & 2 \\ 0 & -5-\lambda & -4 \\ 0 & -2 & -3-\lambda \end{vmatrix} \\ &= (-1-\lambda) \begin{vmatrix} -5-\lambda & -4 \\ -2 & -3-\lambda \end{vmatrix} \\ &= -(1+\lambda)(\lambda^2 + 8\lambda + 7). \end{aligned}$$

Since the solutions of the equation $\lambda^2 + 8\lambda + 7 = 0$ are $\lambda = -1$ and $\lambda = -7$, we conclude that the eigenvalues of A are the following: $\lambda = -1$ (with multiplicity 2) and $\lambda = -7$.

The eigenvectors of A corresponding to the eigenvalue $\lambda = -7$ are the solutions of the following linear system:

$$\begin{pmatrix} 6 & 3 & 2 \\ 0 & 2 & -4 \\ 0 & -2 & 4 \end{pmatrix} \begin{pmatrix} x_1 \\ x_2 \\ x_3 \end{pmatrix} = \begin{pmatrix} 0 \\ 0 \\ 0 \end{pmatrix}$$

that is

$$\begin{cases} 6x_1 + 3x_2 + 2x_3 = 0 \\ 2x_2 - 4x_3 = 0 \\ -2x_2 + 4x_3 = 0 \end{cases}$$

This system has infinitely many solutions given by

$$\begin{cases} x_1 = -\frac{4}{3}\, x_3 \\ x_2 = 2x_3 \end{cases}$$

The eigenspace corresponding to the eigenvalue -7 therefore has dimension 1 and is generated by the vector $v_1 = (-4, 6, 3)$ (obtained by setting $x_3 = 3$).

Now let's look for the eigenvectors of A associated with the eigenvalue $\lambda = -1$:

$$\begin{pmatrix} 0 & 3 & 2 \\ 0 & -4 & -4 \\ 0 & -2 & -2 \end{pmatrix} \begin{pmatrix} x_1 \\ x_2 \\ x_3 \end{pmatrix} = \begin{pmatrix} 0 \\ 0 \\ 0 \end{pmatrix}$$

We get the system

$$\begin{cases} 3x_2 + 2x_3 = 0 \\ -4x_2 - 4x_3 = 0 \\ -2x_2 - 2x_3 = 0 \end{cases}$$

that has infinitely many solutions given by

$$\begin{cases} x_1 \text{ any value} \\ x_2 = 0 \\ x_3 = 0 \end{cases}$$

It thus turns out that the eigenspace corresponding to the eigenvalue -1 has dimension 1 (this is the so-called *geometric multiplicity*) and is generated by the vector $v_2 = (1, 0, 0)$ (obtained by setting $x_1 = 1$). Recalling that the eigenvalue $\lambda = -1$ has multiplicity 2 (this is the *algebraic multiplicity*), we conclude that the matrix A is not diagonalizable, given that for one of its eigenvalues the two multiplicities (algebraic and geometric) do not coincide.

▶ **Exercise 4.3.2.** Given the vectors $v_1 = (2, -3, 1, 0)$ and $v_2 = (0, -1, 1, -1)$, let $f\colon \mathbb{R}^4 \to \mathbb{R}^4$ be a linear function such that $\operatorname{Ker}(f) = \operatorname{Im}(f) = \langle v_1, v_2 \rangle$.

(a) Write the matrix of f with respect to the canonical basis of $\mathbb{R}^4$.

(b) Prove that f has the eigenvalue $\lambda = 0$ with (algebraic) multiplicity 4, but that it is not diagonalizable.

Solution. Note that the vectors v_1 and v_2 are linearly independent, so the kernel of f and the image of f both have dimension 2. Saying that $v_1, v_2 \in \operatorname{Im}(f)$ means that there exist vectors $w_1, w_2 \in \mathbb{R}^4$ such that $f(w_1) = v_1$ and $f(w_2) = v_2$. Of course w_1 and w_2 are not given (indeed, they are not even uniquely determined). We arbitrarily choose as w_1 and w_2 the first two vectors of the canonical basis of $\mathbb{R}^4$:

$$w_1 = e_1 = (1, 0, 0, 0), \qquad w_2 = e_2 = (0, 1, 0, 0).$$

Recall that the columns of the matrix A of f, with respect to the canonical basis, are the images through f of the vectors of the canonical basis of $\mathbb{R}^4$. This means that the first two columns of A are $f(e_1) = v_1$ and $f(e_2) = v_2$. The matrix A therefore has the following form:

$$A = \begin{pmatrix} 2 & 0 & a_1 & b_1 \\ -3 & -1 & a_2 & b_2 \\ 1 & 1 & a_3 & b_3 \\ 0 & -1 & a_4 & b_4 \end{pmatrix}$$

where $f(e_3) = (a_1, a_2, a_3, a_4)$ and $f(e_4) = (b_1, b_2, b_3, b_4)$. Note that we have not yet used the information that the kernel of f must be generated by the vectors v_1 and v_2. Saying that $v_1, v_2 \in \operatorname{Ker}(f)$ is equivalent to saying that

$$\begin{pmatrix} 2 & 0 & a_1 & b_1 \\ -3 & -1 & a_2 & b_2 \\ 1 & 1 & a_3 & b_3 \\ 0 & -1 & a_4 & b_4 \end{pmatrix} \begin{pmatrix} 2 \\ -3 \\ 1 \\ 0 \end{pmatrix} = \begin{pmatrix} 0 \\ 0 \\ 0 \\ 0 \end{pmatrix}, \quad \begin{pmatrix} 2 & 0 & a_1 & b_1 \\ -3 & -1 & a_2 & b_2 \\ 1 & 1 & a_3 & b_3 \\ 0 & -1 & a_4 & b_4 \end{pmatrix} \begin{pmatrix} 0 \\ -1 \\ 1 \\ -1 \end{pmatrix} = \begin{pmatrix} 0 \\ 0 \\ 0 \\ 0 \end{pmatrix}$$

We thus obtain the two systems of linear equations

$$\begin{cases} 4 + a_1 = 0 \\ -3 + a_2 = 0 \\ -1 + a_3 = 0 \\ 3 + a_4 = 0 \end{cases} \qquad \begin{cases} a_1 - b_1 = 0 \\ 1 + a_2 - b_2 = 0 \\ -1 + a_3 - b_3 = 0 \\ 1 + a_4 - b_4 = 0 \end{cases}$$

from which we obtain $a_1 = -4$, $a_2 = 3$, $a_3 = 1$, $a_4 = -3$, and $b_1 = -4$, $b_2 = 4$, $b_3 = 0$, $b_4 = -2$.

The matrix A is therefore

$$A = \begin{pmatrix} 2 & 0 & -4 & -4 \\ -3 & -1 & 3 & 4 \\ 1 & 1 & 1 & 0 \\ 0 & -1 & -3 & -2 \end{pmatrix}$$

The characteristic polynomial of A is given by

$$\det \begin{pmatrix} 2-\lambda & 0 & -4 & -4 \\ -3 & -1-\lambda & 3 & 4 \\ 1 & 1 & 1-\lambda & 0 \\ 0 & -1 & -3 & -2-\lambda \end{pmatrix} = \lambda^4.$$

This means that A has a unique eigenvalue $\lambda = 0$, with multiplicity 4. The eigenvectors corresponding to this eigenvalue are the solutions of

$$\begin{pmatrix} 2 & 0 & -4 & -4 \\ -3 & -1 & 3 & 4 \\ 1 & 1 & 1 & 0 \\ 0 & -1 & -3 & -2 \end{pmatrix} \begin{pmatrix} x_1 \\ x_2 \\ x_3 \\ x_4 \end{pmatrix} = \begin{pmatrix} 0 \\ 0 \\ 0 \\ 0 \end{pmatrix}$$

that is, they are the vectors of the kernel of A (i.e., of f). Since we already know that the kernel of f has dimension two, we conclude that the geometric multiplicity (i.e., $\dim \operatorname{Ker}(f)$) is different from the algebraic multiplicity, so the matrix A is not diagonalizable.

► **Exercise 4.3.3.** Let V be a vector space with basis $\{v_1, v_2, v_3, v_4\}$ and $f\colon V \to V$ be the linear function defined by $f(v_1) = 2v_1 + 3v_2$, $f(v_2) = 3v_1 + 2v_2$, $f(v_3) = v_1 + 3v_3 + 2v_4$, $f(v_4) = 2v_1 - v_2 + 2v_3 + 3v_4$. Find the eigenvalues and eigenvectors of f and determine whether f is diagonalizable.

Solution. The matrix of f with respect to the given basis is

$$A = \begin{pmatrix} 2 & 3 & 1 & 2 \\ 3 & 2 & 0 & -1 \\ 0 & 0 & 3 & 2 \\ 0 & 0 & 2 & 3 \end{pmatrix}$$

Let's find the characteristic polynomial:

$$\det(A - x\cdot\mathbf{1}) = \begin{vmatrix} 2-x & 3 & 1 & 2 \\ 3 & 2-x & 0 & -1 \\ 0 & 0 & 3-x & 2 \\ 0 & 0 & 2 & 3-x \end{vmatrix}$$
$$= \begin{vmatrix} 2-x & 3 \\ 3 & 2-x \end{vmatrix} \cdot \begin{vmatrix} 3-x & 2 \\ 2 & 3-x \end{vmatrix}$$
$$= (x^2 - 4x - 5)(x^2 - 6x + 5).$$

The polynomial $x^2 - 4x - 5$ vanishes for $x = -1$ and for $x = 5$, while the polynomial x^2-6x+5 vanishes for $x = 1$ and for $x = 5$. The eigenvalues of the matrix A (i.e., the eigenvalues of f) are therefore the following: $\lambda_1 = -1$ (with multiplicity 1), $\lambda_2 = 1$ (with multiplicity 1) and $\lambda_3 = 5$ (with multiplicity 2).

We now look for the eigenvectors. For the eigenvalue $\lambda_1 = -1$ we have:

$$\begin{pmatrix} 3 & 3 & 1 & 2 \\ 3 & 3 & 0 & -1 \\ 0 & 0 & 4 & 2 \\ 0 & 0 & 2 & 4 \end{pmatrix} \begin{pmatrix} x_1 \\ x_2 \\ x_3 \\ x_4 \end{pmatrix} = \begin{pmatrix} 0 \\ 0 \\ 0 \\ 0 \end{pmatrix}$$

which gives the following linear system

$$\begin{cases} 3x_1 + 3x_2 + x_3 + 2x_4 = 0 \\ 3x_1 + 3x_2 - x_4 = 0 \\ 4x_3 + 2x_4 = 0 \\ 2x_3 + 4x_4 = 0 \end{cases}$$

whose solutions are given by

$$\begin{cases} x_1 = -x_2 \\ x_3 = 0 \\ x_4 = 0. \end{cases}$$

The eigenspace corresponding to the eigenvalue $\lambda_1 = -1$ therefore has dimension 1 and is generated by the vector $w_1 = v_1 - v_2$ (the eigenvector w_1 has coordinates $(1, -1, 0, 0)$ with respect to the basis v_1, v_2, v_3, v_4 of V).

For the eigenvalue $\lambda_2 = 1$ we have:

$$\begin{pmatrix} 1 & 3 & 1 & 2 \\ 3 & 1 & 0 & -1 \\ 0 & 0 & 2 & 2 \\ 0 & 0 & 2 & 2 \end{pmatrix} \begin{pmatrix} x_1 \\ x_2 \\ x_3 \\ x_4 \end{pmatrix} = \begin{pmatrix} 0 \\ 0 \\ 0 \\ 0 \end{pmatrix}$$

which gives the following linear system

$$\begin{cases} x_1 + 3x_2 + x_3 + 2x_4 = 0 \\ 3x_1 + x_2 - x_4 = 0 \\ 2x_3 + 2x_4 = 0 \\ 2x_3 + 2x_4 = 0 \end{cases}$$

whose solutions are given by

$$\begin{cases} x_1 = -x_2 \\ x_3 = 2x_2 \\ x_4 = -2x_2. \end{cases}$$

The eigenspace corresponding to the eigenvalue $\lambda_2 = 1$ has dimension 1 and is generated by the vector $w_2 = v_1 - v_2 - 2v_3 + 2v_4$ (the eigenvector w_2 has coordinates $(1, -1, -2, 2)$ with respect to the basis v_1, v_2, v_3, v_4 of V).

Finally, for the eigenvalue $\lambda_3 = 5$ (which has multiplicity 2) we have:

$$\begin{pmatrix} -3 & 3 & 1 & 2 \\ 3 & -3 & 0 & -1 \\ 0 & 0 & -2 & 2 \\ 0 & 0 & 2 & -2 \end{pmatrix} \begin{pmatrix} x_1 \\ x_2 \\ x_3 \\ x_4 \end{pmatrix} = \begin{pmatrix} 0 \\ 0 \\ 0 \\ 0 \end{pmatrix}$$

which gives the following linear system

$$\begin{cases} -3x_1 + 3x_2 + x_3 + 2x_4 = 0 \\ 3x_1 - 3x_2 - x_4 = 0 \\ -2x_3 + 2x_4 = 0 \\ 2x_3 - 2x_4 = 0 \end{cases}$$

whose solutions are given by

$$\begin{cases} x_1 = x_2 \\ x_3 = 0 \\ x_4 = 0. \end{cases}$$

The eigenspace corresponding to the eigenvalue $\lambda_3 = 5$ therefore has dimension 1 and is generated by the vector $w_3 = v_1 + v_2$ (the eigenvector w_3 has coordinates $(1, 1, 0, 0)$ with respect to the basis v_1, v_2, v_3, v_4 of V). Thus we conclude that the matrix A (i.e., the linear function f) is not diagonalizable, since for the eigenvalue $\lambda_3 = 5$ the geometric multiplicity (i.e., the dimension of the corresponding eigenspace) is different from the algebraic multiplicity.

▶ **Exercise 4.3.4.** Let $f\colon \mathbb{R}^3 \to \mathbb{R}^3$ be the linear function defined by $f(v_1) = w_1$, $f(v_2) = w_2$ and $f(v_3) = w_3$, where $v_1 = (1,2,3)$, $v_2 = (2,-1,0)$, $v_3 = (0,-1,-1)$, $w_1 = (6,4,10)$, $w_2 = (5,-1,4)$, $w_3 = (-1,-2,-3)$.

(a) Write the matrix of f with respect to the canonical basis of $\mathbb{R}^3$.

(b) Find a basis of the kernel and a basis of the image of f.

(c) Determine the eigenvalues and eigenvectors of f and find out if f is diagonalizable.

Solution. To write the matrix of f with respect to the canonical basis of $\mathbb{R}^3$ we must compute $f(e_1)$, $f(e_2)$ and $f(e_3)$, where $e_1 = (1,0,0)$, $e_2 = (0,1,0)$, $e_3 = (0,0,1)$. To this end we try to express e_1 as a linear combination of the vectors v_1, v_2 and v_3:

$$\lambda_1 v_1 + \lambda_2 v_2 + \lambda_3 v_3 = e_1,$$

which gives the following system of linear equations

$$\begin{cases} \lambda_1 + 2\lambda_2 = 1 \\ 2\lambda_1 - \lambda_2 - \lambda_3 = 0 \\ 3\lambda_1 - \lambda_3 = 0 \end{cases}$$

whose solution is

$$\begin{cases} \lambda_1 = -1 \\ \lambda_2 = 1 \\ \lambda_3 = -3. \end{cases}$$

Therefore we have $e_1 = -v_1 + v_2 - 3v_3$, hence

$$f(e_1) = -f(v_1) + f(v_2) - 3f(v_3) = -w_1 + w_2 - 3w_3 = (2,1,3).$$

The procedure for computing $f(e_2)$ and $f(e_3)$ is completely analogous. We write e_2 as a linear combination of the vectors v_1, v_2 and v_3:

$$\lambda_1 v_1 + \lambda_2 v_2 + \lambda_3 v_3 = e_2.$$

We obtain the following system

$$\begin{cases} \lambda_1 + 2\lambda_2 = 0 \\ 2\lambda_1 - \lambda_2 - \lambda_3 = 1 \\ 3\lambda_1 - \lambda_3 = 0 \end{cases}$$

whose solution is

$$\begin{cases} \lambda_1 = -2 \\ \lambda_2 = 1 \\ \lambda_3 = -6. \end{cases}$$

Therefore we have $e_2 = -2v_1 + v_2 - 6v_3$, hence

$$f(e_2) = -2f(v_1) + f(v_2) - 6f(v_3) = -2w_1 + w_2 - 6w_3 = (-1, 3, 2).$$

Finally, we express e_3 as a linear combination of the vectors v_1, v_2 and v_3:

$$\lambda_1 v_1 + \lambda_2 v_2 + \lambda_3 v_3 = e_3.$$

We get the following system of linear equations

$$\begin{cases} \lambda_1 + 2\lambda_2 = 0 \\ 2\lambda_1 - \lambda_2 - \lambda_3 = 0 \\ 3\lambda_1 - \lambda_3 = 1 \end{cases}$$

whose solution is

$$\begin{cases} \lambda_1 = 2 \\ \lambda_2 = -1 \\ \lambda_3 = 5. \end{cases}$$

Therefore we have $e_3 = 2v_1 - v_2 + 5v_3$, hence

$$f(e_3) = 2f(v_1) - f(v_2) + 5f(v_3) = 2w_1 - w_2 + 5w_3 = (2, -1, 1).$$

The matrix of f with respect to the canonical basis of $\mathbb{R}^3$ is therefore

$$A = \begin{pmatrix} 2 & -1 & 2 \\ 1 & 3 & -1 \\ 3 & 2 & 1 \end{pmatrix}$$

The matrix A could also be determined in another way. In fact, if A is the matrix of f with respect to the canonical basis of $\mathbb{R}^3$, we must have

$$Av_1 = w_1, \quad Av_2 = w_2, \quad Av_3 = w_3,$$

which is equivalent to

$$A \begin{pmatrix} 1 & 2 & 0 \\ 2 & -1 & -1 \\ 3 & 0 & -1 \end{pmatrix} = \begin{pmatrix} 6 & 5 & -1 \\ 4 & -1 & -2 \\ 10 & 4 & -3 \end{pmatrix}$$

From this equality it follows that

$$A = \begin{pmatrix} 6 & 5 & -1 \\ 4 & -1 & -2 \\ 10 & 4 & -3 \end{pmatrix} \begin{pmatrix} 1 & 2 & 0 \\ 2 & -1 & -1 \\ 3 & 0 & -1 \end{pmatrix}^{-1}$$

We therefore need to compute the inverse of the matrix

$$P = \begin{pmatrix} 1 & 2 & 0 \\ 2 & -1 & -1 \\ 3 & 0 & -1 \end{pmatrix}$$

We proceed as follows. We write the matrix

$$\left(\begin{array}{ccc|ccc} 1 & 2 & 0 & 1 & 0 & 0 \\ 2 & -1 & -1 & 0 & 1 & 0 \\ 3 & 0 & -1 & 0 & 0 & 1 \end{array}\right)$$

Now we carry out the following elementary operations *on the rows*: from the second line we subtract twice the first and from the third line we subtract three times the first:

$$\left(\begin{array}{ccc|ccc} 1 & 2 & 0 & 1 & 0 & 0 \\ 0 & -5 & -1 & -2 & 1 & 0 \\ 0 & -6 & -1 & -3 & 0 & 1 \end{array}\right)$$

Now let's divide the second row by -5:

$$\left(\begin{array}{ccc|ccc} 1 & 2 & 0 & 1 & 0 & 0 \\ 0 & 1 & \frac{1}{5} & \frac{2}{5} & -\frac{1}{5} & 0 \\ 0 & -6 & -1 & -3 & 0 & 1 \end{array}\right)$$

To the third row we add the second multiplied by 6:

$$\left(\begin{array}{ccc|ccc} 1 & 2 & 0 & 1 & 0 & 0 \\ 0 & 1 & \frac{1}{5} & \frac{2}{5} & -\frac{1}{5} & 0 \\ 0 & 0 & \frac{1}{5} & -\frac{3}{5} & -\frac{6}{5} & 1 \end{array}\right)$$

Now let's multiply the third row by 5:

$$\left(\begin{array}{ccc|ccc} 1 & 2 & 0 & 1 & 0 & 0 \\ 0 & 1 & \frac{1}{5} & \frac{2}{5} & -\frac{1}{5} & 0 \\ 0 & 0 & 1 & -3 & -6 & 5 \end{array}\right)$$

Now from the second row we subtract the third row divided by 5:

$$\left(\begin{array}{ccc|ccc} 1 & 2 & 0 & 1 & 0 & 0 \\ 0 & 1 & 0 & 1 & 1 & -1 \\ 0 & 0 & 1 & -3 & -6 & 5 \end{array}\right)$$

Finally, we subtract twice the second from the first row:

$$\left(\begin{array}{ccc|ccc} 1 & 0 & 0 & -1 & -2 & 2 \\ 0 & 1 & 0 & 1 & 1 & -1 \\ 0 & 0 & 1 & -3 & -6 & 5 \end{array}\right)$$

We conclude that the inverse of the matrix P is

$$P^{-1} = \begin{pmatrix} -1 & -2 & 2 \\ 1 & 1 & -1 \\ -3 & -6 & 5 \end{pmatrix}$$

Therefore we have

$$A = \begin{pmatrix} 6 & 5 & -1 \\ 4 & -1 & -2 \\ 10 & 4 & -3 \end{pmatrix} \begin{pmatrix} -1 & -2 & 2 \\ 1 & 1 & -1 \\ -3 & -6 & 5 \end{pmatrix} = \begin{pmatrix} 2 & -1 & 2 \\ 1 & 3 & -1 \\ 3 & 2 & 1 \end{pmatrix}$$

The vectors of the kernel of f are the vectors (x_1, x_2, x_3) such that

$$\begin{pmatrix} 2 & -1 & 2 \\ 1 & 3 & -1 \\ 3 & 2 & 1 \end{pmatrix} \begin{pmatrix} x_1 \\ x_2 \\ x_3 \end{pmatrix} = \begin{pmatrix} 0 \\ 0 \\ 0 \end{pmatrix}$$

i.e., the solutions of the following system:

$$\begin{cases} 2x_1 - x_2 + 2x_3 = 0 \\ x_1 + 3x_2 - x_3 = 0 \\ 3x_1 + 2x_2 + x_3 = 0. \end{cases}$$

The solutions are given by

$$\begin{cases} x_1 = -\frac{5}{7}x_3 \\ x_2 = \frac{4}{7}x_3 \\ x_3 \text{ any value,} \end{cases}$$

from which we deduce that $\mathrm{Ker}(f)$ has dimension 1 and its basis is given by the vector $v = (-5, 4, 7)$.

Since $\dim(\operatorname{Ker} f) + \dim(\operatorname{Im} f) = \dim \mathbb{R}^3 = 3$, we have $\dim(\operatorname{Im} f) = 2$, hence to find a basis of the image of f it suffices to find two linearly independent vectors belonging to $\operatorname{Im} f$. We can then choose the first two columns of A, or the vectors w_1 and w_2, or w_2 and w_3, etc.

To compute the eigenvalues of f (i.e., of A) we compute its characteristic polynomial:

$$\det(A - \lambda \cdot \mathbf{1}) = \begin{vmatrix} 2-\lambda & -1 & 2 \\ 1 & 3-\lambda & -1 \\ 3 & 2 & 1-\lambda \end{vmatrix} = -\lambda(\lambda^2 - 6\lambda + 8).$$

The zeros of this polynomial (i.e., the eigenvalues of A) are $\lambda_1 = 0$, $\lambda_2 = 2$ and $\lambda_3 = 4$. Since the matrix A has three distinct eigenvalues it will also have three linearly independent eigenvectors. There will therefore exist a basis of $\mathbb{R}^3$ formed by eigenvectors of A (i.e., of f), which is equivalent to saying that A (i.e., f) is diagonalizable.

Now all we have to do is determine the eigenvectors of the matrix A. For the eigenvalue $\lambda_1 = 0$ we need to solve the system

$$(A - 0 \cdot \mathbf{1}) \begin{pmatrix} x_1 \\ x_2 \\ x_3 \end{pmatrix} = \begin{pmatrix} 0 \\ 0 \\ 0 \end{pmatrix}$$

that is

$$\begin{cases} 2x_1 - x_2 + 2x_3 = 0 \\ x_1 + 3x_2 - x_3 = 0 \\ 3x_1 + 2x_2 + x_3 = 0. \end{cases}$$

The eigenvectors corresponding to the eigenvalue $\lambda_1 = 0$ are therefore the vectors of the kernel of f, which we have already computed.

For the eigenvalue $\lambda_2 = 2$ we need to solve the system

$$(A - 2 \cdot \mathbf{1}) \begin{pmatrix} x_1 \\ x_2 \\ x_3 \end{pmatrix} = \begin{pmatrix} 0 \\ 0 \\ 0 \end{pmatrix}$$

that is

$$\begin{cases} -x_2 + 2x_3 = 0 \\ x_1 + x_2 - x_3 = 0 \\ 3x_1 + 2x_2 - x_3 = 0. \end{cases}$$

The solutions are given by

$$\begin{cases} x_1 = -x_3 \\ x_2 = 2x_3 \\ x_3 \text{ any value} \end{cases}$$

therefore the eigenspace corresponding to the eigenvalue $\lambda_2 = 2$ has dimension 1 and is generated by the vector $(-1, 2, 1)$.

Finally, for the eigenvalue $\lambda_3 = 4$, we need to solve the system

$$(A - 4 \cdot \mathbf{1}) \begin{pmatrix} x_1 \\ x_2 \\ x_3 \end{pmatrix} = \begin{pmatrix} 0 \\ 0 \\ 0 \end{pmatrix}$$

that is

$$\begin{cases} -2x_1 - x_2 + 2x_3 = 0 \\ x_1 - x_2 - x_3 = 0 \\ 3x_1 + 2x_2 - 3x_3 = 0. \end{cases}$$

The solutions are

$$\begin{cases} x_1 = x_3 \\ x_2 = 0 \\ x_3 \text{ any value} \end{cases}$$

therefore the eigenspace corresponding to the eigenvalue $\lambda_3 = 4$ has dimension 1 and is generated by the vector $(1, 0, 1)$.

► **Exercise 4.3.5.** Let $v_1 = (1, 1, 1)$, $v_2 = (2, 0, 1)$, $v_3 = (1, 1, 3) \in \mathbb{R}^3$ and $f \colon \mathbb{R}^3 \to \mathbb{R}^3$ be a linear function such that $f(v_1) = 3v_1$, $f(v_2) = 2v_2$, $f(v_3) = 2v_3 + 2v_2$.

(a) Write the matrix of f with respect to the canonical bases.

(b) Determine the eigenvalues and eigenvectors of f.

(c) Verify that the eigenspaces of f are in direct sum.

Solution. If A is the matrix of f with respect to the canonical bases, we must have

$$Av_1 = 3v_1, \qquad Av_2 = 2v_2, \qquad Av_3 = 2v_3 + 2v_2,$$

which is equivalent to

$$A\begin{pmatrix}1&2&1\\1&0&1\\1&1&3\end{pmatrix}=\begin{pmatrix}1&2&1\\1&0&1\\1&1&3\end{pmatrix}\begin{pmatrix}3&0&0\\0&2&2\\0&0&2\end{pmatrix}$$

Setting

$$P=\begin{pmatrix}1&2&1\\1&0&1\\1&1&3\end{pmatrix}\qquad D=\begin{pmatrix}3&0&0\\0&2&2\\0&0&2\end{pmatrix}$$

we have

$$AP=PD,\qquad \text{i.e.}\qquad A=PDP^{-1}.$$

To determine the matrix A we must therefore compute the inverse of P. We proceed as follows. We write the matrix

$$\left(\begin{array}{ccc|ccc}1&2&1&1&0&0\\1&0&1&0&1&0\\1&1&3&0&0&1\end{array}\right)$$

Now we carry out the following elementary operations *on the rows*: from the second row we subtract the first and from the third row we subtract the second:

$$\left(\begin{array}{ccc|ccc}1&2&1&1&0&0\\0&-2&0&-1&1&0\\0&1&2&0&-1&1\end{array}\right)$$

Now let's swap the second row with the third:

$$\left(\begin{array}{ccc|ccc}1&2&1&1&0&0\\0&1&2&0&-1&1\\0&-2&0&-1&1&0\end{array}\right)$$

To the third row we add twice the second:

$$\left(\begin{array}{ccc|ccc}1&2&1&1&0&0\\0&1&2&0&-1&1\\0&0&4&-1&-1&2\end{array}\right)$$

Now let's divide the third row by 4:

$$\left(\begin{array}{ccc|ccc}1&2&1&1&0&0\\0&1&2&0&-1&1\\0&0&1&-\frac{1}{4}&-\frac{1}{4}&\frac{1}{2}\end{array}\right)$$

Now from the second row we subtract twice the third and from the first row we subtract the third:

$$\left(\begin{array}{ccc|ccc} 1 & 2 & 0 & \frac{5}{4} & \frac{1}{4} & -\frac{1}{2} \\ 0 & 1 & 0 & \frac{1}{2} & -\frac{1}{2} & 0 \\ 0 & 0 & 1 & -\frac{1}{4} & -\frac{1}{4} & \frac{1}{2} \end{array}\right)$$

Finally, we subtract twice the second row from the first:

$$\left(\begin{array}{ccc|ccc} 1 & 0 & 0 & \frac{1}{4} & \frac{5}{4} & -\frac{1}{2} \\ 0 & 1 & 0 & \frac{1}{2} & -\frac{1}{2} & 0 \\ 0 & 0 & 1 & -\frac{1}{4} & -\frac{1}{4} & \frac{1}{2} \end{array}\right)$$

We conclude that the inverse of the matrix P is

$$P^{-1} = \begin{pmatrix} \frac{1}{4} & \frac{5}{4} & -\frac{1}{2} \\ \frac{1}{2} & -\frac{1}{2} & 0 \\ -\frac{1}{4} & -\frac{1}{4} & \frac{1}{2} \end{pmatrix}$$

Therefore we have

$$A = \begin{pmatrix} 1 & 2 & 1 \\ 1 & 0 & 1 \\ 1 & 1 & 3 \end{pmatrix} \begin{pmatrix} 3 & 0 & 0 \\ 0 & 2 & 2 \\ 0 & 0 & 2 \end{pmatrix} \begin{pmatrix} \frac{1}{4} & \frac{5}{4} & -\frac{1}{2} \\ \frac{1}{2} & -\frac{1}{2} & 0 \\ -\frac{1}{4} & -\frac{1}{4} & \frac{1}{2} \end{pmatrix} = \begin{pmatrix} \frac{5}{4} & \frac{1}{4} & \frac{3}{2} \\ \frac{1}{4} & \frac{13}{4} & -\frac{1}{2} \\ -\frac{1}{4} & \frac{3}{4} & \frac{5}{2} \end{pmatrix}$$

The eigenvalues of f are, of course, the eigenvalues of the matrix A, but these coincide with the eigenvalues of the matrix D, since A and D are similar (A is the matrix of f with respect to the canonical basis, while D is the matrix of f with respect to the basis formed by the assigned vectors v_1, v_2 and v_3). Since D is an upper triangular matrix, it is easy to see that its eigenvalues are the elements on the diagonal. We can therefore state that the eigenvalues of f are $\lambda_1 = 2$, with multiplicity 2, and $\lambda_2 = 3$ (with multiplicity 1). The eigenspace corresponding to the eigenvalue $\lambda_1 = 2$ is generated by the vector v_2 (remember that we have $f(v_2) = 2v_2$), therefore it has dimension 1; from this we deduce that f is not diagonalizable. The eigenspace corresponding to the eigenvalue $\lambda_2 = 3$ is generated by the vector v_1 (remember that we have $f(v_1) = 3v_1$). Note that v_3 is not an eigenvector of f, since $f(v_3) = 2v_3 + 2v_2$.

Finally, we observe that the vectors $v_1 = (1, 1, 1)$ and $v_2 = (2, 0, 1)$ are linearly independent and therefore the intersection between the subspaces they generate is zero: $\langle v_1 \rangle \cap \langle v_2 \rangle = \{\mathbf{0}\}$. This means that the eigenspaces of f are in direct sum.

▶ **Exercise 4.3.6.** Given the matrix

$$A = \begin{pmatrix} -t & 3+t & -t \\ 3-t & t & 3-t \\ 6 & -6 & 6 \end{pmatrix}$$

determine if there are any values of $t \in \mathbb{R}$ for which A is invertible. Also determine for which value of t the matrix A is diagonalizable. For this value of t find a diagonal matrix D and an invertible matrix P such that $A = PDP^{-1}$.

Solution. To determine whether A is invertible, we compute its determinant. If we subtract the second from the first row, we have

$$\det A = \begin{vmatrix} -3 & 3 & -3 \\ 3-t & t & 3-t \\ 6 & -6 & 6 \end{vmatrix}$$

Now to the third row we add twice the first, thus obtaining

$$\det A = \begin{vmatrix} -3 & 3 & -3 \\ 3-t & t & 3-t \\ 0 & 0 & 0 \end{vmatrix} = 0.$$

Since the determinant of A is zero for any value of t, we conclude that A is not invertible for any value of the parameter t. (Of course, to conclude that $\det A = 0$ it was enough to observe that the matrix A has two equal columns, the first and the third).

We now look for the eigenvalues of A. The characteristic polynomial is

$$\det(A - x \cdot \mathbf{1}) = \begin{vmatrix} -t-x & 3+t & -t \\ 3-t & t-x & 3-t \\ 6 & -6 & 6-x \end{vmatrix}$$

To compute this determinant from the first row we subtract the second, obtaining

$$\det(A - x \cdot \mathbf{1}) = \begin{vmatrix} -3-x & 3+x & -3 \\ 3-t & t-x & 3-t \\ 6 & -6 & 6-x \end{vmatrix}$$

Now add the second to the first column:

$$\det(A - x \cdot \mathbf{1}) = \begin{vmatrix} 0 & 3+x & -3 \\ 3-x & t-x & 3-t \\ 0 & -6 & 6-x \end{vmatrix}$$

Now we can use Laplace formula along the first column:

$$\begin{aligned}\det(A - x \cdot \mathbf{1}) &= -(3-x)\begin{vmatrix} 3+x & -3 \\ -6 & 6-x \end{vmatrix} \\ &= -(3-x)(3x - x^2) \\ &= -x(3-x)^2.\end{aligned}$$

The eigenvalues of the matrix A are therefore $\lambda_1 = 0$, with multiplicity 1, and $\lambda_2 = 3$, with multiplicity 2.

The eigenvectors corresponding to the eigenvalue $\lambda_1 = 0$ are the solutions of the following linear system

$$\begin{pmatrix} -t & 3+t & -t \\ 3-t & t & 3-t \\ 6 & -6 & 6 \end{pmatrix}\begin{pmatrix} x_1 \\ x_2 \\ x_3 \end{pmatrix} = \begin{pmatrix} 0 \\ 0 \\ 0 \end{pmatrix}$$

that is, they are the vectors that belong to the kernel of A. Solving the system, it turns out that it has infinitely many solutions, given by

$$\begin{cases} x_1 = -x_3 \\ x_2 = 0. \end{cases}$$

This means that the eigenspace for the eigenvalue $\lambda_1 = 0$ (i.e., the kernel of A) has dimension 1 and is generated by the vector $v_1 = (1, 0, -1)$.

We now pass to the eigenvectors corresponding to the eigenvalue $\lambda_2 = 3$. They are the solutions of the following linear system

$$\begin{pmatrix} -t-3 & 3+t & -t \\ 3-t & t-3 & 3-t \\ 6 & -6 & 3 \end{pmatrix}\begin{pmatrix} x_1 \\ x_2 \\ x_3 \end{pmatrix} = \begin{pmatrix} 0 \\ 0 \\ 0 \end{pmatrix}$$

The solutions of this system are given by

$$\begin{cases} x_3 = 2x_2 - 2x_1 \\ (t-3)(x_1 - x_2) = 0. \end{cases}$$

We therefore distinguish two cases. If $t \neq 3$, the previous system becomes

$$\begin{cases} x_3 = 2x_2 - 2x_1 \\ x_1 - x_2 = 0 \end{cases}$$

whose solutions are given by

$$\begin{cases} x_1 = x_2 \\ x_3 = 0. \end{cases}$$

In this case the eigenspace corresponding to the eigenvalue $\lambda_2 = 3$ has dimension 1 and is generated by the vector $v_2 = (1, 1, 0)$. The matrix A is therefore not diagonalizable, given that the algebraic multiplicity of the eigenvalue $\lambda_2 = 3$ is different from its geometric multiplicity.

If instead $t = 3$, the previous system reduces to the single equation

$$x_3 = 2x_2 - 2x_1.$$

In this case there are infinitely many solutions, depending on *two* parameters (x_1 and x_2). The eigenspace corresponding to the eigenvalue $\lambda_2 = 3$ now has dimension 2 and is generated by the vectors $v_2 = (1, 0, -2)$ and $v_3 = (0, 1, 2)$. Thus we conclude that, for $t = 3$, the matrix A is diagonalizable. The matrices P and D are

$$P = \begin{pmatrix} 1 & 1 & 0 \\ 0 & 0 & 1 \\ -1 & -2 & 2 \end{pmatrix}, \qquad D = \begin{pmatrix} 0 & 0 & 0 \\ 0 & 3 & 0 \\ 0 & 0 & 3 \end{pmatrix}.$$

▶ **Exercise 4.3.7.** Let $v_1 = (1, -1, 2)$, $v_2 = (-1, 2, 0)$, $v_3 = (0, 1, 1) \in \mathbb{R}^3$ and $f\colon \mathbb{R}^3 \to \mathbb{R}^3$ be the linear function whose kernel is spanned by v_1 and such that $f(v_2) = 2v_2$ and $f(v_3) = v_3$.

(a) Using the change-of-basis formula, write the matrix of f with respect to the canonical bases.

(b) Determine if there exists a linear function $g\colon \mathbb{R}^3 \to \mathbb{R}^3$ such that the composition $g \circ f$ is invertible.

Solution. Since $\mathrm{Ker}(f)$ is spanned by v_1, we have $f(v_1) = \mathbf{0} = 0v_1$, so v_1 is an eigenvector of f for the eigenvalue 0. Knowing that $f(v_2) = 2v_2$ and $f(v_3) = v_3$, we conclude that v_2 is an eigenvector for the eigenvalue 2 and v_3 is an eigenvector for the eigenvalue 1. If we set

$$P = \begin{pmatrix} 1 & -1 & 0 \\ -1 & 2 & 1 \\ 2 & 0 & 1 \end{pmatrix} \quad \text{and} \quad D = \begin{pmatrix} 0 & 0 & 0 \\ 0 & 2 & 0 \\ 0 & 0 & 1 \end{pmatrix}$$

we have $AP = PD$, where A is the matrix of f with respect to the canonical bases. Since the vectors v_1, v_2 and v_3 are linearly independent, the matrix P is invertible and therefore the matrix A can be obtained using the formula

$$A = PDP^{-1}.$$

Now all that remains is to determine the inverse of P. We proceed as follows. We write the matrix

$$\left(\begin{array}{ccc|ccc} 1 & -1 & 0 & 1 & 0 & 0 \\ -1 & 2 & 1 & 0 & 1 & 0 \\ 2 & 0 & 1 & 0 & 0 & 1 \end{array}\right)$$

Now we carry out the following elementary operations *on the rows*: we add the first to the second row and we subtract twice the first from the third row:

$$\left(\begin{array}{ccc|ccc} 1 & -1 & 0 & 1 & 0 & 0 \\ 0 & 1 & 1 & 1 & 1 & 0 \\ 0 & 2 & 1 & -2 & 0 & 1 \end{array}\right)$$

Now we subtract twice the second from the third row:

$$\left(\begin{array}{ccc|ccc} 1 & -1 & 0 & 1 & 0 & 0 \\ 0 & 1 & 1 & 1 & 1 & 0 \\ 0 & 0 & -1 & -4 & -2 & 1 \end{array}\right)$$

We multiply the third row by -1 to get:

$$\left(\begin{array}{ccc|ccc} 1 & -1 & 0 & 1 & 0 & 0 \\ 0 & 1 & 1 & 1 & 1 & 0 \\ 0 & 0 & 1 & 4 & 2 & -1 \end{array}\right)$$

Now we subtract the third from the second row:

$$\left(\begin{array}{ccc|ccc} 1 & -1 & 0 & 1 & 0 & 0 \\ 0 & 1 & 0 & -3 & -1 & 1 \\ 0 & 0 & 1 & 4 & 2 & -1 \end{array}\right)$$

Finally, we add the second to the first row:

$$\left(\begin{array}{ccc|ccc} 1 & 0 & 0 & -2 & -1 & 1 \\ 0 & 1 & 0 & -3 & -1 & 1 \\ 0 & 0 & 1 & 4 & 2 & -1 \end{array}\right)$$

We conclude that the inverse of the matrix P is

$$P^{-1} = \begin{pmatrix} -2 & -1 & 1 \\ -3 & -1 & 1 \\ 4 & 2 & -1 \end{pmatrix}$$

Therefore we have

$$A = \begin{pmatrix} 1 & -1 & 0 \\ -1 & 2 & 1 \\ 2 & 0 & 1 \end{pmatrix} \begin{pmatrix} 0 & 0 & 0 \\ 0 & 2 & 0 \\ 0 & 0 & 1 \end{pmatrix} \begin{pmatrix} -2 & -1 & 1 \\ -3 & -1 & 1 \\ 4 & 2 & -1 \end{pmatrix} = \begin{pmatrix} 6 & 2 & -2 \\ -8 & -2 & 3 \\ 4 & 2 & -1 \end{pmatrix}$$

To answer the second question, just remember that $f(v_1) = \mathbf{0}$. Therefore, for every linear function g, we have

$$(g \circ f)(v_1) = g(f(v_1)) = g(\mathbf{0}) = \mathbf{0},$$

hence $v_1 \in \operatorname{Ker}(g \circ f)$. This means that $\operatorname{Ker}(g \circ f) \neq \{\mathbf{0}\}$ and therefore $g \circ f$ is not injective. Not being injective it cannot be invertible!

► **Exercise 4.3.8.** Let V be a vector space with basis v_1, v_2, v_3, v_4. Let us consider the linear function $f\colon V \to V$ defined by $f(v_1) = v_1 + 3v_3$, $f(v_2) = v_2 + v_4$, $f(v_3) = 2v_1 + 2v_3$, $f(v_4) = -2v_2 - 2v_4$.

(a) Determine if f is surjective and find a basis of $\operatorname{Im} f$.

(b) Determine, if possible, a basis w_1, w_2, w_3, w_4 of V with respect to which the matrix of f is diagonal.

(c) Determine if there are two linearly independent vectors $u_1, u_2 \in V$ such that $f(u_1) = f(u_2)$.

Solution. From the definition of f we immediately deduce that its matrix with respect to the basis v_1, v_2, v_3, v_4 is the following:

$$A = \begin{pmatrix} 1 & 0 & 2 & 0 \\ 0 & 1 & 0 & -2 \\ 3 & 0 & 2 & 0 \\ 0 & 1 & 0 & -2 \end{pmatrix}$$

It is easy to verify that the rank of A is 3 (note that there are two equal rows), so f is not surjective, $\dim \operatorname{Im}(f) = 3$ and a basis of $\operatorname{Im}(f)$ is given by the vectors whose coordinates are represented by the first three columns of

A (that are linearly independent). Therefore a basis of $\mathrm{Im}(f)$ is given by the vectors

$$v_1 + 3v_3, \quad v_2 + v_4, \quad 2v_1 + 2v_3.$$

To find a basis of V with respect to which the matrix of f is diagonal, it suffices to compute the eigenvalues and eigenvectors of f (i.e., of A) and check whether A is diagonalizable or not.

The characteristic polynomial of A is

$$\det(A - x \cdot \mathbf{1}) = \begin{vmatrix} 1-x & 0 & 2 & 0 \\ 0 & 1-x & 0 & -2 \\ 3 & 0 & 2-x & 0 \\ 0 & 1 & 0 & -2-x \end{vmatrix}$$

By swapping the second row with the third, we have

$$\det(A - x \cdot \mathbf{1}) = -\begin{vmatrix} 1-x & 0 & 2 & 0 \\ 3 & 0 & 2-x & 0 \\ 0 & 1-x & 0 & -2 \\ 0 & 1 & 0 & -2-x \end{vmatrix}$$

Now exchanging the second column with the third, we obtain

$$\det(A - x \cdot \mathbf{1}) = \begin{vmatrix} 1-x & 2 & 0 & 0 \\ 3 & 2-x & 0 & 0 \\ 0 & 0 & 1-x & -2 \\ 0 & 0 & 1 & -2-x \end{vmatrix}$$

From the properties of the determinant, we have

$$\begin{vmatrix} 1-x & 2 & 0 & 0 \\ 3 & 2-x & 0 & 0 \\ 0 & 0 & 1-x & -2 \\ 0 & 0 & 1 & -2-x \end{vmatrix} = \begin{vmatrix} 1-x & 2 \\ 3 & 2-x \end{vmatrix} \cdot \begin{vmatrix} 1-x & -2 \\ 1 & -2-x \end{vmatrix}$$
$$= (x^2 - 3x - 4)(x^2 + x).$$

The eigenvalues of A are therefore the solutions of the equations

$$x^2 - 3x - 4 = 0 \quad \text{and} \quad x^2 + x = 0,$$

from which it follows that the eigenvalues are $x_1 = 0$, $x_2 = 4$ (both with multiplicity 1) and $x_3 = x_4 = -1$ (with multiplicity 2).

The eigenvectors corresponding to the eigenvalue $x_1 = 0$ are the solutions of the system

$$\begin{cases} x_1 + 2x_3 = 0 \\ x_2 - 2x_4 = 0 \\ 3x_1 + 2x_3 = 0 \\ x_2 - 2x_4 = 0 \end{cases}$$

which is equivalent to

$$\begin{cases} x_1 = 0 \\ x_2 = 2x_4 \\ x_3 = 0 \end{cases}$$

from which it follows that the eigenspace corresponding to the eigenvalue 0 (i.e., the kernel of f) has dimension 1 and is generated by the vector $w_1 = 2v_2 + v_4$.

The eigenvectors corresponding to the eigenvalue $x_2 = 4$ are the solutions of the system

$$\begin{cases} -3x_1 + 2x_3 = 0 \\ -3x_2 - 2x_4 = 0 \\ 3x_1 - 2x_3 = 0 \\ x_2 - 6x_4 = 0 \end{cases}$$

which is equivalent to

$$\begin{cases} x_1 = \frac{2}{3}x_3 \\ x_2 = 0 \\ x_4 = 0 \end{cases}$$

from which it follows that the eigenspace corresponding to the eigenvalue 4 has dimension 1 and is generated by the vector $w_2 = 2v_1 + 3v_3$.

Finally, the eigenvectors corresponding to the eigenvalue -1 are the solutions of the system

$$\begin{cases} 2x_1 + 2x_3 = 0 \\ 2x_2 - 2x_4 = 0 \\ 3x_1 + 3x_3 = 0 \\ x_2 - x_4 = 0 \end{cases}$$

which is equivalent to

$$\begin{cases} x_1 = -x_3 \\ x_2 = x_4 \end{cases}$$

from which it follows that the eigenspace corresponding to the eigenvalue -1 has dimension 2 and is generated by the vectors $w_3 = -v_1 + v_3$ and $w_4 = v_2 + v_4$.

From what we have seen we conclude that the matrix A is diagonalizable, that a basis of V with respect to which the matrix of f is diagonal is given by the eigenvectors w_1, w_2, w_3, w_4, and that the matrix of f with respect to this basis is the following:

$$A' = \begin{pmatrix} 0 & 0 & 0 & 0 \\ 0 & 4 & 0 & 0 \\ 0 & 0 & -1 & 0 \\ 0 & 0 & 0 & -1 \end{pmatrix}$$

Finally, given that f is not injective (because $\mathrm{Ker}(f) \neq \{\mathbf{0}\}$), there certainly exist two linearly independent vectors $u_1, u_2 \in V$ such that $f(u_1) = f(u_2)$. Just take, for example, $u_1 = v_1$ and $u_2 = u_1 + w_1 = v_1 + 2v_2 + v_4$. u_1 and u_2 are linearly independent and we have $f(u_2) = f(u_1 + w_1) = f(u_1) + f(w_1) = f(u_1)$, because $f(w_1) = \mathbf{0}$, since $w_1 \in \mathrm{Ker}(f)$.

► **Exercise 4.3.9.** Given $h \in \mathbb{R}$, let us consider the matrix

$$A_h = \begin{pmatrix} 0 & 1 & 0 \\ h & 0 & 1 \\ 0 & 1 & 0 \end{pmatrix}$$

(a) Determine for which values of h the matrix A_h is diagonalizable.

(b) For each $\bar{h}$ such that $A_{\bar{h}}$ has eigenvectors with multiplicity > 1 find all the eigenspaces.

(c) For $h = 3$ find a matrix P such that $P^{-1}A_3P$ is diagonal.

(d) For the $\bar{h}$ of point (b) show that $(A_{\bar{h}})^3 = 0$. For such $\bar{h}$, is it true that $A_{\bar{h}}$ is similar to the matrix $\begin{pmatrix} 0 & 1 & 0 \\ 0 & 0 & 0 \\ 0 & 0 & 0 \end{pmatrix}$?

Solution. The characteristic polynomial of A_h is

$$\det(A_h - x \cdot \mathbf{1}) = \begin{vmatrix} -x & 1 & 0 \\ h & -x & 1 \\ 0 & 1 & -x \end{vmatrix} = -x(x^2 - 1 - h).$$

From the equation $x(x^2 - 1 - h) = 0$ we get $x = 0$, $x^2 = 1 + h$. We can then distinguish the following three cases:

- If $h > -1$ the matrix A_h has three distinct real eigenvalues: 0, $-\sqrt{1+h}$, $\sqrt{1+h}$. In this case A_h is diagonalizable.
- If $h < -1$ the matrix A_h has only one real eigenvalue, 0, and two complex conjugate eigenvalues. In this case A_h is not diagonalizable over the field of real numbers (but it is over the field of complex numbers).
- If $h = -1$ the matrix A_h has a real eigenvalue, 0, with multiplicity 3. In this case, to find out if A_h is diagonalizable we need to compute the dimension of the eigenspace corresponding to the eigenvalue 0 (the geometric multiplicity).

We then analyze the only doubtful case, namely the case $h = -1$. The matrix then becomes

$$A_{-1} = \begin{pmatrix} 0 & 1 & 0 \\ -1 & 0 & 1 \\ 0 & 1 & 0 \end{pmatrix}$$

The eigenspace corresponding to the eigenvalue 0 (i.e., the kernel of A_{-1}) is given by the solutions of the following system

$$\begin{cases} x_2 = 0 \\ -x_1 + x_3 = 0 \end{cases}$$

from which we deduce that this eigenspace has dimension 1 and is generated by the eigenvector $(1, 0, 1)$. This means that the geometric multiplicity is equal to 1 and is therefore different from the algebraic multiplicity of the eigenvalue (which is 3). The matrix A_{-1} is therefore not diagonalizable.

Remark: actually it was not necessary to determine the eigenspace corresponding to the eigenvalue 0 to discover that the matrix A_{-1} is not diagonalizable. Indeed A_{-1} has only the eigenvalue 0 with multiplicity 3 therefore, if it were diagonalizable, it would be similar to the zero matrix. However the only matrix similar to the zero matrix is the zero matrix itself!

The only $\bar{h}$ for which $A_{\bar{h}}$ has eigenvalues with multiplicity > 1 is $\bar{h} = -1$. As we have already seen, the only eigenspace of the matrix A_{-1} is the one corresponding to the eigenvalue 0, of which we have already found a basis.

Let us now consider the case $h = 3$. The matrix becomes

$$A_3 = \begin{pmatrix} 0 & 1 & 0 \\ 3 & 0 & 1 \\ 0 & 1 & 0 \end{pmatrix}$$

As seen in point (a), the matrix A_3 is diagonalizable and its eigenvalues are 0, -2 and 2. The eigenvectors corresponding to the eigenvalue 0 are given by the solutions of the system

$$\begin{cases} x_2 = 0 \\ 3x_1 + x_3 = 0 \end{cases}$$

A basis of the eigenspace corresponding to the eigenvalue 0 is therefore given by the vector $(1, 0, -3)$.

The eigenvectors corresponding to the eigenvalue -2 are the solutions of the system

$$\begin{cases} 2x_1 + x_2 = 0 \\ 3x_1 + 2x_2 + x_3 = 0 \\ x_2 + 2x_3 = 0 \end{cases}$$

A basis of the eigenspace corresponding to the eigenvalue -2 is therefore given by the vector $(1, -2, 1)$.

Finally, the eigenvectors for the eigenvalue 2 are obtained by solving the system

$$\begin{cases} -2x_1 + x_2 = 0 \\ 3x_1 - 2x_2 + x_3 = 0 \\ x_2 - 2x_3 = 0 \end{cases}$$

A basis of the eigenspace corresponding to the eigenvalue 2 is therefore given by the vector $(1, 2, 1)$.

The matrix P is the matrix whose columns are the three eigenvectors just found

$$P = \begin{pmatrix} 1 & 1 & 1 \\ 0 & -2 & 2 \\ -3 & 1 & 1 \end{pmatrix}$$

If we set

$$D = \begin{pmatrix} 0 & 0 & 0 \\ 0 & -2 & 0 \\ 0 & 0 & 2 \end{pmatrix}$$

we have $A_3 P = PD$, hence $D = P^{-1} A_3 P$.

Finally, since the $\bar{h}$ of point (b) is $\bar{h} = -1$, we have to compute $(A_{-1})^3$. We have

$$(A_{-1})^2 = \begin{pmatrix} 0 & 1 & 0 \\ -1 & 0 & 1 \\ 0 & 1 & 0 \end{pmatrix} \begin{pmatrix} 0 & 1 & 0 \\ -1 & 0 & 1 \\ 0 & 1 & 0 \end{pmatrix} = \begin{pmatrix} -1 & 0 & 1 \\ 0 & 0 & 0 \\ -1 & 0 & 1 \end{pmatrix}$$

hence

$$(A_{-1})^3 = A_{-1}\,(A_{-1})^2 = \begin{pmatrix} 0 & 1 & 0 \\ -1 & 0 & 1 \\ 0 & 1 & 0 \end{pmatrix} \begin{pmatrix} -1 & 0 & 1 \\ 0 & 0 & 0 \\ -1 & 0 & 1 \end{pmatrix} = \begin{pmatrix} 0 & 0 & 0 \\ 0 & 0 & 0 \\ 0 & 0 & 0 \end{pmatrix}$$

as claimed. The matrices

$$A_{-1} = \begin{pmatrix} 0 & 1 & 0 \\ -1 & 0 & 1 \\ 0 & 1 & 0 \end{pmatrix} \quad \text{and} \quad B = \begin{pmatrix} 0 & 1 & 0 \\ 0 & 0 & 0 \\ 0 & 0 & 0 \end{pmatrix}$$

are not similar. To prove it, just note that $(A_{-1})^2 \neq 0$, while $B^2 = 0$. More simply, just note that $\mathrm{rk}(A_{-1}) = 2$ while $\mathrm{rk}(B) = 1$ therefore, having different ranks, the matrices A_{-1} and B cannot correspond to the same linear function.

► **Exercise 4.3.10.** Let $f\colon \mathbb{R}^3 \to \mathbb{R}^3$ be the linear function defined by setting

$$f(x, y, z) = (2x + y - 3z, -x + 2z, x - 2y - 4z).$$

(a) Write the matrix A of f with respect to the canonical bases.

(b) Find bases of $\mathrm{Ker}(f)$ and $\mathrm{Im}(f)$ and determine if $\mathrm{Ker}(f)$ and $\mathrm{Im}(f)$ are in direct sum.

(c) Write the matrix B of f with respect to the canonical basis of the domain and the basis $w_1 = (0, -1, 1)$, $w_2 = (1, 0, 1)$, $w_3 = (1, -1, 0)$ of the codomain.

(d) Determine a matrix S such that $B = SA$.

(e) Are the matrices A and B similar? Why?

Solution. The columns of the matrix A are given by the components of the images, via f, of the vectors of the canonical basis of $\mathbb{R}^3$. We therefore have

$$A = \begin{pmatrix} 2 & 1 & -3 \\ -1 & 0 & 2 \\ 1 & -2 & -4 \end{pmatrix}$$

The vectors of the kernel of f are given by the solutions of the system

$$\begin{pmatrix} 2 & 1 & -3 \\ -1 & 0 & 2 \\ 1 & -2 & -4 \end{pmatrix} \begin{pmatrix} x_1 \\ x_2 \\ x_3 \end{pmatrix} = \begin{pmatrix} 0 \\ 0 \\ 0 \end{pmatrix}$$

that is

$$\begin{cases} 2x_1 + x_2 - 3x_3 = 0 \\ -x_1 + 2x_3 = 0 \\ x_1 - 2x_2 - 4x_3 = 0 \end{cases}$$

We find

$$\begin{cases} x_1 = 2x_3 \\ x_2 = -x_3 \end{cases}$$

from which it follows that $\mathrm{Ker}(f)$ has dimension 1 and a basis is given by the vector $(2, -1, 1)$.

Since $\dim \mathrm{Ker}(f) + \dim \mathrm{Im}(f) = 3$, we have $\dim \mathrm{Im}(f) = 2$ and therefore a basis of $\mathrm{Im}(f)$ is given by two linearly independent columns of A, for example $(2, -1, 1)$ and $(1, 0, -2)$. Since the vector $(2, -1, 1)$ belongs to both the kernel and the image of f, we conclude that $\mathrm{Ker}(f)$ and $\mathrm{Im}(f)$ are not in direct sum.

Let us now consider the basis $w_1 = (0, -1, 1)$, $w_2 = (1, 0, 1)$, $w_3 = (1, -1, 0)$ of the codomain of f. Since we have maintained the canonical basis in the domain of f, to determine the columns of the matrix B we must do as follows. We start from the first vector of the (canonical) basis of the domain, $e_1 = (1, 0, 0)$. Its image via f is $f(e_1) = (2, -1, 1)$. Now we need to express the vector $f(e_1)$ as a linear combination of the vectors of the basis chosen for the codomain, i.e., of the vectors w_1, w_2 and w_3:

$$f(e_1) = \begin{pmatrix} 2 \\ -1 \\ 1 \end{pmatrix} = a_1 \begin{pmatrix} 0 \\ -1 \\ 1 \end{pmatrix} + a_2 \begin{pmatrix} 1 \\ 0 \\ 1 \end{pmatrix} + a_3 \begin{pmatrix} 1 \\ -1 \\ 0 \end{pmatrix}$$

The three coefficients a_1, a_2 and a_3 make up the first column of the matrix B. By solving the system we find

$$\begin{cases} a_1 = 0 \\ a_2 = 1 \\ a_3 = 1 \end{cases}$$

To determine the second column of B we need to repeat what we have just done starting from the second vector $e_2 = (0, 1, 0)$ of the canonical basis of the domain:

$$f(e_2) = \begin{pmatrix} 1 \\ 0 \\ -2 \end{pmatrix} = b_1 \begin{pmatrix} 0 \\ -1 \\ 1 \end{pmatrix} + b_2 \begin{pmatrix} 1 \\ 0 \\ 1 \end{pmatrix} + b_3 \begin{pmatrix} 1 \\ -1 \\ 0 \end{pmatrix}$$

The coefficients b_1, b_2 and b_3 make up the second column of B. By solving the system we find

$$\begin{cases} b_1 = -3/2 \\ b_2 = -1/2 \\ b_3 = 3/2 \end{cases}$$

Finally, to find the third column of B we consider the third vector $e_3 = (0,0,1)$ of the canonical basis of the domain:

$$f(e_3) = \begin{pmatrix} -3 \\ 2 \\ -4 \end{pmatrix} = c_1 \begin{pmatrix} 0 \\ -1 \\ 1 \end{pmatrix} + c_2 \begin{pmatrix} 1 \\ 0 \\ 1 \end{pmatrix} + c_3 \begin{pmatrix} 1 \\ -1 \\ 0 \end{pmatrix}$$

The coefficients c_1, c_2 and c_3 make up the third column of B. By solving the system we find

$$\begin{cases} c_1 = -3/2 \\ c_2 = -5/2 \\ c_3 = -1/2 \end{cases}$$

Therefore the matrix B is

$$B = \begin{pmatrix} 0 & -3/2 & -3/2 \\ 1 & -1/2 & -5/2 \\ 1 & 3/2 & -1/2 \end{pmatrix}$$

Now, if we denote by

$$P = \begin{pmatrix} 0 & 1 & 1 \\ -1 & 0 & -1 \\ 1 & 1 & 0 \end{pmatrix}$$

the matrix whose columns are the components of the vectors of the new basis w_1, w_2, w_3, with respect to the canonical basis, from the change-of-basis formula we know that $PB = A$ (this equality can also be verified by a direct computation), from which we get $B = P^{-1}A$ (note that the matrix P is invertible, because w_1, w_2, w_3 is a basis). Computing the inverse of P we find

$$P^{-1} = \begin{pmatrix} -1/2 & -1/2 & 1/2 \\ 1/2 & 1/2 & 1/2 \\ 1/2 & -1/2 & -1/2 \end{pmatrix}$$

This is the matrix S.

Warning: from the formula $B = SA$ the matrix S cannot be obtained by simply writing $S = BA^{-1}$. In fact, remember that $\mathrm{Ker}(f) \neq \{\mathbf{0}\}$, therefore the function f and, consequently, also the matrix A, is not invertible!

Finally, to establish whether A and B are similar, let us compute their characteristic polynomials. We have

$$\det(A - x \cdot \mathbf{1}) = \begin{vmatrix} 2-x & 1 & -3 \\ -1 & -x & 2 \\ 1 & -2 & -4-x \end{vmatrix} = -x^3 - 2x^2,$$

while

$$\det(B - x \cdot \mathbf{1}) = \begin{vmatrix} -x & -3/2 & -3/2 \\ 1 & -1/2-x & -5/2 \\ 1 & 3/2 & -1/2-x \end{vmatrix} = -x^3 - x^2 - 7x.$$

The matrices A and B have different characteristic polynomials, therefore they cannot be similar.

▶ **Exercise 4.3.11.** Let us consider the matrix

$$A = \begin{pmatrix} 8 & -1 & 1 \\ 0 & t & 1 \\ -2 & 2 & 6 \end{pmatrix}$$

(a) Determine the value of t such that the vector $v = (1, 1, -1)$ is an eigenvector of A.

(b) For the value of t found in point (a), determine the eigenvalues and eigenvectors of A and determine if A is diagonalizable.

(c) Does there exist an orthonormal basis of $\mathbb{R}^3$ formed by eigenvectors of A? Why?

(d) Does there exist a *non-diagonal* matrix similar to A?

Solution. Saying that v is an eigenvector of A means that $Av = \lambda v$, for some number λ. We have

$$\begin{pmatrix} 8 & -1 & 1 \\ 0 & t & 1 \\ -2 & 2 & 6 \end{pmatrix} \begin{pmatrix} 1 \\ 1 \\ -1 \end{pmatrix} = \begin{pmatrix} 6 \\ t-1 \\ -6 \end{pmatrix} = \lambda \begin{pmatrix} 1 \\ 1 \\ -1 \end{pmatrix}$$

from which it follows that $\lambda = 6$ and $t = 7$.

For $t = 7$ the matrix A becomes

$$\begin{pmatrix} 8 & -1 & 1 \\ 0 & 7 & 1 \\ -2 & 2 & 6 \end{pmatrix}$$

and computing its characteristic polynomial we find

$$\det(A - x\cdot\mathbf{1}) = \begin{vmatrix} 8-x & -1 & 1 \\ 0 & 7-x & 1 \\ -2 & 2 & 6-x \end{vmatrix} = -x^3 + 21x^2 - 146x + 336.$$

We have already seen that A has an eigenvalue equal to 6, so the characteristic polynomial of A must be divisible by $x - 6$. Indeed we have

$$-x^3 + 21x^2 - 146x + 336 = -(x-6)(x^2 - 15x + 56).$$

The other two eigenvalues of A are therefore the solutions of the equation

$$x^2 - 15x + 56 = 0,$$

from which we get $x = 7$, $x = 8$. The matrix A therefore has three distinct eigenvalues, given by $\lambda_1 = 6$, $\lambda_2 = 7$ and $\lambda_3 = 8$. Since all eigenvalues are real and distinct, the matrix A is diagonalizable.

We have already seen that an eigenvector for the eigenvalue $\lambda_1 = 6$ is $v = (1, 1, -1)$.

The eigenvectors corresponding to the eigenvalue $\lambda_2 = 7$ are found by solving the system

$$\begin{pmatrix} 1 & -1 & 1 \\ 0 & 0 & 1 \\ -2 & 2 & -1 \end{pmatrix} \begin{pmatrix} x_1 \\ x_2 \\ x_3 \end{pmatrix} = \begin{pmatrix} 0 \\ 0 \\ 0 \end{pmatrix}$$

from which we obtain

$$\begin{cases} x_1 = x_2 \\ x_3 = 0 \end{cases}$$

The eigenspace corresponding to the eigenvalue 7 therefore has dimension 1 and a basis is given by the vector $(1, 1, 0)$.

Finally, the eigenvectors corresponding to the eigenvalue $\lambda_3 = 8$ are given by the solutions of the system

$$\begin{pmatrix} 0 & -1 & 1 \\ 0 & -1 & 1 \\ -2 & 2 & -2 \end{pmatrix} \begin{pmatrix} x_1 \\ x_2 \\ x_3 \end{pmatrix} = \begin{pmatrix} 0 \\ 0 \\ 0 \end{pmatrix}$$

from which we obtain

$$\begin{cases} x_1 = 0 \\ x_2 = x_3 \end{cases}$$

The eigenspace corresponding to the eigenvalue 8 therefore has dimension 1 and a basis is given by the vector $(0, 1, 1)$.

The eigenspaces of A are generated by the vectors $(1, 1, -1)$, $(1, 1, 0)$ and $(0, 1, 1)$, respectively; these three vectors form a basis of $\mathbb{R}^3$ but are not pairwise orthogonal. Therefore there is no orthogonal basis (and therefore not even an orthonormal basis) of $\mathbb{R}^3$ formed by eigenvectors of A.

In fact, this conclusion could be reached immediately, simply by remembering that an orthonormal basis of $\mathbb{R}^3$ formed by eigenvectors of a matrix of order 3 exists if and only if the matrix is symmetric (and our matrix A is not symmetric!).

The last question has a trivial answer: there certainly exists a non-diagonal matrix similar to A, for example the matrix A itself.

► **Exercise 4.3.12.** Write the matrix, with respect to the canonical bases, of an endomorphism f of $\mathbb{R}^2$ having eigenvalues $\lambda_1 = 2$ and $\lambda_2 = -3$ and such that $f(1, 2) = (2, 1)$.

Solution. Since, by hypothesis, f has two distinct eigenvalues, it is diagonalizable. If we denote by V_{λ_1} and V_{λ_2} the eigenspaces of f, we have

$$V_{\lambda_1} \oplus V_{\lambda_2} = \mathbb{R}^2.$$

This means that there exist two vectors $v_1 \in V_{\lambda_1}$ and $v_2 \in V_{\lambda_2}$ such that $(1, 2) = v_1 + v_2$. Applying f we obtain $f(1, 2) = f(v_1) + f(v_2)$, but we know that $f(v_1) = 2v_1$ and $f(v_2) = -3v_2$ (in fact v_1 and v_2 are eigenvectors of f corresponding to the eigenvalues $\lambda_1 = 2$ and $\lambda_2 = -3$, respectively).

Recalling that, by hypothesis, $f(1, 2) = (2, 1)$, we obtain the following system

$$\begin{cases} v_1 + v_2 = (1, 2) \\ 2v_1 - 3v_2 = (2, 1) \end{cases}$$

from which we obtain $v_1 = (1, 7/5)$ and $v_2 = (0, 3/5)$. We have thus determined the eigenvectors of f.

Now we recall that, if A is the matrix of f with respect to the canonical basis of $\mathbb{R}^2$, we have

$$A = PDP^{-1},$$

where

$$D = \begin{pmatrix} 2 & 0 \\ 0 & -3 \end{pmatrix} \qquad P = \begin{pmatrix} 1 & 0 \\ 7/5 & 3/5 \end{pmatrix}$$

Computing the inverse of P, we obtain

$$P^{-1} = \begin{pmatrix} 1 & 0 \\ -7/3 & 5/3 \end{pmatrix}$$

from which we get

$$A = \begin{pmatrix} 1 & 0 \\ 7/5 & 3/5 \end{pmatrix} \begin{pmatrix} 2 & 0 \\ 0 & -3 \end{pmatrix} \begin{pmatrix} 1 & 0 \\ -7/3 & 5/3 \end{pmatrix} = \begin{pmatrix} 2 & 0 \\ 7 & -3 \end{pmatrix}.$$

Another method to solve this problem is as follows. Let us denote by

$$A = \begin{pmatrix} a & b \\ c & d \end{pmatrix}$$

the matrix of f with respect to the canonical bases. The equality $f(1,2) = (2,1)$ is equivalent to

$$\begin{pmatrix} a & b \\ c & d \end{pmatrix} \begin{pmatrix} 1 \\ 2 \end{pmatrix} = \begin{pmatrix} 2 \\ 1 \end{pmatrix}$$

which gives the equations

$$a + 2b = 2, \qquad c + 2d = 1.$$

We now recall that the trace of the matrix A is equal to the sum of the eigenvalues λ_1 and λ_2, therefore $a+d = \lambda_1 + \lambda_2 = -1$, while the determinant of A is equal to the product of the eigenvalues, i.e., $ad - bc = \lambda_1\lambda_2 = -6$. Thus we get the system

$$\begin{cases} a + 2b = 2 \\ c + 2d = 1 \\ a + d = -1 \\ ad - bc = -6 \end{cases}$$

whose solution is

$$\begin{cases} a = 2 \\ b = 0 \\ c = 7 \\ d = -3. \end{cases}$$

► **Exercise 4.3.13.** Let f be an endomorphism of $\mathbb{R}^3$ with the following properties: (*i*) the characteristic polynomial of f is $x^2(x-2)$, (*ii*) the vector $v_1 = (1,-1,2)$ is a basis of the kernel of f, (*iii*) $v_2 = (0,2,-1)$ is the generator of the eigenspace corresponding to the unique non-zero eigenvalue of f, (*iv*) $f(v_3) = v_1$, where $v_3 = (1,0,2)$.

(a) Write the matrix A of f with respect to the basis $\{v_1, v_2, v_3\}$.

(b) Write the matrix B of f with respect to the canonical basis.

(c) Determine whether the matrix B is diagonalizable.

(d) For $t \in \mathbb{R}$ determine $f^{-1}(t, 3, 0)$.

Solution. Since the characteristic polynomial of f is $x^2(x-2)$, the eigenvalues of f are $\lambda_1 = 0$ (with multiplicity 2) and $\lambda_2 = 2$ (with multiplicity 1). The vector v_2 is therefore an eigenvector for the eigenvalue 2, this means that $f(v_2) = 2v_2$.

We have $f(v_1) = \mathbf{0}$, $f(v_2) = 2v_2$, $f(v_3) = v_1$, therefore the matrix of f with respect to the basis $\{v_1, v_2, v_3\}$ is

$$A = \begin{pmatrix} 0 & 0 & 1 \\ 0 & 2 & 0 \\ 0 & 0 & 0 \end{pmatrix}$$

To determine the matrix B we have to compute the images through f of the vectors e_1, e_2, e_3 of the canonical basis of $\mathbb{R}^3$. Looking at the given vectors, we notice that $v_3 - v_1 = e_2$, therefore $f(e_2) = f(v_3) - f(v_1) = v_1 - \mathbf{0} = (1, -1, 2)$.

Then we have $e_3 = 2e_2 - v_2$, hence $f(e_3) = 2f(e_2) - f(v_2) = 2(1, -1, 2) - 2(0, 2, -1) = (2, -6, 6)$. Finally, we have $e_1 = v_1 + e_2 - 2e_3$, hence $f(e_1) = f(v_1) + f(e_2) - 2f(e_3) = (-3, 11, -10)$.

Therefore the matrix B is

$$B = \begin{pmatrix} -3 & 1 & 2 \\ 11 & -1 & -6 \\ -10 & 2 & 6 \end{pmatrix}$$

The matrices A and B are similar (representing the same endomorphism f with respect to different bases), so B is diagonalizable if and only if A is. However, we know that $\lambda_1 = 0$ is an eigenvalue of f with algebraic multiplicity 2, while the corresponding eigenspace (i.e., the kernel of f) has dimension 1, since its basis is given by the vector v_1. Thus, for the eigenvalue 0, the algebraic and geometric multiplicities are different. This means that neither A nor B is diagonalizable.

To compute $f^{-1}(t, 3, 0)$ we need to solve the system

$$\begin{pmatrix} -3 & 1 & 2 \\ 11 & -1 & -6 \\ -10 & 2 & 6 \end{pmatrix} \begin{pmatrix} x_1 \\ x_2 \\ x_3 \end{pmatrix} = \begin{pmatrix} t \\ 3 \\ 0 \end{pmatrix}$$

that is

$$\begin{cases} -3x_1 + x_2 + 2x_3 = t \\ 11x_1 - x_2 - 6x_3 = 3 \\ -10x_1 + 2x_2 + 6x_3 = 0 \end{cases}$$

This system is equivalent to

$$\begin{cases} x_2 = 3 - x_1 \\ x_3 = 2x_1 - 1 \\ t = 1 \end{cases}$$

Therefore, for $t = 1$ the system has infinitely many solutions and we have

$$f^{-1}(1,3,0) = \{(a, 3-a, 2a-1) \mid \forall a \in \mathbb{R}\} = (0,3,-1) + \langle(1,-1,2)\rangle,$$

while, if $t \neq -1$, the system has no solutions and therefore $f^{-1}(t,3,0) = \varnothing$.

▶ **Exercise 4.3.14.** Let's consider the matrix

$$A_{(t)} = \begin{pmatrix} 1 & 0 & 2 \\ 0 & -3 & 0 \\ t+1 & 0 & 1 \end{pmatrix}$$

where t is a real parameter.

(a) Determine for which values of t the eigenvalues of $A_{(t)}$ are real.

(b) Find for which values of t the matrix has eigenvalues with multiplicity greater than 1. For each of the values of t found, determine whether the corresponding matrix is diagonalizable.

(c) Determine the value of t for which there is an *orthonormal* basis of $\mathbb{R}^3$ formed by eigenvectors of the matrix $A_{(t)}$.

Solution. We compute the characteristic polynomial of the matrix $A_{(t)}$:

$$\det(A_{(t)} - x \cdot \mathbf{1}) = \begin{vmatrix} 1-x & 0 & 2 \\ 0 & -3-x & 0 \\ t+1 & 0 & 1-x \end{vmatrix} = (-3-x)(x^2 - 2x - 2t - 1).$$

The eigenvalues of $A_{(t)}$ are therefore $\lambda_1 = -3$, $\lambda_2 = 1 + \sqrt{2t+2}$, $\lambda_3 = 1 - \sqrt{2t+2}$. From this it follows that these eigenvalues are real if and only if $2t + 2 \geq 0$, i.e., for $t \geq -1$.

To have eigenvalues with multiplicity greater than 1, two of the three eigenvalues must coincide. The possibilities are as follows:

(1) $1+\sqrt{2t+2}=1-\sqrt{2t+2}$,

(2) $-3=1+\sqrt{2t+2}$,

(3) $-3=1-\sqrt{2t+2}$.

In the first case the equation $1+\sqrt{2t+2}=1-\sqrt{2t+2}$ has a unique solution $t=-1$. For this value of t the eigenvalues are $\lambda_1=-3$, $\lambda_2=\lambda_3=1$.

In the second case the equation $1+\sqrt{2t+2}=-3$ has no solutions, while in the third case the equation $1-\sqrt{2t+2}=-3$ has a unique solution $t=7$. For this value of t the eigenvalues are $\lambda_1=-3$, $\lambda_2=5$, $\lambda_3=-3$.

We analyze the two cases separately, $t=-1$ and $t=7$. If $t=-1$ the matrix becomes

$$\begin{pmatrix}1 & 0 & 2\\ 0 & -3 & 0\\ 0 & 0 & 1\end{pmatrix}$$

We determine the eigenspace corresponding to the eigenvalue $\lambda=1$ (which has algebraic multiplicity 2). The eigenvectors corresponding to the eigenvalue $\lambda=1$ are given by the solutions of the system

$$\begin{pmatrix}0 & 0 & 2\\ 0 & -4 & 0\\ 0 & 0 & 0\end{pmatrix}\begin{pmatrix}x_1\\ x_2\\ x_3\end{pmatrix}=\begin{pmatrix}0\\ 0\\ 0\end{pmatrix}$$

from which we get

$$\begin{cases}x_2=0\\ x_3=0\end{cases}$$

The eigenspace for the eigenvalue 1 therefore has dimension 1 and a basis is given by the vector $(1,0,0)$. Since the dimension of this eigenspace (the so-called geometric multiplicity) is smaller than the algebraic multiplicity of the corresponding eigenvalue, the matrix obtained for $t=-1$ is not diagonalizable.

We now pass to the case $t=7$. For this value of t the matrix becomes

$$\begin{pmatrix}1 & 0 & 2\\ 0 & -3 & 0\\ 8 & 0 & 1\end{pmatrix}$$

We determine the eigenspace corresponding to the eigenvalue $\lambda=-3$ (which has algebraic multiplicity 2). The eigenvectors for the eigenvalue $\lambda=-3$ are given by the solutions of the system

$$\begin{pmatrix} 4 & 0 & 2 \\ 0 & 0 & 0 \\ 8 & 0 & 4 \end{pmatrix} \begin{pmatrix} x_1 \\ x_2 \\ x_3 \end{pmatrix} = \begin{pmatrix} 0 \\ 0 \\ 0 \end{pmatrix}$$

from which we obtain $x_3 = -2x_1$, while the unknowns x_1 and x_2 remain undetermined. This means that the eigenspace corresponding to the eigenvalue -3 has dimensions 2 and a basis is given by the vectors $(1, 0, -2)$ and $(0, 1, 0)$. Since the dimension of this eigenspace is equal to the algebraic multiplicity of the corresponding eigenvalue, the matrix obtained for $t = 7$ is diagonalizable (we note that it is useless to calculate the dimension of the eigenspace corresponding to the eigenvalue $\lambda_2 = 5$ because this eigenvalue has multiplicity 1, therefore also its geometric multiplicity is necessarily equal to 1).

Finally, we recall that there exists an orthonormal basis of $\mathbb{R}^3$ of eigenvectors of the matrix $A_{(t)}$ if and only if this matrix is symmetric, which occurs if and only if $t + 1 = 2$, i.e., if and only if $t = 1$.

► **Exercise 4.3.15.** Given $v_1 = (1, 0, 0)$, $v_2 = (1, 1, 2)$, $v_3 = (0, 1, 1) \in \mathbb{R}^3$, let f be an endomorphism of $\mathbb{R}^3$ having 1 as the only eigenvalue, with algebraic multiplicity 3, and such that $f(v_1) = v_2$, $f(v_2) = v_3$.

(a) Write the matrix A of f with respect to the basis $\{v_1, v_2, v_3\}$.

(b) Write the matrix B of f with respect to the canonical basis.

(c) Find the eigenspace corresponding to the eigenvalue 1 and determine whether f is diagonalizable.

(d) Find a basis of $\mathbb{R}^3$ with respect to which the matrix of f is upper triangular.

Solution. Since $f(v_1) = v_2$ and $f(v_2) = v_3$, the first two columns of the matrix A of f with respect to the basis $\{v_1, v_2, v_3\}$ are, respectively, $(0, 1, 0)$ and $(0, 0, 1)$. We can therefore write the matrix A in the form

$$A = \begin{pmatrix} 0 & 0 & a \\ 1 & 0 & b \\ 0 & 1 & c \end{pmatrix}$$

Let us now compute the characteristic polynomial of this matrix:

$$\det(A - x \cdot \mathbf{1}) = \begin{vmatrix} -x & 0 & a \\ 1 & -x & b \\ 0 & 1 & c - x \end{vmatrix} = -x^3 + cx^2 + bx + a.$$

The requirement that f have 1 as the only eigenvalue, with algebraic multiplicity 3, means that its characteristic polynomial must be

$$(1-x)^3 = -x^3 + 3x^2 - 3x + 1.$$

From the equality $-x^3 + cx^2 + bx + a = -x^3 + 3x^2 - 3x + 1$ it follows that $a = 1$, $b = -3$, $c = 3$, hence the matrix A is

$$A = \begin{pmatrix} 0 & 0 & 1 \\ 1 & 0 & -3 \\ 0 & 1 & 3 \end{pmatrix}$$

Remark. For those familiar with Hamilton–Cayley's theorem another way to determine the matrix A is as follows. Knowing that the characteristic polynomial of f is $(1-x)^3$, it follows from Hamilton–Cayley's theorem that $(\mathrm{id} - f)^3 = \mathrm{id} - 3f + 3f^2 - f^3 = 0$, hence $f^3 = \mathrm{id} - 3f + 3f^2$. By applying this equality to the vector v_1, we obtain $f^3(v_1) = v_1 - 3f(v_1) + 3f^2(v_1)$. But $f(v_1) = v_2$, $f^2(v_1) = f(f(v_1)) = f(v_2) = v_3$ and finally $f^3(v_1) = f(f^2(v_1)) = f(v_3)$, hence the previous equality allows us to compute $f(v_3)$. We get $f(v_3) = v_1 - 3v_2 + 3v_3$, and this determines the third column of A.

The matrix B of f with respect to the canonical basis is obtained from the matrix A through the formula $B = PAP^{-1}$, where the change-of-basis matrix P is the matrix whose columns contain the components of the vectors v_1, v_2 and v_3 with respect to the canonical basis. So we have

$$P = \begin{pmatrix} 1 & 1 & 0 \\ 0 & 1 & 1 \\ 0 & 2 & 1 \end{pmatrix}$$

Computing the inverse matrix of P we obtain

$$P^{-1} = \begin{pmatrix} 1 & 1 & -1 \\ 0 & -1 & 1 \\ 0 & 2 & -1 \end{pmatrix}$$

Now we can determine the matrix B:

$$B = PAP^{-1} = \begin{pmatrix} 1 & -3 & 1 \\ 1 & 0 & 0 \\ 2 & -5 & 2 \end{pmatrix}$$

Using the matrix B we can now easily determine the eigenspace corresponding to the eigenvalue 1. The eigenvectors of f corresponding to the eigenvalue 1 (expressed with respect to the canonical basis) are given by the solutions of the system

$$\begin{pmatrix} 0 & -3 & 1 \\ 1 & -1 & 0 \\ 2 & -5 & 1 \end{pmatrix} \begin{pmatrix} x_1 \\ x_2 \\ x_3 \end{pmatrix} = \begin{pmatrix} 0 \\ 0 \\ 0 \end{pmatrix}$$

from which we obtain

$$\begin{cases} x_1 = x_2 \\ x_3 = 3x_2 \end{cases}$$

The eigenspace corresponding to the eigenvalue 1 therefore has dimension 1 and a basis is given by the vector $(1, 1, 3)$. Since the dimension of this eigenspace is smaller than the algebraic multiplicity of the corresponding eigenvalue, f (and therefore both the matrices A and B) are not diagonalizable.

Let $\{w_1, w_2, w_3\}$ be a basis of $\mathbb{R}^3$ with respect to which the matrix of f is upper triangular; let's denote by C this matrix. Since C is a matrix representing the function f, it must have 1 as its only eigenvalue, with algebraic multiplicity 3. Being upper triangular, this means that all elements on the main diagonal must be 1. The matrix C therefore has the following form:

$$C = \begin{pmatrix} 1 & ? & ? \\ 0 & 1 & ? \\ 0 & 0 & 1 \end{pmatrix}$$

The fact that the first column is $(1, 0, 0)$ means that the first vector of the basis, w_1, must be an eigenvector for the eigenvalue 1; we can therefore set $w_1 = (1, 1, 3)$.

The second column cannot be $(0, 1, 0)$, otherwise w_2 would also be an eigenvector for the eigenvalue 1, but this is impossible since the eigenspace corresponding to the eigenvalue 1 has dimension 1. The first element of the second column must therefore be non-zero. Let's try to see if we can find a vector w_2 such that the second column of C is $(1, 1, 0)$. This means that we must have $f(w_2) = w_1 + w_2$, i.e., setting $w_2 = (x_1, x_2, x_3)$,

$$\begin{pmatrix} 1 & -3 & 1 \\ 1 & 0 & 0 \\ 2 & -5 & 2 \end{pmatrix} \begin{pmatrix} x_1 \\ x_2 \\ x_3 \end{pmatrix} = \begin{pmatrix} 1 \\ 1 \\ 3 \end{pmatrix} + \begin{pmatrix} x_1 \\ x_2 \\ x_3 \end{pmatrix}$$

We obtain the system

$$\begin{cases} -3x_2 + x_3 = 1 \\ x_1 - x_2 = 1 \\ 2x_1 - 5x_2 + x_3 = 3 \end{cases}$$

which is equivalent to

$$\begin{cases} x_1 = x_2 + 1 \\ x_3 = 3x_2 + 1 \end{cases}$$

A solution is given by the vector $w_2 = (1, 0, 1)$.

Similarly, we deduce that the third column of C cannot be $(0, 0, 1)$ (otherwise w_3 would also be an eigenvector for the eigenvalue 1). Let's try to see if we can find a vector w_3 such that the third column is $(0, 1, 1)$. This means that we must have $f(w_3) = w_2 + w_3$, i.e., setting $w_3 = (x_1, x_2, x_3)$,

$$\begin{pmatrix} 1 & -3 & 1 \\ 1 & 0 & 0 \\ 2 & -5 & 2 \end{pmatrix} \begin{pmatrix} x_1 \\ x_2 \\ x_3 \end{pmatrix} = \begin{pmatrix} 1 \\ 0 \\ 1 \end{pmatrix} + \begin{pmatrix} x_1 \\ x_2 \\ x_3 \end{pmatrix}$$

We obtain the system

$$\begin{cases} -3x_2 + x_3 = 1 \\ x_1 - x_2 = 0 \\ 2x_1 - 5x_2 + x_3 = 1 \end{cases}$$

which is equivalent to

$$\begin{cases} x_1 = x_2 \\ x_3 = 3x_2 + 1 \end{cases}$$

A solution is given by the vector $w_3 = (0, 0, 1)$.

We can therefore conclude that, with respect to the basis formed by the vectors $w_1 = (1, 1, 3)$, $w_2 = (1, 0, 1)$, $w_3 = (0, 0, 1)$, the matrix of f is

$$C = \begin{pmatrix} 1 & 1 & 0 \\ 0 & 1 & 1 \\ 0 & 0 & 1 \end{pmatrix}$$

► **Exercise 4.3.16.** Let us consider the matrix

$$A = \begin{pmatrix} -5 & 0 & 0 & 0 \\ 3 & -1 & 0 & 2 \\ 2 & 0 & 1 & 0 \\ -1 & 4 & 0 & -3 \end{pmatrix}$$

(a) Determine the eigenvalues and the eigenvectors of A, and establish whether A is diagonalizable.

(b) Let $U \subset \mathbb{R}^4$ be the subspace spanned by $u_1 = (0, 1, 0, -1)$, $u_2 = (0, 3, 0, 1)$. Check that, for any $v \in U$, we have $Av \in U$.

(c) Write the matrix B of the linear function $f: U \to U$ defined by $f(v) = Av$, with respect to the basis $\{u_1, u_2\}$.

Solution. We compute the characteristic polynomial of A:

$$\begin{aligned}
\det(A - x \cdot \mathbf{1}) &= \begin{vmatrix} -5-x & 0 & 0 & 0 \\ 3 & -1-x & 0 & 2 \\ 2 & 0 & 1-x & 0 \\ -1 & 4 & 0 & -3-x \end{vmatrix} \\
&= (-5-x) \begin{vmatrix} -1-x & 0 & 2 \\ 0 & 1-x & 0 \\ 4 & 0 & -3-x \end{vmatrix} \\
&= (-5-x)(1-x) \begin{vmatrix} -1-x & 2 \\ 4 & -3-x \end{vmatrix} \\
&= (-5-x)(1-x)(x^2 + 4x - 5)
\end{aligned}$$

The eigenvalues of A are therefore -5, 1 and the two solutions of the equation $x^2 + 4x - 5 = 0$ which, in turn, are -5 and 1. Thus we conclude that the matrix A has two eigenvalues, $\lambda_1 = 1$ and $\lambda_2 = -5$, both with algebraic multiplicity 2.

We determine the eigenspace corresponding to the eigenvalue $\lambda_1 = 1$. The eigenvectors for this eigenvalue are given by the solutions of the system

$$\begin{pmatrix} -6 & 0 & 0 & 0 \\ 3 & -2 & 0 & 2 \\ 2 & 0 & 0 & 0 \\ -1 & 4 & 0 & -4 \end{pmatrix} \begin{pmatrix} x_1 \\ x_2 \\ x_3 \\ x_4 \end{pmatrix} = \begin{pmatrix} 0 \\ 0 \\ 0 \\ 0 \end{pmatrix}$$

This system is equivalent to

$$\begin{cases} x_1 = 0 \\ x_2 = x_4 \end{cases}$$

that has infinitely many solutions, depending on two parameters (x_3 and x_4). This means that the eigenspace corresponding to the eigenvalue 1 has dimension 2 and a basis is given by the vectors $(0, 0, 1, 0)$ and $(0, 1, 0, 1)$.

Let us now determine the eigenspace corresponding to the eigenvalue $\lambda_2 = -5$. The eigenvectors for this eigenvalue are given by the solutions of

the system

$$\begin{pmatrix} 0 & 0 & 0 & 0 \\ 3 & 4 & 0 & 2 \\ 2 & 0 & 6 & 0 \\ -1 & 4 & 0 & 2 \end{pmatrix} \begin{pmatrix} x_1 \\ x_2 \\ x_3 \\ x_4 \end{pmatrix} = \begin{pmatrix} 0 \\ 0 \\ 0 \\ 0 \end{pmatrix}$$

This system is equivalent to

$$\begin{cases} x_1 = 0 \\ x_3 = 0 \\ x_4 = -2x_2 \end{cases}$$

that has infinitely many solutions, for each value of x_2. This means that the eigenspace corresponding to the eigenvalue -5 has dimension 1 and a basis is given by the vector $(0, 1, 0, -2)$. Since this eigenspace has dimension 1 while the corresponding eigenvalue has algebraic multiplicity 2, we conclude that the matrix A is not diagonalizable.

To verify that $Av \in U$ for each $v \in U$, it suffices to check it for the two generators $u_1 = (0, 1, 0, -1)$ and $u_2 = (0, 3, 0, 1)$ of U (every other vector of U is a linear combination of u_1 and u_2).

We have $Au_1 = (0, -3, 0, 7)$ and $Au_2 = (0, -1, 0, 9)$; we must therefore verify that the vectors $(0, -3, 0, 7)$ and $(0, -1, 0, 9)$ can be written as a linear combination of u_1 and u_2.

Writing $\alpha_1 u_1 + \alpha_2 u_2 = (0, -3, 0, 7)$ we find $\alpha_1 = -6$ and $\alpha_2 = 1$, while writing $\beta_1 u_1 + \beta_2 u_2 = (0, -1, 0, 9)$ gives $\beta_1 = -7$ and $\beta_2 = 2$. Thus we have found the following equalities:

$$Au_1 = -6u_1 + u_2, \qquad Au_2 = -7u_1 + 2u_2,$$

which confirm that $Au_1 \in U$ and $Au_2 \in U$.

The matrix B of the function $f\colon U \to U$ defined by $f(v) = Av$, with respect to the basis $\{u_1, u_2\}$, has as columns the components of the vectors Au_1 and Au_2 with respect to the basis $\{u_1, u_2\}$. From the previous equalities $Au_1 = -6u_1 + u_2$ and $Au_2 = -7u_1 + 2u_2$, we deduce that B is

$$B = \begin{pmatrix} -6 & -7 \\ 1 & 2 \end{pmatrix}$$

▶ **Exercise 4.3.17.** Let f be the endomorphism of $\mathbb{R}^3$ whose matrix, with respect to the canonical basis, is

$$A = \begin{pmatrix} 2 & 1 & 1 \\ 1 & 2 & -1 \\ 1 & -1 & 2 \end{pmatrix}$$

(a) Is it possible to determine whether A is similar to a diagonal matrix, without computing its eigenvalues and eigenvectors?

(b) Determine the eigenvalues and the eigenspaces of A.

(c) Find an orthonormal basis of $\mathbb{R}^3$ with respect to which the matrix of f is diagonal.

Solution. In order to state that the matrix A is similar to a diagonal matrix, it is enough to observe that it is symmetric (and to remember that all symmetric matrices are diagonalizable).

We now compute the characteristic polynomial of A:

$$\det(A - x \cdot \mathbf{1}) = \begin{vmatrix} 2-x & 1 & 1 \\ 1 & 2-x & -1 \\ 1 & -1 & 2-x \end{vmatrix} = -x(x-3)^2$$

The eigenvalues of A are 0 (with multiplicity 1) and 3 (with multiplicity 2).

We determine the eigenspace corresponding to the eigenvalue 0. The eigenvectors corresponding to this eigenvalue are given by the solutions of the system

$$\begin{pmatrix} 2 & 1 & 1 \\ 1 & 2 & -1 \\ 1 & -1 & 2 \end{pmatrix} \begin{pmatrix} x_1 \\ x_2 \\ x_3 \end{pmatrix} = \begin{pmatrix} 0 \\ 0 \\ 0 \end{pmatrix}$$

This system is equivalent to

$$\begin{cases} x_2 = -x_1 \\ x_3 = -x_1 \end{cases}$$

which has infinitely many solutions, for any value of x_1. This means that the eigenspace corresponding to the eigenvalue 0 has dimension 1 and a basis is given by the vector $(1, -1, -1)$.

Now we determine the eigenspace corresponding to the eigenvalue 3. We get the system

$$\begin{pmatrix} -1 & 1 & 1 \\ 1 & -1 & -1 \\ 1 & -1 & -1 \end{pmatrix} \begin{pmatrix} x_1 \\ x_2 \\ x_3 \end{pmatrix} = \begin{pmatrix} 0 \\ 0 \\ 0 \end{pmatrix}$$

which is equivalent to the equation $x_1 = x_2+x_3$. It follows that the eigenspace corresponding to the eigenvalue 3 has dimension 2 and a basis is given by the vectors $(1, 1, 0)$ and $(1, 0, 1)$.

To determine an orthonormal basis of $\mathbb{R}^3$ with respect to which the matrix of f is diagonal we need to determine an orthonormal basis consisting of eigenvectors of A.

Note that the vectors $(1, 1, 0)$ and $(1, 0, 1)$ are not orthogonal. Using these two vectors it is easy to construct an orthonormal basis of the eigenspace corresponding to the eigenvalue 3.

We begin by setting $w_1 = (1, 1, 0)$ and $w_2 = (1, 0, 1)+\alpha(1, 1, 0)$. Imposing the orthogonality condition $w_1 \cdot w_2 = 0$, we find $\alpha = -1/2$, from which we obtain $w_2 = (1/2, -1/2, 1)$. Finally, we normalize the vectors w_1 and w_2 by dividing each one by its norm. Thus we get the vectors

$$w_1' = \left(\tfrac{1}{\sqrt{2}}, \tfrac{1}{\sqrt{2}}, 0\right), \qquad w_2' = \left(\tfrac{1}{\sqrt{6}}, -\tfrac{1}{\sqrt{6}}, \tfrac{2}{\sqrt{6}}\right)$$

The eigenvector $w_3 = (1, -1, -1)$ is orthogonal to the vectors w_1 and w_2, since it belongs to an eigenspace corresponding to a different eigenvalue. It is therefore sufficient to normalize it, dividing it by its norm, thus obtaining the vector

$$w_3' = \left(\tfrac{1}{\sqrt{3}}, -\tfrac{1}{\sqrt{3}}, -\tfrac{1}{\sqrt{3}}\right)$$

The orthonormal basis is formed by the vectors w_1', w_2' and w_3'.

► **Exercise 4.3.18.** Let f be an endomorphism of $\mathbb{R}^3$ such that

$$f^{-1}(-1, 0, 1) = (-1, 0, 1) + \langle(1, 2, 1), (2, 1, 2)\rangle$$

(a) Determine the matrix A of f, with respect to the canonical basis of $\mathbb{R}^3$.

(b) Find bases of the kernel and of the image of f.

(c) Determine an orthonormal basis of $\mathbb{R}^3$ with respect to which the matrix of f is

$$D = \begin{pmatrix} 0 & 0 & 0 \\ 0 & 0 & 0 \\ 0 & 0 & 1 \end{pmatrix}$$

(d) Find the values of $t \in \mathbb{R}$ for which there is an endomorphism f_t of $\mathbb{R}^3$ such that

$$f_t^{-1}(1, 0, -1) = (1, 0, -1) + \langle(1, 0, 1), (t, 1, 1)\rangle,$$

and f_t is diagonalizable, but not orthogonally diagonalizable.

Solution. Recall that if f is a linear function and v and w are two vectors such that $f(v) = w$, then $f^{-1}(w) = v + \mathrm{Ker}(f)$.

From the equality $f^{-1}(-1,0,1) = (-1,0,1)+\langle(1,2,1),(2,1,2)\rangle$ we deduce that $f(-1,0,1) = (-1,0,1)$ and that $\mathrm{Ker}(f) = \langle(1,2,1),(2,1,2)\rangle$. The equality $f(-1,0,1) = (-1,0,1)$ means that the vector $(-1,0,1)$ is an eigenvector of f for the eigenvalue 1, while the fact that $\dim \mathrm{Ker}(f) = 2$ means that f has an eigenvalue equal to 0 whose eigenspace (i.e., $\mathrm{Ker}(f)$) has dimension 2.

We therefore conclude that f has the eigenvalues 1 (with multiplicity 1) and 0 (with multiplicity 2), that the eigenspace corresponding to the eigenvalue 1 is generated by the vector $(-1,0,1)$, while the eigenspace corresponding to the eigenvalue 0 is generated by the vectors $(1,2,1)$ and $(2,1,2)$. Therefore f is diagonalizable and a basis $\{w_1, w_2, w_3\}$ of $\mathbb{R}^3$ with respect to which f has matrix D is obtained by taking as w_1 and w_2 the vectors of a basis of $\mathrm{Ker}(f)$ and as w_3 the eigenvector $(-1,0,1)$.

Note that $(0,1,0) = \frac{1}{3}\big(2(1,2,1) - (2,1,2)\big) \in \mathrm{Ker}(f)$ and $(1,0,1) = (1,2,1) - 2(0,1,0) \in \mathrm{Ker}(f)$, hence

$$\mathrm{Ker}(f) = \langle(1,2,1),(2,1,2)\rangle = \langle(0,1,0),(1,0,1)\rangle.$$

Since the vectors $(0,1,0)$ and $(1,0,1)$ are orthogonal, they form an orthogonal basis of $\mathrm{Ker}(f)$. Since they are also orthogonal to the vector $(-1,0,1)$, the vectors $(0,1,0)$, $(1,0,1)$ and $(-1,0,1)$ form an orthogonal basis of $\mathbb{R}^3$ with respect to which the matrix of f is the matrix D. Now there is nothing left to do but normalize these vectors, dividing each of them by its norm; the orthonormal basis required at point (c) is therefore formed by the vectors

$$w_1 = (0,1,0), \quad w_2 = (1/\sqrt{2}, 0, 1/\sqrt{2}), \quad w_3 = (-1/\sqrt{2}, 0, 1/\sqrt{2}).$$

We now determine the matrix A of f, with respect to the canonical basis.

We have seen that $\mathrm{Ker}(f)$ is spanned by the vectors $(0,1,0)$ and $(1,0,1)$, so we know that $f(0,1,0) = f(1,0,1) = (0,0,0)$, moreover $f(-1,0,1) = (-1,0,1)$. From the linearity of f it follows that $f(1,0,1) + f(-1,0,1) = f(0,0,2) = (-1,0,1)$, and $f(1,0,1) - f(-1,0,1) = f(2,0,0) = (1,0,-1)$. Thus we conclude that

$$f(1,0,0) = (1/2, 0, -1/2), \quad f(0,1,0) = (0,0,0), \quad f(0,0,1) = (-1/2, 0, 1/2)$$

hence the matrix is

$$A = \begin{pmatrix} 1/2 & 0 & -1/2 \\ 0 & 0 & 0 \\ -1/2 & 0 & 1/2 \end{pmatrix}$$

We have already determined a basis of the kernel of f, a basis of $\mathrm{Im}(f)$ is given by one of the non-zero columns of A: $\mathrm{Im}(f) = \langle(1/2, 0, -1/2)\rangle$.

We now turn to the last point. Reasoning analogously to what was done previously, from the equality

$$f_t^{-1}(1,0,-1) = (1,0,-1) + \langle(1,0,1),(t,1,1)\rangle$$

it follows that $\mathrm{Ker}(f_t) = \langle(1,0,1),(t,1,1)\rangle$ and $f_t(1,0,-1) = (1,0,-1)$.

Note that the vectors $(1,0,1)$, $(t,1,1)$ and $(1,0,-1)$ are linearly independent for any value of t. Exactly as before, we deduce that such a function f_t exists and is diagonalizable for every $t \in \mathbb{R}$.

By determining the matrix A_t of f_t with respect to the canonical basis (with a procedure analogous to that used to determine the matrix A), we find

$$A_t = \begin{pmatrix} \frac{1}{2} & \frac{1}{2} - \frac{1}{2}t & -\frac{1}{2} \\ 0 & 0 & 0 \\ -\frac{1}{2} & \frac{1}{2}t - \frac{1}{2} & \frac{1}{2} \end{pmatrix}$$

As stated, this matrix is diagonalizable for every t; it is orthogonally diagonalizable only when it is symmetric, i.e., for $t = 1$. In conclusion, the function f_t is not orthogonally diagonalizable for $t \neq 1$.

► **Exercise 4.3.19.** Let us consider the following matrices:

$$A = \begin{pmatrix} 1 & -1 & 2 \\ -1 & 1 & -2 \\ 2 & -2 & 4 \end{pmatrix}, \qquad B = \begin{pmatrix} a & b & 0 \\ 1 & 0 & 0 \\ 1 & 0 & 0 \end{pmatrix}$$

(a) Determine the values of a and b for which the matrices A and B are similar.

(b) For the values of a and b found in point (a), find an invertible matrix P such that $A = PBP^{-1}$.

Solution. To start, let's see if the matrix A is diagonalizable. The characteristic polynomial of A is

$$\det(A - x \cdot \mathbf{1}) = \begin{vmatrix} 1-x & -1 & 2 \\ -1 & 1-x & -2 \\ 2 & -2 & 4-x \end{vmatrix} = -x^2(x-6)$$

The eigenvalues of A are therefore 0 (with multiplicity 2) and 6.

We now determine the eigenspace of A corresponding to the eigenvalue 0. The eigenvectors corresponding to this eigenvalue are given by the solutions of the system

$$\begin{pmatrix} 1 & -1 & 2 \\ -1 & 1 & -2 \\ 2 & -2 & 4 \end{pmatrix} \begin{pmatrix} x_1 \\ x_2 \\ x_3 \end{pmatrix} = \begin{pmatrix} 0 \\ 0 \\ 0 \end{pmatrix}$$

This system is equivalent to the equation $x_1 = x_2 - 2x_3$, which has infinitely many solutions, for any x_2 and x_3. This means that the eigenspace corresponding to the eigenvalue 0 has dimension 2 and a basis is given by the vectors $v_1 = (1, 1, 0)$ and $v_2 = (-2, 0, 1)$.

Now we determine the eigenspace of A corresponding to the eigenvalue 6. We get the system

$$\begin{pmatrix} -5 & -1 & 2 \\ -1 & -5 & -2 \\ 2 & -2 & -2 \end{pmatrix} \begin{pmatrix} x_1 \\ x_2 \\ x_3 \end{pmatrix} = \begin{pmatrix} 0 \\ 0 \\ 0 \end{pmatrix}$$

which is equivalent to

$$\begin{cases} -5x_1 - x_2 + 2x_3 = 0 \\ 2x_1 - 2x_2 - 2x_3 = 0 \end{cases}$$

whose solutions are given by

$$\begin{cases} x_2 = -x_1 \\ x_3 = 2x_1 \end{cases}$$

This means that the eigenspace corresponding to the eigenvalue 6 has dimension 1 and a basis is given by the vector $v_3 = (1, -1, 2)$.

Thus we conclude that A is similar to the diagonal matrix

$$D = \begin{pmatrix} 0 & 0 & 0 \\ 0 & 0 & 0 \\ 0 & 0 & 6 \end{pmatrix}$$

and, by setting

$$R = \begin{pmatrix} 1 & -2 & 1 \\ 1 & 0 & -1 \\ 0 & 1 & 2 \end{pmatrix}$$

we have $AR = RD$, i.e., $A = RDR^{-1}$.

We now turn to the matrix B. The characteristic polynomial of B is

$$\det(B - x \cdot \mathbf{1}) = \begin{vmatrix} a-x & b & 0 \\ 1 & -x & 0 \\ 1 & 0 & -x \end{vmatrix} = -x(x^2 - ax - b)$$

Since A and B must be similar, they must have the same characteristic polynomial, so we must have

$$-x^2(x-6) = -x(x^2 - ax - b)$$

from which it follows that $a = 6$ and $b = 0$.

Of course, having the same characteristic polynomial is only a necessary but not a sufficient condition for two matrices to be similar. We must therefore check whether the matrix

$$B = \begin{pmatrix} 6 & 0 & 0 \\ 1 & 0 & 0 \\ 1 & 0 & 0 \end{pmatrix}$$

is similar to A or, equivalently, to the diagonal matrix D (since A is similar to D).

We know that the eigenvalues of B are the same as those of A, namely 0 (with multiplicity 2) and 6 (with multiplicity 1).

We determine the eigenspace of B for the eigenvalue 0. The eigenvectors corresponding to this eigenvalue are given by the solutions of the system

$$\begin{pmatrix} 6 & 0 & 0 \\ 1 & 0 & 0 \\ 1 & 0 & 0 \end{pmatrix} \begin{pmatrix} x_1 \\ x_2 \\ x_3 \end{pmatrix} = \begin{pmatrix} 0 \\ 0 \\ 0 \end{pmatrix}$$

This system is equivalent to the equation $x_1 = 0$, which has infinitely many solutions, for any x_2 and x_3. This means that the eigenspace of the matrix B corresponding to the eigenvalue 0 has dimension 2 and a basis is given by the vectors $w_1 = (0, 1, 0)$ and $w_2 = (0, 0, 1)$.

Now we determine the eigenspace of B corresponding to the eigenvalue 6. We get the system

$$\begin{pmatrix} 0 & 0 & 0 \\ 1 & -6 & 0 \\ 1 & 0 & -6 \end{pmatrix} \begin{pmatrix} x_1 \\ x_2 \\ x_3 \end{pmatrix} = \begin{pmatrix} 0 \\ 0 \\ 0 \end{pmatrix}$$

which is equivalent to

$$\begin{cases} x_1 - 6x_2 = 0 \\ x_1 - 6x_3 = 0 \end{cases}$$

whose solutions are given by

$$\begin{cases} x_1 = 6x_2 \\ x_3 = x_2 \end{cases}$$

This means that the eigenspace corresponding to the eigenvalue 6 has dimension 1 and a basis is given by the vector $w_3 = (6, 1, 1)$.

Thus we conclude that the matrix B is also similar to the diagonal matrix D and, if we set

$$S = \begin{pmatrix} 0 & 0 & 6 \\ 1 & 0 & 1 \\ 0 & 1 & 1 \end{pmatrix}$$

we have $BS = SD$, i.e., $D = S^{-1}BS$.

Recalling that $A = RDR^{-1}$, we have

$$\begin{aligned} A &= RDR^{-1} \\ &= R(S^{-1}BS)R^{-1} \\ &= (RS^{-1})B(SR^{-1}) \\ &= (RS^{-1})B(RS^{-1})^{-1} \end{aligned}$$

from which it follows that setting $P = RS^{-1}$ we have $A = PBP^{-1}$, as required.

Now all we have to do is compute the inverse of S and multiply it by R. With easy calculations we get

$$S^{-1} = \begin{pmatrix} -1/6 & 1 & 0 \\ -1/6 & 0 & 1 \\ 1/6 & 0 & 0 \end{pmatrix}$$

hence

$$P = RS^{-1} = \begin{pmatrix} 1/3 & 1 & -2 \\ -1/3 & 1 & 0 \\ 1/6 & 0 & 1 \end{pmatrix}.$$

▶ **Exercise 4.3.20.** Let $u = (1, -1, 0)$, $v = (0, -2, -1)$, $w = (4, -1, 1) \in \mathbb{R}^3$. Let $f\colon \mathbb{R}^3 \to \mathbb{R}^3$ be a linear function such that $f(u) = w$, $f(v) = 2w$ and $\operatorname{Im}(f) \subset \operatorname{Ker}(f)$.

(a) Write the matrix A of f with respect to the canonical basis and determine a basis of the kernel of f.

(b) Find a basis of $\mathbb{R}^3$ with respect to which the matrix of f is $B = \begin{pmatrix} 0 & 0 & 1 \\ 0 & 0 & 0 \\ 0 & 0 & 0 \end{pmatrix}$.

(c) Determine an invertible matrix S such that $A = SBS^{-1}$. Is such a matrix S unique?

(d) Determine whether there exists an invertible matrix R such that $A = RCR^{-1}$, where C is the matrix $C = \begin{pmatrix} 0 & 0 & 1 \\ 0 & 1 & 0 \\ 0 & 0 & 0 \end{pmatrix}$.

Solution. Since $f(u) = w$, we have $w \in \operatorname{Im}(f)$. Knowing that $\operatorname{Im}(f) \subset \operatorname{Ker}(f)$ we also have $w \in \operatorname{Ker}(f)$, hence $f(w) = \mathbf{0}$. So let's try to write the vectors of the canonical basis of $\mathbb{R}^3$ as a linear combination of u, v and w. Note that $v + w = (4, -3, 0)$ hence $v + w - 3u = (1, 0, 0) = e_1$. Therefore $f(e_1) = f(v) + f(w) - 3f(u) = 2w - 3w = -w = (-4, 1, -1)$.

Then we have $e_2 = -u + e_1$, hence $f(e_2) = -f(u) + f(e_1) = -2w = (-8, 2, -2)$. Finally, $e_3 = -v - 2e_2$, hence $f(e_3) = -f(v) - 2f(e_2) = 2w = (8, -2, 2)$. Therefore the matrix A is

$$A = \begin{pmatrix} -4 & -8 & 8 \\ 1 & 2 & -2 \\ -1 & -2 & 2 \end{pmatrix}$$

The kernel of f is the set of solutions of the system

$$\begin{cases} -4x_1 - 8x_2 + 8x_3 = 0 \\ x_1 + 2x_2 - 2x_3 = 0 \\ -x_1 - 2x_2 + 2x_3 = 0 \end{cases}$$

which is equivalent to the equation $x_1 + 2x_2 - 2x_3 = 0$, from which we get $x_1 = -2x_2 + 2x_3$. $\operatorname{Ker}(f)$ has therefore dimension 2 and a basis is given by the vectors $(-2, 1, 0)$ and $(2, 0, 1)$.

Let $\{v_1, v_2, v_3\}$ be a basis with respect to which the matrix of f is B. This means that $f(v_1) = \mathbf{0}$, $f(v_2) = \mathbf{0}$ and $f(v_3) = v_1$. Comparing the requirement $f(v_3) = v_1$ with the hypothesis $f(u) = w$, we see that we can take $v_1 = w$ and $v_3 = u$. To determine v_2 we note that we must have $f(v_2) = \mathbf{0}$, therefore v_2 must be a vector of $\operatorname{Ker}(f)$, linearly independent of $v_1 = w$ and $v_3 = u$. One can take, for example, $v_2 = (-2, 1, 0)$, or even $v_2 = (2, 0, 1)$ (the verification that the vectors v_1, v_2, v_3 are linearly independent is immediate). We therefore note that the basis $\{v_1, v_2, v_3\}$ with respect to which the matrix of f is B is not unique.

The invertible matrix S such that $A = SBS^{-1}$ (or, equivalently, $AS = SB$) is the change-of-basis matrix, whose columns are formed by the components of the vectors v_1, v_2, v_3, with respect to the canonical basis. Therefore we have

$$S = \begin{pmatrix} 4 & -2 & 1 \\ -1 & 1 & -1 \\ 1 & 0 & 0 \end{pmatrix}$$

This matrix is not unique, because the basis $\{v_1, v_2, v_3\}$ is not unique.

Finally, we observe that there does not exist an invertible matrix R such that $A = RCR^{-1}$. If it existed, the matrices A and C would be similar and therefore the matrix C would also be similar to the matrix B (indeed in point (b) we proved that the matrix A is similar to the matrix B). But B and C are certainly not similar since B has rank 1 while C has rank 2.

▶ **Exercise 4.3.21.** Let $V = M_2(\mathbb{R})$ be the vector space of square matrices of order 2 with real entries. Given a real number k, let $f_k \colon V \to V$ be the linear function defined by setting $f_k(A) = 3\,A + k\,A^T$, for each matrix $A \in V$.

(a) For $k \in \mathbb{R}$, find the dimension and a basis of $\operatorname{Ker}(f_k)$.

(b) Write the matrix of f_k, with respect to the canonical basis of V.

(c) Now we set $k = 3$. Find a basis of the image of f_3.

(d) Now we set $k = -2$. Determine the eigenvalues and eigenspaces of f_{-2} and state whether it is diagonalizable.

Solution. Setting $A = \left(\begin{smallmatrix} a & b \\ c & d \end{smallmatrix}\right)$, we have

$$f_k(A) = 3\begin{pmatrix} a & b \\ c & d \end{pmatrix} + k\begin{pmatrix} a & c \\ b & d \end{pmatrix} = \begin{pmatrix} (3+k)a & 3b + kc \\ 3c + kb & (3+k)d \end{pmatrix}.$$

The matrix A belongs to the kernel of f_k if and only if

$$\begin{cases} (3+k)a = 0 \\ 3b + kc = 0 \\ 3c + kb = 0 \\ (3+k)d = 0 \end{cases}$$

By solving this system, we distinguish the following cases: (i) if $k \neq \pm 3$, the unique solution is $a = b = c = d = 0$, hence $\operatorname{Ker}(f_k) = \{\mathbf{0}\}$. (ii) if $k = 3$, we have $a = d = 0$, $b = -c$ and c can take any value. In this case $\operatorname{Ker}(f_3)$ has dimension 1 and a basis is given by the matrix $\left(\begin{smallmatrix} 0 & 1 \\ -1 & 0 \end{smallmatrix}\right)$. We note that the kernel of f_3 is the set of antisymmetric matrices.

(iii) Finally, if $k = -3$, we have $b = c$, while a, c and d can take any value. Therefore $\operatorname{Ker}(f_{-3})$ has dimension 3 and a basis is given by the matrices $\left(\begin{smallmatrix} 1 & 0 \\ 0 & 0 \end{smallmatrix}\right)$, $\left(\begin{smallmatrix} 0 & 1 \\ 1 & 0 \end{smallmatrix}\right)$, $\left(\begin{smallmatrix} 0 & 0 \\ 0 & 1 \end{smallmatrix}\right)$. We note that the kernel of f_{-3} coincides with the set of symmetric matrices.

The canonical basis of $V = M_2(\mathbb{R})$ is given by the matrices $E_1 = \left(\begin{smallmatrix} 1 & 0 \\ 0 & 0 \end{smallmatrix}\right)$, $E_2 = \left(\begin{smallmatrix} 0 & 1 \\ 0 & 0 \end{smallmatrix}\right)$, $E_3 = \left(\begin{smallmatrix} 0 & 0 \\ 1 & 0 \end{smallmatrix}\right)$, $E_4 = \left(\begin{smallmatrix} 0 & 0 \\ 0 & 1 \end{smallmatrix}\right)$. We have:

$$f_k(E_1) = \begin{pmatrix} 3+k & 0 \\ 0 & 0 \end{pmatrix} = (3+k)E_1, \quad f_k(E_2) = \begin{pmatrix} 0 & 3 \\ k & 0 \end{pmatrix} = 3E_2 + kE_3,$$

$$f_k(E_3) = \begin{pmatrix} 0 & k \\ 3 & 0 \end{pmatrix} = kE_2 + 3E_3, \quad f_k(E_4) = \begin{pmatrix} 0 & 0 \\ 0 & 3+k \end{pmatrix} = (3+k)E_4.$$

It follows that the matrix of f_k with respect to the canonical basis is

$$F_k = \begin{pmatrix} 3+k & 0 & 0 & 0 \\ 0 & 3 & k & 0 \\ 0 & k & 3 & 0 \\ 0 & 0 & 0 & 3+k \end{pmatrix}$$

For $k = 3$, the matrix of f_k becomes

$$F_3 = \begin{pmatrix} 6 & 0 & 0 & 0 \\ 0 & 3 & 3 & 0 \\ 0 & 3 & 3 & 0 \\ 0 & 0 & 0 & 6 \end{pmatrix}$$

whose rank is 3. Thus we have $\dim \operatorname{Im}(f_3) = 3$ and the columns of F_3 represent the coordinates of the vectors that generate the image of f_3. Thus we conclude that a basis of $\mathrm{Im}(f_3)$ is given by the matrices $\left(\begin{smallmatrix}6&0\\0&0\end{smallmatrix}\right)$, $\left(\begin{smallmatrix}0&3\\3&0\end{smallmatrix}\right)$, $\left(\begin{smallmatrix}0&0\\0&6\end{smallmatrix}\right)$. We can note that the image of f_3 coincides with the set of symmetric matrices.

For $k = -2$, the matrix of f_k becomes

$$F_{-2} = \begin{pmatrix} 1 & 0 & 0 & 0 \\ 0 & 3 & -2 & 0 \\ 0 & -2 & 3 & 0 \\ 0 & 0 & 0 & 1 \end{pmatrix}$$

We compute the characteristic polynomial of F_{-2}:

$$\det(F_{-2} - x \cdot \mathbf{1}) = \begin{vmatrix} 1-x & 0 & 0 & 0 \\ 0 & 3-x & -2 & 0 \\ 0 & -2 & 3-x & 0 \\ 0 & 0 & 0 & 1-x \end{vmatrix} = (1-x)^3(5-x).$$

The eigenvalues of F_{-2} are 1 (with multiplicity 3) and 5 (with multiplicity 1). We determine the eigenspace corresponding to the eigenvalue 1. The eigenvectors corresponding to this eigenvalue are given by the solutions of the system

$$\begin{pmatrix} 0 & 0 & 0 & 0 \\ 0 & 2 & -2 & 0 \\ 0 & -2 & 2 & 0 \\ 0 & 0 & 0 & 0 \end{pmatrix} \begin{pmatrix} x_1 \\ x_2 \\ x_3 \\ x_4 \end{pmatrix} = \begin{pmatrix} 0 \\ 0 \\ 0 \\ 0 \end{pmatrix}$$

This system is equivalent to the equation $x_2 = x_3$. This means that the eigenspace corresponding to the eigenvalue 1 has dimension 3 and a basis is given by the matrices $\left(\begin{smallmatrix}1&0\\0&0\end{smallmatrix}\right)$, $\left(\begin{smallmatrix}0&1\\1&0\end{smallmatrix}\right)$, $\left(\begin{smallmatrix}0&0\\0&1\end{smallmatrix}\right)$. The eigenspace corresponding to the eigenvalue 1 therefore coincides with the set of symmetric matrices.

Now we determine the eigenspace corresponding to the eigenvalue 5. We get the system

$$\begin{pmatrix} -4 & 0 & 0 & 0 \\ 0 & -2 & -2 & 0 \\ 0 & -2 & -2 & 0 \\ 0 & 0 & 0 & -4 \end{pmatrix} \begin{pmatrix} x_1 \\ x_2 \\ x_3 \\ x_4 \end{pmatrix} = \begin{pmatrix} 0 \\ 0 \\ 0 \\ 0 \end{pmatrix}$$

which is equivalent to

$$\begin{cases} x_1 = x_4 = 0 \\ x_2 = -x_3 \end{cases}$$

The eigenspace corresponding to the eigenvalue 5 therefore has dimension 1 and a basis is given by the matrix $\left(\begin{smallmatrix} 0 & 1 \\ -1 & 0 \end{smallmatrix}\right)$. This eigenspace therefore coincides with the set of antisymmetric matrices. From what we have seen, it follows that f_{-2} is diagonalizable.

► **Exercise 4.3.22.** Let $f\colon \mathbb{R}^3 \to \mathbb{R}^3$ be the endomorphism whose matrix, with respect to the canonical bases, is

$$A = \begin{pmatrix} 1 & 0 & 0 \\ 0 & t & 1-t \\ 2t-1 & t & t \end{pmatrix}$$

(a) Determine for which $t \in \mathbb{R}$ the eigenvalues of f are real, specifying the cases in which these eigenvalues have multiplicity greater than 1.

(b) Determine for which values of t the matrix A is diagonalizable over the field of real numbers.

(c) Determine the value of t for which there is an orthonormal basis of eigenvectors of f and write such a basis.

Solution. The characteristic polynomial of A is

$$\det(A - x \cdot \mathbf{1}) = \begin{vmatrix} 1-x & 0 & 0 \\ 0 & t-x & 1-t \\ 2t-1 & t & t-x \end{vmatrix} = (1-x)(x^2 - 2tx + 2t^2 - t)$$

The eigenvalues of A are therefore

$$\lambda_1 = 1, \qquad \lambda_2 = t - \sqrt{t - t^2}, \qquad \lambda_3 = t + \sqrt{t - t^2}.$$

These eigenvalues are real when $t - t^2 \geq 0$, i.e., for $0 \leq t \leq 1$.

The cases in which the eigenvalues are not distinct are the following: for $t = 0$ we have $\lambda_2 = \lambda_3 = 0$, so the eigenvalues are 1 (with multiplicity 1) and 0 (with multiplicity 2); for $t = 1$ we have $\lambda_2 = \lambda_3 = 1$, so in this case there is a unique eigenvalue equal to 1, with multiplicity 3. Finally, it is also possible that we have $\lambda_1 = \lambda_3$, which occurs for $t = 1/2$. In this case the eigenvalues are 1 (with multiplicity 2) and 0 (with multiplicity 1).

If $0 < t < 1/2$, or $1/2 < t < 1$, the matrix A has three real and distinct eigenvalues, so it is certainly diagonalizable. We still need to analyze the three cases $t = 0$, $t = 1$ and $t = 1/2$.

If $t = 0$, the matrix A becomes

$$\begin{pmatrix} 1 & 0 & 0 \\ 0 & 0 & 1 \\ -1 & 0 & 0 \end{pmatrix}$$

The eigenvectors corresponding to the eigenvalue 0 (which has multiplicity 2) are given by the solutions of the following system:

$$\begin{cases} x_1 = 0 \\ x_3 = 0 \end{cases}$$

This means that the dimension of this eigenspace is 1, and a basis is given by the vector $(0, 1, 0)$. In this case the matrix A is therefore not diagonalizable.

If $t = 1$, the matrix A becomes

$$\begin{pmatrix} 1 & 0 & 0 \\ 0 & 1 & 0 \\ 1 & 1 & 1 \end{pmatrix}$$

The eigenvectors corresponding to the eigenvalue 1 (which has multiplicity 3) are given by the solutions of the equation $x_1 + x_2 = 0$. This means that the dimension of this eigenspace is 2 and therefore, also in this case, the matrix A is not diagonalizable.

Finally, if $t = 1/2$, the matrix A becomes

$$\begin{pmatrix} 1 & 0 & 0 \\ 0 & 1/2 & 1/2 \\ 0 & 1/2 & 1/2 \end{pmatrix}$$

We note that this matrix is symmetric and therefore it is diagonalizable.

Recall that a matrix admits an orthonormal basis of eigenvectors if and only if it is symmetric. Since the matrix A is symmetric only for $t = 1/2$, we can conclude that an orthonormal basis of eigenvectors of f exists if only if $t = 1/2$. To find such a basis, we determine the eigenvectors of the matrix

$$\begin{pmatrix} 1 & 0 & 0 \\ 0 & 1/2 & 1/2 \\ 0 & 1/2 & 1/2 \end{pmatrix}$$

For the eigenvalue 0 we find the eigenvector $v_1 = (0, 1, -1)$, while for the eigenvalue 1 (which has multiplicity 2), the corresponding eigenspace has dimension 2 and a basis is given by the eigenvectors $v_2 = (1,0,0)$ and $v_3 = (0,1,1)$. It is straightforward to check that the vectors v_1, v_2, v_3 are pairwise orthogonal. To find an orthonormal basis of eigenvectors, all that remains is to normalize them, dividing each vector by its norm. Thus we obtain the orthonormal basis formed by the following vectors: $(0, 1/\sqrt{2}, -1/\sqrt{2})$, $(1,0,0)$, $(0, 1/\sqrt{2}, 1/\sqrt{2})$.

▶ **Exercise 4.3.23.** Let's consider the matrix

$$A = \begin{pmatrix} 2 & 4 & -2 \\ 1 & 3 & -2 \\ 3 & 8 & -5 \end{pmatrix}.$$

(a) Let $B = (2, k, 1)$. Using the Rouché–Capelli theorem, state for which value of $k \in \mathbb{R}$ the system $AX = B$ has solutions.

(b) Find the eigenvalues of A and determine if A is diagonalizable.

(c) Determine if A is similar to the matrix $\begin{pmatrix} 0 & 1 & 0 \\ 0 & 1 & 1 \\ 0 & 0 & -1 \end{pmatrix}$.

(d) Is A similar to A^2? Is A similar to A^3?

Solution. (a) We need to compute the rank of the matrix

$$(A \mid B) = \begin{pmatrix} 2 & 4 & -2 & 2 \\ 1 & 3 & -2 & k \\ 3 & 8 & -5 & 1 \end{pmatrix}$$

Reducing this matrix to the echelon form, we find that A has rank 2 while the complete matrix $(A \mid B)$ has rank 2 only if $k = 0$. This means that the system $AX = B$ has solutions only for $k = 0$.

(b) We have

$$\det \begin{pmatrix} 2-\lambda & 4 & -2 \\ 1 & 3-\lambda & -2 \\ 3 & 8 & -5-\lambda \end{pmatrix} = \lambda(1-\lambda)(1+\lambda)$$

so the eigenvalues of A are 0, 1 and -1. Since the eigenvalues are distinct, the matrix is diagonalizable (note that it is not necessary to compute the eigenvectors).

(c) We have seen that A is diagonalizable, so it is similar to the matrix

$$D = \begin{pmatrix} 0 & 0 & 0 \\ 0 & 1 & 0 \\ 0 & 0 & -1 \end{pmatrix}$$

Also the matrix

$$\begin{pmatrix} 0 & 1 & 0 \\ 0 & 1 & 1 \\ 0 & 0 & -1 \end{pmatrix}$$

has eigenvalues 0, 1 and -1, so it is similar to the matrix D. Since the two matrices are similar to the same matrix D, then they are also similar to each other.

(d) We have seen that A is similar to D, so we have $A = PDP^{-1}$ (where P is the matrix whose columns are the eigenvectors of A). Then we have

$$A^2 = PDP^{-1}PDP^{-1} = PD^2P^{-1}$$

and

$$D^2 = \begin{pmatrix} 0 & 0 & 0 \\ 0 & 1 & 0 \\ 0 & 0 & 1 \end{pmatrix}$$

hence the eigenvalues of A^2 are 0, 1 and 1, and are therefore different from those of A. This means that A and A^2 are not similar, because they have different eigenvalues.

Similarly we have

$$A^3 = PDP^{-1}PDP^{-1}PDP^{-1} = PD^3P^{-1}$$

and

$$D^3 = \begin{pmatrix} 0 & 0 & 0 \\ 0 & 1 & 0 \\ 0 & 0 & -1 \end{pmatrix} = D$$

hence $A^3 = PD^3P^{-1} = PDP^{-1} = A$. The matrices A^3 and A are equal, so they are also similar.

5

Euclidean Vector Spaces

5.1 Symmetric bilinear forms

► **Exercise 5.1.1.** Let V be a vector space over the field $\mathbb{C}$ of complex numbers with basis $\{v_1, v_2, v_3\}$ and let g be the symmetric bilinear form whose matrix, with respect to the given basis, is

$$G = \begin{pmatrix} 2 & 3 & -1 \\ 3 & 4 & 0 \\ -1 & 0 & 1 \end{pmatrix}$$

Find an invertible matrix P such that ${}^tPGP = \mathbf{1}$.

Solution. Since ${}^tPGP$ is the matrix of the bilinear form g with respect to a new basis $\{v_1', v_2', v_3'\}$, obtained from the basis $\{v_1, v_2, v_3\}$ by means of the change-of-basis matrix P, i.e.,

$$(v_1', v_2', v_3') = (v_1, v_2, v_3)P,$$

the problem reduces to the construction of a basis $\{v_1', v_2', v_3'\}$ with respect to which the matrix of g is the identity, i.e., to the determination of an orthonormal basis. For this we use the Gram–Schmidt procedure. Since $g(v_1, v_1) = 2$, we set $v_1' = v_1/\sqrt{2}$. Therefore we have $g(v_1', v_1') = 1$. A vector v_2'' orthogonal to v_1' is found in the following way: we set $v_2'' = av_1' + v_2$ and require that $g(v_1', v_2'') = 0$. We obtain $a = -g(v_1', v_2)$, hence we have:

$$\begin{aligned} v_2'' &= v_2 - g(v_1', v_2)v_1' \\ &= v_2 - g(v_1/\sqrt{2}, v_2)v_1/\sqrt{2} \\ &= v_2 - \tfrac{3}{2}\, v_1. \end{aligned}$$

Now we need to normalize this vector. We have $g(v_2'', v_2'') = g(v_2 - \frac{3}{2}\, v_1, v_2 - \frac{3}{2}\, v_1) = -\frac{1}{2}$, therefore we set $v_2' = v_2''/\sqrt{-\frac{1}{2}} = -i\sqrt{2}\, v_2'' = \frac{3}{2}\, i\sqrt{2}\, v_1 - i\sqrt{2}\, v_2$.

It only remains to determine a vector v_3'' orthogonal to v_1' and v_2'. Repeating the reasoning done previously, we have

$$\begin{aligned} v_3'' &= v_3 - g(v_1', v_3)v_1' - g(v_2', v_3)v_2' \\ &= v_3 - g(v_1/\sqrt{2}, v_3)v_1/\sqrt{2} - g\big(\tfrac{3}{2}\, i\sqrt{2}\, v_1 - i\sqrt{2}\, v_2, v_3\big)\big(\tfrac{3}{2}\, i\sqrt{2}\, v_1 - i\sqrt{2}\, v_2\big) \\ &= v_3 - 4v_1 + 3v_2. \end{aligned}$$

The latter vector must also be normalized. We have $g(v_3'', v_3'') = g(v_3 - 4v_1 + 3v_2, v_3 - 4v_1 + 3v_2) = 5$, hence we set $v_3' = v_3''/\sqrt{5} = -\frac{4}{\sqrt{5}}\, v_1 + \frac{3}{\sqrt{5}}\, v_2 + \frac{1}{\sqrt{5}}\, v_3$.

The change-of-basis matrix is

$$P = \begin{pmatrix} \frac{1}{\sqrt{2}} & \frac{3}{2} i\sqrt{2} & -\frac{4}{\sqrt{5}} \\ 0 & -i\sqrt{2} & \frac{3}{\sqrt{5}} \\ 0 & 0 & \frac{1}{\sqrt{5}} \end{pmatrix}$$

As an exercise one can verify that ${}^tPGP = \mathbf{1}$.

► **Exercise 5.1.2.** Let V be a vector space over the field $\mathbb{B}$ of real numbers with basis $\{v_1, v_2, v_3\}$ and let g be the symmetric bilinear form whose matrix, with respect to the given basis, is

$$G = \begin{pmatrix} 0 & 2 & 1 \\ 2 & 3 & -1 \\ 1 & -1 & 2 \end{pmatrix}$$

Find a basis of V with respect to which the matrix of g is diagonal, with only 1 and -1 on the diagonal.

Solution. We use the Gram–Schmidt procedure. Since the vector v_1 is isotropic, we cannot take a multiple of it as the first vector v_1' of the new basis. To solve this problem it is sufficient to exchange the vectors v_1 and v_2 of the given basis and apply the Gram–Schmidt procedure starting from v_2. Since $g(v_2, v_2) = 3$, we set $v_1' = v_2/\sqrt{3}$. Therefore we have $g(v_1', v_1') = 1$. Now we set $v_2'' = av_1' + v_1$ and we require that $g(v_1', v_2'') = 0$. We get $a = -g(v_1', v_1) = -2/\sqrt{3}$, hence $v_2'' = -2/\sqrt{3}\, v_1' + v_1 = -\frac{2}{3}\, v_2 + v_1$. Since $g(v_2'', v_2'') = -\frac{4}{3}$ but $\sqrt{-\frac{4}{3}}$ does not exist in $\mathbb{R}$, we set $v_2' = v_2''/\sqrt{\frac{4}{3}} = \frac{\sqrt{3}}{2}\, v_2'' = -\frac{\sqrt{3}}{3}\, v_2 + \frac{\sqrt{3}}{2}\, v_1$, hence we get $g(v_2', v_2') = -1$.

To find the last vector of the basis, we set $v_3'' = av_1' + bv_2' + v_3$ and require that $g(v_1', v_3'') = 0$ and $g(v_2', v_3'') = 0$. We get $a = -g(v_1', v_3) = 1/\sqrt{3}$ and

$b = g(v'_2, v_3) = 5\sqrt{3}/6$ (recall that in this case we have $g(v'_2, v'_2) = -1$). Therefore we get $v''_3 = \frac{1}{\sqrt{3}} v'_1 + \frac{5\sqrt{3}}{6} v'_2 + v_3 = -\frac{1}{2} v_2 + \frac{5}{4} v_1 + v_3$. We obtain $g(v''_3, v''_3) = 15/4$, hence we set $v'_3 = \frac{2}{\sqrt{15}} v''_3$, in order to have $g(v'_3, v'_3) = 1$.

The matrix of g in the basis $\{v'_1, v'_2, v'_3\}$ is

$$G' = \begin{pmatrix} 1 & 0 & 0 \\ 0 & -1 & 0 \\ 0 & 0 & 1 \end{pmatrix}.$$

► **Exercise 5.1.3.** Let V be a vector space over the field $\mathbb{R}$ of real numbers with basis $\{v_1, v_2, v_3\}$ and let g be the symmetric bilinear form whose matrix, with respect to the given basis, is

$$G = \begin{pmatrix} 0 & 2 & -1 \\ 2 & 0 & 3 \\ -1 & 3 & 0 \end{pmatrix}$$

Find a basis of V with respect to which the matrix of g is diagonal, with only 1 and -1 on the diagonal.

Solution. Since all the vectors of the given basis are isotropic, it is not possible to take a multiple of them as a vector of the new basis. In this case, instead, we will have to consider, for example, the vectors $v''_1 = v_1 + v_2$ and $v''_2 = v_1 - v_2$. These two vectors are orthogonal, since $g(v''_1, v''_2) = g(v_1, v_1) - g(v_2, v_2) = 0$, and they are not isotropic. In fact we have $g(v''_1, v''_1) = 4$ and $g(v''_2, v''_2) = -4$. Therefore for the new basis we take the vectors $v'_1 = \frac{1}{2} v''_1$ and $v'_2 = \frac{1}{2} v''_2$, getting $g(v'_1, v'_1) = 1$ and $g(v'_2, v'_2) = -1$. Now we can find v'_3 in the usual way. We set $v''_3 = av'_1 + bv'_2 + v_3$ and require that $g(v'_1, v''_3) = 0$ and $g(v'_2, v''_3) = 0$. We get $a = -g(v'_1, v_3) = -1$ and $b = g(v'_2, v_3) = -2$. Therefore we have $v''_3 = -v'_1 - 2v'_2 + v_3$ and, since $g(v''_3, v''_3) = 3$, we set $v'_3 = v''_3/\sqrt{3}$, thus getting $g(v'_3, v'_3) = 1$.

The matrix of g in the basis $\{v'_1, v'_2, v'_3\}$ is

$$G' = \begin{pmatrix} 1 & 0 & 0 \\ 0 & -1 & 0 \\ 0 & 0 & 1 \end{pmatrix}.$$

► **Exercise 5.1.4.** Let V be a real vector space of dimension 3 with basis $\{v_1, v_2, v_3\}$. Let g be the symmetric bilinear form whose matrix, with respect to the given basis, is

$$G = \begin{pmatrix} -3 & 1 & 0 \\ 1 & 2 & -1 \\ 0 & -1 & -1 \end{pmatrix}.$$

(a) Verify that g is non-degenerate and find an orthogonal basis of V.

(b) Calculate the inertia index $i(g)$.

(c) Determine whether there are non-zero isotropic vectors and, if so, determine a totally isotropic subspace of maximum dimension.

Solution. (a) We have

$$\det G = -3 \begin{vmatrix} 2 & -1 \\ -1 & -1 \end{vmatrix} - \begin{vmatrix} 1 & -1 \\ 0 & -1 \end{vmatrix} = 10,$$

hence g is non-degenerate.

Now we determine an orthogonal basis. We set $w_1 = v_1$ and we have $g(w_1, w_1) = -3$. Note that it is not necessary to normalize the vectors of the new basis, because an orthogonal basis is required and not an orthonormal basis.

Observing the matrix G we notice that the vector v_3 is orthogonal to w_1, therefore as the second vector of the orthogonal basis we can take $w_2 = v_3$; then we have $g(w_2, w_2) = -1$.

It now remains only to determine a third vector w_3 orthogonal to the previous two. We set $w_3 = \lambda_1 v_1 + \lambda_2 v_2 + \lambda_3 v_3$ and require that w_3 is orthogonal to w_1 and w_2. We obtain the following system:

$$\begin{cases} g(w_1, w_3) = -3\lambda_1 + \lambda_2 = 0 \\ g(w_2, w_3) = -\lambda_2 - \lambda_3 = 0, \end{cases}$$

from which we get

$$\begin{cases} \lambda_1 = -\frac{1}{3}\lambda_3 \\ \lambda_2 = -\lambda_3. \end{cases}$$

Setting, for example, $\lambda_3 = 3$, we obtain $w_3 = -v_1 - 3v_2 + 3v_3$. We now find $g(w_3, w_3) = 30$, from which we deduce that the matrix of g with respect to the orthogonal basis $\{w_1, w_2, w_3\}$ is

$$\begin{pmatrix} -3 & 0 & 0 \\ 0 & -1 & 0 \\ 0 & 0 & 30 \end{pmatrix}$$

(b) Looking at the matrix just found, we discover that the restriction of g to the subspace $L\{w_1, w_2\}$, generated by w_1 and w_2, is negative definite, while the restriction of g to the subspace generated by w_3 is positive definite. We therefore deduce that the inertia index of g is given by $i(g) = 1 - 2 = -1$.

(c) Since g is not defined, there are surely some non-zero isotropic vectors. For example, if we consider a vector $w = \lambda_1 w_1 + \lambda_2 w_2 + \lambda_3 w_3$, the condition that w is isotropic is

$$g(w, w) = -3\lambda_1^2 - \lambda_2^2 + 30\lambda_3^2 = 0,$$

which obviously has nonzero real solutions.

So let U be a totally isotropic subspace. Since, by definition, we have $U \subset U^\perp$, and $\dim U^\perp = 3 - \dim U$, we deduce that U can have, at most, dimension 1. Therefore to determine a totally isotropic subspace of maximum dimension it is sufficient to determine a non-zero isotropic vector; for example the vector $w = \sqrt{30}\, w_2 + w_3$. The subspace $U = L\{w\}$ is a totally isotropic subspace of maximum dimension (obviously such a subspace is not unique).

► **Exercise 5.1.5.** Let $V = \mathbb{R}^4$ and g be the symmetric bilinear form with matrix

$$G = \begin{pmatrix} 3 & -2 & 1 & 1 \\ -2 & 1 & -1 & -1 \\ 1 & -1 & 1 & 4 \\ 1 & -1 & 4 & 2 \end{pmatrix}$$

with respect to the canonical basis.

(a) Verify that g is non-degenerate, compute its inertia index and determine an orthogonal basis of V.

(b) Is $(\mathbb{R}^4, g)$ isometric to $\mathbb{R}^4$ with the usual scalar product? And to Minkowski's space?

Solution. (a) To prove that g is non-degenerate, it suffices to prove that $\det G \neq 0$, or that there is an orthogonal basis of V. Since, in any case, we have to determine an orthogonal basis, we can avoid calculating the determinant of G.

Let $\{e_1, \ldots, e_4\}$ be the canonical basis of $\mathbb{R}^4$. As the first vector of an orthogonal basis we can take $v_1 = e_1$; we have $g(v_1, v_1) = 3$. We now look for a second vector v_2 orthogonal to v_1: we can take v_2 of the following form

$$v_2 = \lambda_1 v_1 + e_2,$$

and require that $g(v_1, v_2) = 0$. We get $\lambda_1 = \frac{2}{3}$, hence

$$v_2 = \tfrac{2}{3}\, v_1 + e_2 = \tfrac{2}{3}\, e_1 + e_2.$$

Now it is easy to compute $g(v_2, v_2) = -\frac{1}{3}$.

To determine the third vector v_3 of the orthogonal basis the procedure is analogous. Let

$$v_3 = \lambda_1 v_1 + \lambda_2 v_2 + e_3,$$

and require that $g(v_1, v_3) = 0$ and $g(v_2, v_3) = 0$. We obtain the following two equations:

$$\begin{cases} g(v_1, v_3) = 3\lambda_1 + 1 = 0 \\ g(v_2, v_3) = -\frac{1}{3}\,\lambda_2 - \frac{1}{3} = 0, \end{cases}$$

whose solution is $\lambda_1 = -\frac{1}{3}$ and $\lambda_2 = -1$. Therefore we have

$$v_3 = -\tfrac{1}{3}\, v_1 - v_2 + e_3 = -e_1 - e_2 + e_3,$$

and $g(v_3, v_3) = 1$. Similarly, setting

$$v_4 = \lambda_1 v_1 + \lambda_2 v_2 + \lambda_3 v_3 + e_4,$$

and requiring that $g(v_1, v_4) = 0$, $g(v_2, v_4) = 0$ and $g(v_3, v_4) = 0$, we obtain $\lambda_1 = -\frac{1}{3}$, $\lambda_2 = -1$ and $\lambda_3 = -4$, therefore

$$v_4 = -\tfrac{1}{3}\, v_1 - v_2 - 4v_3 + e_4 = 3e_1 + 3e_2 - 4e_3 + e_4,$$

and $g(v_4, v_4) = -14$.

In conclusion, we have seen that there is an orthogonal basis, given by $\{v_1, \ldots, v_4\}$; this proves that g is non-degenerate. Furthermore, we have seen that the matrix of g with respect to the basis $\{v_1, \ldots, v_4\}$ is

$$\begin{pmatrix} 3 & 0 & 0 & 0 \\ 0 & -\frac{1}{3} & 0 & 0 \\ 0 & 0 & 1 & 0 \\ 0 & 0 & 0 & -14 \end{pmatrix}$$

from which we deduce that the inertia index of g is $i(g) = 0$.

(b) The usual scalar product in $\mathbb{R}^4$ has inertia index 4. Therefore $\mathbb{R}^4$ endowed with the usual scalar product is not isometric to the space V endowed with the symmetric bilinear form g, which has inertia index 0 (remember Sylvester's theorem). Analogously, (V, g) is not isometric to Minkowski's space, because the latter is $\mathbb{R}^4$ endowed with the bilinear form given by the matrix

$$\begin{pmatrix} -1 & 0 & 0 & 0 \\ 0 & 1 & 0 & 0 \\ 0 & 0 & 1 & 0 \\ 0 & 0 & 0 & 1 \end{pmatrix}$$

whose inertia index is 2.

▶ **Exercise 5.1.6.** In the vector space $\mathbb{R}^4$ consider the symmetric bilinear form g of matrix

$$G = \begin{pmatrix} 0 & 2 & -2 & 0 \\ 2 & 0 & 1 & -3 \\ -2 & 1 & 0 & -1 \\ 0 & -3 & -1 & 0 \end{pmatrix}$$

with respect to the canonical basis.

(a) Determine the inertia index of g.

(b) Find, if it exists, a basis of $\mathbb{R}^4$ with respect to which the matrix of g is

$$\begin{pmatrix} 1 & 0 & 0 & 0 \\ 0 & 1 & 0 & 0 \\ 0 & 0 & 0 & 1 \\ 0 & 0 & 1 & 0 \end{pmatrix}$$

(c) Find, if it exists, a basis of $\mathbb{R}^4$ with respect to which the matrix of g is

$$\begin{pmatrix} 0 & 1 & 0 & 0 \\ 1 & 0 & 0 & 0 \\ 0 & 0 & 0 & 1 \\ 0 & 0 & 1 & 0 \end{pmatrix}$$

(d) Find, if it exists, a basis of $\mathbb{R}^4$ with respect to which the matrix of g is

$$\begin{pmatrix} 0 & 1 & 0 & 0 \\ 1 & 0 & 0 & 0 \\ 0 & 0 & -1 & 0 \\ 0 & 0 & 0 & -1 \end{pmatrix}$$

Solution. (a) To compute the inertia index we first look for an orthogonal basis (since it will also be useful later on). Let $\{e_1, \ldots, e_4\}$ be the canonical basis of $\mathbb{R}^4$. In this case all vectors of the canonical basis are isotropic, so we cannot take any of them as the first vector of an orthogonal basis. We then take $v_1 = e_1 + e_2$; this vector is not isotropic, since we have $g(v_1, v_1) = 2g(e_1, e_2) = 4$.

As second vector it is easy to see that we can take $v_2 = e_1 - e_2$, given that it is orthogonal to v_1: $g(v_1, v_2) = 0$. Then we have $g(v_2, v_2) = -4$.

Now we can set

$$v_3 = \lambda_1 v_1 + \lambda_2 v_2 + e_3,$$

and require that $g(v_1, v_3) = g(v_2, v_3) = 0$. We get $\lambda_1 = \frac{1}{4}$ and $\lambda_2 = -\frac{3}{4}$. Then

$$v_3 = \tfrac{1}{4}\, v_1 - \tfrac{3}{4}\, v_2 + e_3 = -\tfrac{1}{2}\, e_1 + e_2 + e_3,$$

from which we get $g(v_3, v_3) = 2$. Similarly, setting

$$v_4 = \lambda_1 v_1 + \lambda_2 v_2 + \lambda_3 v_3 + e_4,$$

and requiring that $g(v_1, v_4) = g(v_2, v_4) = g(v_3, v_4) = 0$, we get $\lambda_1 = \frac{3}{4}$, $\lambda_2 = \frac{3}{4}$ and $\lambda_3 = 2$, hence

$$v_4 = \tfrac{3}{4}\, v_1 + \tfrac{3}{4}\, v_2 + 2v_3 + e_4 = \tfrac{1}{2}\, e_1 + 2e_2 + 2e_3 + e_4,$$

and $g(v_4, v_4) = -8$. The matrix of g with respect to the basis $\{v_1, \ldots, v_4\}$ is

$$\begin{pmatrix} 4 & 0 & 0 & 0 \\ 0 & -4 & 0 & 0 \\ 0 & 0 & 2 & 0 \\ 0 & 0 & 0 & -8 \end{pmatrix}$$

from which we deduce that the inertia index of g is $i(g) = 0$.

(b) Given that, as is easily verified, the matrix $\begin{pmatrix} 0 & 1 \\ 1 & 0 \end{pmatrix}$ has inertia index 0, we deduce that the matrix

$$\begin{pmatrix} 1 & 0 & 0 & 0 \\ 0 & 1 & 0 & 0 \\ 0 & 0 & 0 & 1 \\ 0 & 0 & 1 & 0 \end{pmatrix}$$

has inertia index $2 + 0 = 2$, which is different from the inertia index of g. From Sylvester's theorem it then follows that there does not exist a basis of $\mathbb{R}^4$ with respect to which g has this matrix.

(c) From the above it follows that the matrix

$$\begin{pmatrix} 0 & 1 & 0 & 0 \\ 1 & 0 & 0 & 0 \\ 0 & 0 & 0 & 1 \\ 0 & 0 & 1 & 0 \end{pmatrix}$$

has inertia index $0+0=0$, which is equal to the inertia index of g. Sylvester's theorem then states that there exists a basis of $\mathbb{R}^4$ with respect to which g has this matrix. Recalling the matrix of g with respect to the basis $\{v_1,\dots,v_4\}$ found previously, it is easy to see that by setting, for example, $w_1 = v_1 + v_2$, $w_2 = \frac{1}{4}(v_1 - v_2)$, $w_3 = v_3 + \frac{1}{2}v_4$ and $w_4 = \frac{1}{4}(v_3 - \frac{1}{2}v_4)$, we obtain a basis with respect to which the matrix of g coincides with this matrix.

(d) From the above it follows that the matrix

$$\begin{pmatrix} 0 & 1 & 0 & 0 \\ 1 & 0 & 0 & 0 \\ 0 & 0 & -1 & 0 \\ 0 & 0 & 0 & -1 \end{pmatrix}$$

has inertia index $0-2=-2$, which is different from the inertia index of g. From Sylvester's theorem it then follows that there does not exist a basis of $\mathbb{R}^4$ with respect to which g has this matrix.

5.2 Alternating bilinear forms

► **Exercise 5.2.1.** Let V be a real vector space and $\{v_1,\dots,v_4\}$ be its basis. Let $g\colon V\times V\to\mathbb{R}$ be the alternating bilinear form with matrix

$$G = \begin{pmatrix} 0 & 3 & 4 & -1 \\ -3 & 0 & 5 & 2 \\ -4 & -5 & 0 & -6 \\ 1 & -2 & 6 & 0 \end{pmatrix}$$

with respect to the given basis.

Find a basis $\{w_1,\dots,w_4\}$ of V such that the matrix of g with respect to this new basis is

$$G' = \begin{pmatrix} 0 & 1 & 0 & 0 \\ -1 & 0 & 0 & 0 \\ 0 & 0 & 0 & 1 \\ 0 & 0 & -1 & 0 \end{pmatrix}$$

Solution. As the first vector we can take $w_1 = v_1$. Now we have to find a vector w_2 such that $g(w_1, w_2) = 1$ (the fact that $g(w_2, w_2) = 0$ is obvious, since g is alternating). Noting that $g(w_1, v_2) = g(v_1, v_2) = 3$, we see that it suffices to set $w_2 = \frac{1}{3} v_2$ to get $g(w_1, w_2) = 1$. To find w_3 we set $w_3 = aw_1 + bw_2 + cv_3$. Requiring that $g(w_1, w_3) = 0$ and $g(w_2, w_3) = 0$, we obtain the system

$$\begin{cases} b + 4c = 0 \\ -a + \frac{5}{3} c = 0. \end{cases}$$

Setting, for example, $c = 3$, we get $b = -12$ and $a = 5$. Therefore we have $w_3 = 5w_1 - 12w_2 + 3v_3 = 5v_1 - 4v_2 + 3v_3$.

Finally, to determine w_4 we set $w_4 = aw_1 + bw_2 + cw_3 + dv_4$ and require that $g(w_1, w_4) = 0$, $g(w_2, w_4) = 0$ and $g(w_3, w_4) = 1$. We obtain the system

$$\begin{cases} b - d = 0 \\ -a + \frac{2}{3} d = 0 \\ -31d = 1, \end{cases}$$

whose solution is $a = -\frac{2}{93}$, $b = d = -\frac{1}{31}$ (and any c). Setting $c = 0$ we obtain $w_4 = -\frac{2}{93} w_1 - \frac{1}{31} w_2 - \frac{1}{31} v_4 = -\frac{2}{93} v_1 - \frac{1}{93} v_2 - \frac{1}{31} v_4$.

Remark: now we can write the change-of-basis matrix P. In fact we have

$$(w_1, \ldots, w_4) = (v_1, \ldots, v_4) \begin{pmatrix} 1 & 0 & 5 & -2/93 \\ 0 & 1/3 & -4 & -1/93 \\ 0 & 0 & 3 & 0 \\ 0 & 0 & 0 & -1/31 \end{pmatrix}$$

hence

$$P = \begin{pmatrix} 1 & 0 & 5 & -2/93 \\ 0 & 1/3 & -4 & -1/93 \\ 0 & 0 & 3 & 0 \\ 0 & 0 & 0 & -1/31 \end{pmatrix}.$$

It can be verified by a direct computation that $G' = {}^tPGP$.

5.3 Isometries

► **Exercise 5.3.1.** Let V be a real vector space of finite dimension and $g\colon V \times V \to \mathbb{R}$ be a positive definite symmetric bilinear form. If $\phi\colon V \to V$ is an isometry, prove that we have

$$\operatorname{Im}(\phi - \mathrm{id}) = \operatorname{Ker}(\phi - \mathrm{id})^{\perp}.$$

Solution. We prove that $\mathrm{Im}(\phi - \mathrm{id}) \subset \mathrm{Ker}(\phi - \mathrm{id})^\perp$. Let $v \in V$ and $w \in \mathrm{Ker}(\phi - \mathrm{id})$; we must prove that $(\phi - \mathrm{id})(v)$ is orthogonal to w, i.e., that $g((\phi - \mathrm{id})(v), w) = 0$. Note that $\phi(w) = w$, because $(\phi - \mathrm{id})(w) = 0$. Therefore we get:

$$\begin{aligned} g((\phi - \mathrm{id})(v), w) &= g(\phi(v), w) - g(v, w) \\ &= g(\phi(v), \phi(w)) - g(v, w) \\ &= 0, \end{aligned}$$

where we used the fact that $g(\phi(v), \phi(w)) = g(v, w)$, because ϕ is an isometry.

Since the previous proof is valid for every v and every w, we have thus proved that $\mathrm{Im}(\phi - \mathrm{id}) \subset \mathrm{Ker}(\phi - \mathrm{id})^\perp$. On the other hand, let us remember that

$$\dim \mathrm{Im}(\phi - \mathrm{id}) = \dim V - \dim \mathrm{Ker}(\phi - \mathrm{id})$$

and

$$\dim \mathrm{Ker}(\phi - \mathrm{id})^\perp = \dim V - \dim \mathrm{Ker}(\phi - \mathrm{id}).$$

Hence $\mathrm{Im}(\phi - \mathrm{id})$ and $\mathrm{Ker}(\phi - \mathrm{id})^\perp$ have the same dimension and therefore are equal.

5.4 Other exercises

► **Exercise 5.4.1.** Consider $\mathbb{C}$ as a vector space over $\mathbb{R}$ and define on it the bilinear form $g\colon \mathbb{C} \times \mathbb{C} \to \mathbb{R}$ given by

$$g(z_1, z_2) = 2\Re(z_1 z_2),$$

where $\Re(z)$ denotes the real part of the complex number z.

(a) Prove that g is bilinear and non-degenerate.

(b) Write the matrix of g with respect to the basis $\{1, i\}$ of $\mathbb{C}$.

(c) Let $z_1, z_2 \in \mathbb{C}$ be linearly independent over $\mathbb{R}$ (i.e., z_1 and z_2 are a basis of $\mathbb{C}$ considered as an $\mathbb{R}$-vector space). Write the matrix G of g with respect to the basis $\{z_1, z_2\}$ and prove that

$$\det G = \left[\det \begin{pmatrix} z_1 & z_2 \\ \bar{z}_1 & \bar{z}_2 \end{pmatrix}\right]^2$$

where $\bar{z}$ denotes the complex conjugate of $z \in \mathbb{C}$.

Solution. (a) The symmetry of g is obvious from the definition. Then to prove that g is bilinear it suffices to prove that it is linear with respect to the first argument. We have:

$$\begin{aligned} g(\lambda_1 z_1 + \lambda_2 z_2, z_3) &= 2\Re((\lambda_1 z_1 + \lambda_2 z_2) z_3) \\ &= 2\Re(\lambda_1 z_1 z_3 + \lambda_2 z_2 z_3) \\ &= 2\lambda_1 \Re(z_1 z_3) + 2\lambda_2 \Re(z_2 z_3) \\ &= \lambda_1 g(z_1 z_3) + \lambda_2 g(z_2 z_3), \end{aligned}$$

for any $z_1, z_2 \in \mathbb{C}$ and any $\lambda_1, \lambda_2 \in \mathbb{R}$.

To prove that g is non-degenerate, we need to prove that, for any $z \neq 0$ there exists $w \in \mathbb{C}$ such that $g(z, w) \neq 0$. Now, if $z = a + ib \neq 0$, we have $a \neq 0$ or $b \neq 0$. In the first case we have $g(z, 1) = 2a \neq 0$, while in the second we have $g(z, i) = -2b \neq 0$. Therefore g is non-degenerate.

(b) The entries of the matrix of g with respect to the basis $\{1, i\}$ are $g(1, 1)$, $g(1, i)$, $g(i, 1)$ and $g(i, i)$. Therefore the matrix is

$$\begin{pmatrix} 2 & 0 \\ 0 & -2 \end{pmatrix}$$

(c) As in the previous case, the (i, j)-th element of the matrix of g with respect to the basis $\{z_1, z_2\}$ is $g(z_i, z_j)$. It is easy to check that $g(z_1, z_1) = 2\Re(z_1^2) = z_1^2 + \bar{z}_1^2$, $g(z_1, z_2) = g(z_2, z_1) = 2\Re(z_1 z_2) = z_1 z_2 + \bar{z}_1 \bar{z}_2$ and $g(z_2, z_2) = 2\Re(z_2^2) = z_2^2 + \bar{z}_2^2$. Therefore we have

$$G = \begin{pmatrix} z_1^2 + \bar{z}_1^2 & z_1 z_2 + \bar{z}_1 \bar{z}_2 \\ z_1 z_2 + \bar{z}_1 \bar{z}_2 & z_2^2 + \bar{z}_2^2 \end{pmatrix}$$

It is now easy to compute the determinant of G and verify that the required equality holds.

► **Exercise 5.4.2.** Let A be a symmetric matrix with real entries. Is it true that the matrix $\mathbf{1} + A^2$ is invertible?

Solution. Since the matrix A is symmetric A^2 is also symmetric, therefore it represents a symmetric bilinear form in the vector space $\mathbb{R}^n$, where n is the order of A^2. The symmetric bilinear form defined by A^2 is positive semidefinite; in fact, for each column vector v, we have:

$${}^t v A^2 v = ({}^t v A)(Av) = {}^t(Av)(Av) = {}^t w w \geq 0,$$

where we set $w = Av$ and we remarked that ${}^t ww$ is the usual scalar product of w by itself.

The matrix $\mathbf{1} + A^2$ therefore represents the sum of a symmetric positive definite bilinear form (of matrix $\mathbf{1}$) with a positive semidefinite one, and is therefore positive definite and therefore non-degenerate. It follows that $\det(\mathbf{1} + A^2) \neq 0$ hence $\mathbf{1} + A^2$ is invertible.

5.5 Recap exercises

► **Exercise 5.5.1.** In the vector space $V = \mathbb{R}^4$ consider the symmetric bilinear form g whose matrix, with respect to the canonical basis, is

$$G = \begin{pmatrix} 4 & 0 & 1 & 0 \\ 0 & 2 & -1 & -2 \\ 1 & -1 & 4 & 0 \\ 0 & -2 & 0 & 4 \end{pmatrix}$$

Prove that g is non-degenerate and find an orthogonal basis of V. Also determine a diagonal matrix D and an invertible matrix P such that $D = {}^tPGP$.

Solution. To prove that g is non-degenerate, it suffices to verify that the determinant of G is different from 0. Indeed we have $\det G = 44$, so g is non-degenerate. To find an orthogonal basis we apply the Gram–Schmidt procedure to the canonical basis $\{e_1, e_2, e_3, e_4\}$ of $\mathbb{R}^4$.

We set $w_1 = e_1$; then $g(w_1, w_1) = g(e_1, e_1) = 4$. Now we set $w_2 = e_2 + \alpha_1 w_1$. Requiring that $g(w_2, w_1) = 0$ we get

$$\alpha_1 = -\frac{g(e_2, w_1)}{g(w_1, w_1)} = -\frac{g(e_2, e_1)}{g(e_1, e_1)} = 0.$$

Therefore $w_2 = e_2$ (this was obvious from the start since, from the matrix of g, we see that e_2 is orthogonal to e_1).

Now we set $w_3 = e_3 + \alpha_1 w_1 + \alpha_2 w_2$. Requiring that $g(w_3, w_1) = 0$ and $g(w_3, w_2) = 0$, we obtain

$$\alpha_1 = -\frac{g(e_3, w_1)}{g(w_1, w_1)} = -\frac{g(e_3, e_1)}{g(e_1, e_1)} = -\frac{1}{4}$$

$$\alpha_2 = -\frac{g(e_3, w_2)}{g(w_2, w_2)} = -\frac{g(e_3, e_2)}{g(e_2, e_2)} = \frac{1}{2},$$

hence

$$w_3 = e_3 - \tfrac{1}{4}\, w_1 + \tfrac{1}{2}\, w_2 = -\tfrac{1}{4}\, e_1 + \tfrac{1}{2}\, e_2 + e_3.$$

We have:

$$\begin{aligned} g(w_3, w_3) &= g\big(-\tfrac{1}{4}\, e_1 + \tfrac{1}{2}\, e_2 + e_3, -\tfrac{1}{4}\, e_1 + \tfrac{1}{2}\, e_2 + e_3\big) \\ &= \tfrac{1}{16}\, g(e_1, e_1) + \tfrac{1}{4}\, g(e_2, e_2) + g(e_3, e_3) \\ &\qquad -\tfrac{1}{4}\, g(e_1, e_2) - \tfrac{1}{2}\, g(e_1, e_3) + g(e_2, e_3) \\ &= \tfrac{13}{4} \end{aligned}$$

and similarly

$$g(e_4, w_3) = -\tfrac{1}{4}\, g(e_4, e_1) + \tfrac{1}{2}\, g(e_4, e_2) + g(e_4, e_3) = -1.$$

Finally we set $w_4 = e_4 + \alpha_1 w_1 + \alpha_2 w_2 + \alpha_3 w_3$. Requiring that $g(w_4, w_1) = 0$, $g(w_4, w_2) = 0$ and $g(w_4, w_3) = 0$, we obtain

$$\begin{aligned} \alpha_1 &= -\frac{g(e_4, w_1)}{g(w_1, w_1)} = 0 \\ \alpha_2 &= -\frac{g(e_4, w_2)}{g(w_2, w_2)} = 1 \\ \alpha_3 &= -\frac{g(e_4, w_3)}{g(w_3, w_3)} = \frac{4}{13} \end{aligned}$$

hence

$$w_4 = e_4 + w_2 + \tfrac{4}{13}\, w_3 = -\tfrac{1}{13}\, e_1 + \tfrac{15}{13}\, e_2 + \tfrac{4}{13}\, e_3 + e_4.$$

Then we compute

$$g(w_4, w_4) = g\big(-\tfrac{1}{13}\, e_1 + \tfrac{15}{13}\, e_2 + \tfrac{4}{13}\, e_3 + e_4, -\tfrac{1}{13}\, e_1 + \tfrac{15}{13}\, e_2 + \tfrac{4}{13}\, e_3 + e_4\big) = \tfrac{22}{13}.$$

The basis $\{w_1, w_2, w_3, w_4\}$ is an orthogonal basis and the matrix of g with respect to this basis is the following diagonal matrix:

$$D = \begin{pmatrix} 4 & 0 & 0 & 0 \\ 0 & 2 & 0 & 0 \\ 0 & 0 & 13/4 & 0 \\ 0 & 0 & 0 & 22/13 \end{pmatrix}$$

which, by the way, proves that g is non-degenerate. The change-of-basis matrix is

$$P = \begin{pmatrix} 1 & 0 & -1/4 & -1/13 \\ 0 & 1 & 1/2 & 15/13 \\ 0 & 0 & 1 & 4/13 \\ 0 & 0 & 0 & 1 \end{pmatrix}.$$

► **Exercise 5.5.2.** In the vector space $\mathbb{R}^3$ consider the symmetric bilinear form g defined by

$$g(v,w) = x_1y_1 - x_1y_2 + 2x_1y_3 - x_2y_1 + 2x_2y_2 - x_2y_3 + 2x_3y_1 - x_3y_2 + 4x_3y_3,$$

where $v = (x_1, x_2, x_3)$ and $w = (y_1, y_2, y_3)$.

(a) Write the matrix of g with respect to the canonical basis.

(b) Determine whether g is non-degenerate and whether it is positive definite, negative definite or indefinite.

(c) Find an orthogonal basis.

(d) Determine whether there are isotropic vectors and, if so, find at least one.

Solution. Let $G = (g_{ij})$ be the matrix of g with respect to the canonical basis. By definition we have $g_{ij} = g(e_i, e_j)$. On the other hand, if $v = (x_1, x_2, x_3)$ and $w = (y_1, y_2, y_3)$, we also have

$$g(v,w) = \sum_{i,j=1}^{3} g_{ij}x_iy_j,$$

hence the coefficient of the term x_iy_j in the expression of g is the element g_{ij} of the matrix of g with respect to the canonical basis. The matrix G is

$$G = \begin{pmatrix} 1 & -1 & 2 \\ -1 & 2 & -1 \\ 2 & -1 & 4 \end{pmatrix}$$

We note that the element $g_{11} = 1$ is positive and that the minor of order 2

$$\begin{vmatrix} 1 & -1 \\ -1 & 2 \end{vmatrix} = 1$$

is also positive. However the determinant of G is

$$\begin{vmatrix} 1 & -1 & 2 \\ -1 & 2 & -1 \\ 2 & -1 & 4 \end{vmatrix} = -1 < 0$$

therefore g is non-degenerate and it is indefinite.

The Gram–Schmidt procedure, applied to the canonical basis $\{e_1, e_2, e_3\}$ of $\mathbb{R}^3$, allows to construct an orthogonal basis (as we shall see, the fact that g is not positive definite has no influence).

We set $w_1 = e_1$; then we have $g(w_1, w_1) = g(e_1, e_1) = 1$. Now we set $w_2 = e_2 + \alpha_1 w_1$. By requiring w_2 to be orthogonal to w_1, i.e., $g(w_2, w_1) = 0$, we get

$$\alpha_1 = -\frac{g(e_2, w_1)}{g(w_1, w_1)} = -\frac{g(e_2, e_1)}{g(e_1, e_1)} = 1.$$

Therefore $w_2 = e_2 + e_1$. Now we compute $g(w_2, w_2)$:

$$g(w_2, w_2) = g\big(e_2 + e_1, e_2 + e_1\big) = 1.$$

Finally we set $w_3 = e_3 + \alpha_1 w_1 + \alpha_2 w_2$. Requiring that $g(w_3, w_1) = 0$ and $g(w_3, w_2) = 0$, we obtain

$$\alpha_1 = -\frac{g(e_3, w_1)}{g(w_1, w_1)} = -\frac{g(e_3, e_1)}{g(e_1, e_1)} = -2$$

$$\alpha_2 = -\frac{g(e_3, w_2)}{g(w_2, w_2)} = -\frac{g\big(e_3, e_2 + e_1\big)}{g(w_2, w_2)} = -1,$$

hence

$$w_3 = e_3 - 2w_1 - w_2 = -3e_1 - e_2 + e_3.$$

Finally we have:

$$g(w_3, w_3) = g(-3e_1 - e_2 + e_3, -3e_1 - e_2 + e_3) = -1.$$

The basis $\{w_1, w_2, w_3\}$ is an orthogonal basis and the matrix of g with respect to this basis is the following diagonal matrix:

$$D = \begin{pmatrix} 1 & 0 & 0 \\ 0 & 1 & 0 \\ 0 & 0 & -1 \end{pmatrix}$$

which, by the way, proves that g is non-degenerate and indefinite (because there are both positive and negative numbers on the diagonal). The change-of-basis matrix is

$$P = \begin{pmatrix} 1 & 1 & -3 \\ 0 & 1 & -1 \\ 0 & 0 & 1 \end{pmatrix}.$$

Now we could check with a direct computation that $D = {}^tPGP$.

Since g is indefinite there are isotropic vectors. To find one we can note that $g(w_1, w_1) = 1$ while $g(w_3, w_3) = -1$. Therefore, setting $u = w_1 + w_3$ we have

$$g(u, u) = g(w_1+w_3, w_1+w_3) = g(w_1, w_1)+2g(w_1, w_3)+g(w_3, w_3) = 1-1 = 0,$$

hence $u = w_1 + w_3 = -2e_1 - e_2 + e_3 = (-2, -1, 1)$ is an isotropic vector.

► **Exercise 5.5.3.** In the vector space $V = \mathbb{R}^3$ consider the symmetric bilinear form g whose matrix, with respect to the canonical basis, is

$$G = \begin{pmatrix} 2 & -1 & 1 \\ -1 & 3 & 1 \\ 1 & 1 & 2 \end{pmatrix}$$

Prove that g is non-degenerate and determine an orthogonal basis of V. Also determine a diagonal matrix D and an invertible matrix P such that $D = {}^tPGP$.

Solution. To prove that g is non-degenerate, it suffices to verify that the determinant of G is different from 0. We have

$$\det G = 2\begin{vmatrix} 3 & 1 \\ 1 & 2 \end{vmatrix} + \begin{vmatrix} -1 & 1 \\ 1 & 2 \end{vmatrix} + \begin{vmatrix} -1 & 3 \\ 1 & 1 \end{vmatrix} = 3,$$

hence g is non-degenerate. To find an orthogonal basis we apply the Gram–Schmidt procedure to the canonical basis $\{e_1, e_2, e_3\}$ of $\mathbb{R}^3$.

Set $w_1 = e_1$; we have $g(w_1, w_1) = g(e_1, e_1) = 2$. Now we set $w_2 = e_2 + \alpha_1 w_1$. By requiring w_2 to be orthogonal to w_1, i.e., $g(w_2, w_1) = 0$, we get

$$\alpha_1 = -\frac{g(e_2, w_1)}{g(w_1, w_1)} = -\frac{g(e_2, e_1)}{g(e_1, e_1)} = \frac{1}{2}.$$

Therefore $w_2 = e_2 + \frac{1}{2}\,e_1$. Now we compute $g(w_2, w_2)$:

$$g(w_2, w_2) = g\big(e_2 + \tfrac{1}{2}\,e_1, e_2 + \tfrac{1}{2}\,e_1\big) = \frac{5}{2}.$$

Finally we set $w_3 = e_3 + \alpha_1 w_1 + \alpha_2 w_2$. Requiring that $g(w_3, w_1) = 0$ and $g(w_3, w_2) = 0$, we obtain

$$\alpha_1 = -\frac{g(e_3, w_1)}{g(w_1, w_1)} = -\frac{g(e_3, e_1)}{g(e_1, e_1)} = -\frac{1}{2}$$

$$\alpha_2 = -\frac{g(e_3, w_2)}{g(w_2, w_2)} = -\frac{g\big(e_3, e_2 + \frac{1}{2}\,e_1\big)}{g(w_2, w_2)} = -\frac{3}{5},$$

hence

$$w_3 = e_3 - \tfrac{1}{2}\,w_1 - \tfrac{3}{5}\,w_2 = -\tfrac{4}{5}\,e_1 - \tfrac{3}{5}\,e_2 + e_3.$$

Finally we have:

$$g(w_3, w_3) = g\big(-\tfrac{4}{5}\,e_1 - \tfrac{3}{5}\,e_2 + e_3, -\tfrac{4}{5}\,e_1 - \tfrac{3}{5}\,e_2 + e_3\big) = \frac{3}{5}.$$

The basis $\{w_1, w_2, w_3\}$ is an orthogonal basis and the matrix of g with respect to this basis is the following diagonal matrix:

$$D = \begin{pmatrix} 2 & 0 & 0 \\ 0 & 5/2 & 0 \\ 0 & 0 & 3/5 \end{pmatrix}$$

which, by the way, proves that g is non-degenerate. The change-of-basis matrix is

$$P = \begin{pmatrix} 1 & 1/2 & -4/5 \\ 0 & 1 & -3/5 \\ 0 & 0 & 1 \end{pmatrix}.$$

► **Exercise 5.5.4.** In the Euclidean vector space $\mathbb{R}^3$, endowed with the usual scalar product, consider the subspace U with equation $x + y - z = 0$. Write the vector $v = (3, -2, 4)$ as a sum $v = v_1 + v_2$, with $v_1 \in U$ and $v_2 \in U^\perp$. Let $f\colon \mathbb{R}^3 \to \mathbb{R}^3$ be the function which associates to a vector $w \in \mathbb{R}^3$ its orthogonal projection $f(w)$ onto the subspace U. Write the matrix Π of f with respect to the canonical basis of $\mathbb{R}^3$.

Solution. From the equation $x+y-z = 0$ of U we get $z = x+y$. The vectors of U therefore depend on two parameters, which means that U has dimension 2 (it is a plane in $\mathbb{R}^3$). A basis of U is given by the vectors $u_1 = (1, 0, 1)$ and $u_2 = (0, 1, 1)$. A vector $w = (a, b, c)$ belongs to the orthogonal subspace of U if and only if

$$\begin{cases} w \cdot u_1 = a + c = 0 \\ w \cdot u_2 = b + c = 0. \end{cases}$$

This system has infinitely many solutions (depending on a parameter), given by

$$\begin{cases} a = -c \\ b = -c \\ c \text{ any value} \end{cases}$$

which means that $U^\perp$ has dimension 1 and is generated by the vector $w = (1, 1, -1)$ (note that the vector w has as its components the coefficients appearing in the equation of U).

Now let's look for two vectors $v_1 \in U$ and $v_2 \in U^\perp$ such that $v_1 + v_2 = v$. Since $v_1 \in U$, we must have $v_1 = \alpha_1 u_1 + \alpha_2 u_2$, while $v_2 \in U^\perp$ implies that $v_2 = \beta w$. Therefore we have $v = v_1 + v_2 = \alpha_1 u_1 + \alpha_2 u_2 + \beta w$. This gives us the following system of linear equations

$$\begin{cases} \alpha_1 + \beta = 3 \\ \alpha_2 + \beta = -2 \\ \alpha_1 + \alpha_2 - \beta = 4 \end{cases}$$

whose solution is

$$\begin{cases} \alpha_1 = 4 \\ \alpha_2 = -1 \\ \beta = -1. \end{cases}$$

Hence we have

$$v_1 = 4u_1 - u_2 = (4, -1, 3), \qquad v_2 = -w = (-1, -1, 1)$$

(the vector v_1 is called the orthogonal projection of v onto the subspace U).

We now describe another method for determining the vectors v_1 and v_2. Since $v_2 \in U^\perp$ we must have $v_2 = \beta w = (\beta, \beta, -\beta)$, because w is a basis of $U^\perp$. From the equality $v = v_1 + v_2$ we can therefore obtain $v_1 = v - v_2 = (3 - \beta, -2 - \beta, 4 + \beta)$. Since $v_1 \in U$, its coordinates must satisfy the equation of U. Substituting these coordinates into the equation of U we obtain $-3 - 3\beta = 0$, from which we get $\beta = -1$. Therefore the vectors we are looking for are $v_2 = (-1, -1, 1)$ and $v_1 = v - v_2 = (4, -1, 3)$.

As we can see, this second method is much more advantageous than the first, as it uses only one unknown β, while in the first method we used the three unknowns α_1, α_2 and β.

Let us now consider the function f which associates to each vector $w \in \mathbb{R}^3$ its orthogonal projection on U. The columns of the matrix of f with respect to the canonical basis of $\mathbb{R}^3$ are therefore the orthogonal projections of the vectors $e_1 = (1, 0, 0)$, $e_2 = (0, 1, 0)$, $e_3 = (0, 0, 1)$ onto the subspace U. The calculations necessary to determine the orthogonal projections of these vectors on U are completely analogous to those performed previously to determine the orthogonal projection of the vector v. For example, to determine the orthogonal projection of the vector e_1 onto U we need to solve

the system

$$\begin{cases} \alpha_1 + \beta = 1 \\ \alpha_2 + \beta = 0 \\ \alpha_1 + \alpha_2 - \beta = 0 \end{cases}$$

which is none other than the previous system in which the column of constant terms is now given by the components of the vector e_1. The solution of this system is

$$\begin{cases} \alpha_1 = 2/3 \\ \alpha_2 = -1/3 \\ \beta = 1/3 \end{cases}$$

from which we obtain the orthogonal projection $f(e_1)$ of the first vector of the canonical basis:

$$f(e_1) = \tfrac{2}{3}\,u_1 - \tfrac{1}{3}\,u_2 = (2/3, -1/3, 1/3);$$

this is the first column of the matrix. The second and third columns are determined in a completely analogous way, repeating the procedure just described for the vectors e_2 and e_3. Thus the following matrix is obtained

$$\Pi = \begin{pmatrix} 2/3 & -1/3 & 1/3 \\ -1/3 & 2/3 & 1/3 \\ 1/3 & 1/3 & 2/3 \end{pmatrix}.$$

There is another way to compute the matrix Π of the orthogonal projection onto the subspace U. We know from the theory that the matrix Π is given by the following formula:

$$\Pi = A\,({}^tAA)^{-1}\,{}^tA,$$

where A is the matrix whose columns are the vectors u_1 and u_2 of a basis of U. In this case we have

$$A = \begin{pmatrix} 1 & 0 \\ 0 & 1 \\ 1 & 1 \end{pmatrix}, \qquad {}^tA = \begin{pmatrix} 1 & 0 & 1 \\ 0 & 1 & 1 \end{pmatrix}$$

hence

$${}^tAA = \begin{pmatrix} 2 & 1 \\ 1 & 2 \end{pmatrix} \qquad \text{and} \qquad ({}^tAA)^{-1} = \frac{1}{3}\begin{pmatrix} 2 & -1 \\ -1 & 2 \end{pmatrix}.$$

Finally we obtain

$$\Pi = A\,({}^tAA)^{-1}\,{}^tA = \frac{1}{3}\begin{pmatrix} 2 & -1 & 1 \\ -1 & 2 & 1 \\ 1 & 1 & 2 \end{pmatrix}.$$

► **Exercise 5.5.5.** In the vector space $V = \mathbb{R}^4$ consider the symmetric bilinear form g whose matrix, with respect to the canonical basis, is

$$G = \begin{pmatrix} 1 & -1 & 0 & 1 \\ -1 & 4 & 0 & -2 \\ 0 & 0 & 5 & 1 \\ 1 & -2 & 1 & 3 \end{pmatrix}$$

Prove that g is non-degenerate and determine an orthogonal basis of V. Also determine a diagonal matrix D and an invertible matrix P such that $D = {}^tPGP$.

Solution. To prove that g is non-degenerate, it suffices to verify that the determinant of G is different from 0. We have $\det G = 22$, hence g is non-degenerate. To find an orthogonal basis we apply the Gram–Schmidt procedure to the canonical basis $\{e_1, e_2, e_3, e_4\}$ of $\mathbb{R}^4$.

Set $w_1 = e_1$; we have $g(w_1, w_1) = g(e_1, e_1) = 1$. Now we set $w_2 = e_2 + \alpha_1 w_1$. Requiring that $g(w_2, w_1) = 0$, we get

$$\alpha_1 = -\frac{g(e_2, w_1)}{g(w_1, w_1)} = -\frac{g(e_2, e_1)}{g(e_1, e_1)} = 1.$$

Therefore $w_2 = e_2 + w_1 = e_2 + e_1$. We get $g(w_2, w_2) = g(e_1 + e_2, e_1 + e_2) = 3$.

Now we set $w_3 = e_3 + \alpha_1 w_1 + \alpha_2 w_2$. Requiring that $g(w_3, w_1) = 0$ and $g(w_3, w_2) = 0$, we obtain

$$\alpha_1 = -\frac{g(e_3, w_1)}{g(w_1, w_1)} = -\frac{g(e_3, e_1)}{g(e_1, e_1)} = 0$$
$$\alpha_2 = -\frac{g(e_3, w_2)}{g(w_2, w_2)} = -\frac{g(e_3, e_1 + e_2)}{g(w_2, w_2)} = 0,$$

hence $w_3 = e_3$.

Finally we set $w_4 = e_4 + \alpha_1 w_1 + \alpha_2 w_2 + \alpha_3 w_3$. Requiring that $g(w_4, w_1) = 0$, $g(w_4, w_2) = 0$ and $g(w_4, w_3) = 0$, we obtain

$$\alpha_1 = -\frac{g(e_4, w_1)}{g(w_1, w_1)} = -1$$
$$\alpha_2 = -\frac{g(e_4, w_2)}{g(w_2, w_2)} = \frac{1}{3}$$
$$\alpha_3 = -\frac{g(e_4, w_3)}{g(w_3, w_3)} = -\frac{1}{5}$$

hence

$$w_4 = e_4 - w_1 + \tfrac{1}{3}\, w_2 - \tfrac{1}{5}\, w_3 = -\tfrac{2}{3}\, e_1 + \tfrac{1}{3}\, e_2 - \tfrac{1}{5}\, e_3 + e_4.$$

Finally we compute

$$g(w_4, w_4) = g\big(-\tfrac{2}{3}\, e_1 + \tfrac{1}{3}\, e_2 - \tfrac{1}{5}\, e_3 + e_4, -\tfrac{2}{3}\, e_1 + \tfrac{1}{3}\, e_2 - \tfrac{1}{5}\, e_3 + e_4\big) = \frac{22}{15}.$$

The basis $\{w_1, w_2, w_3, w_4\}$ is an orthogonal basis and the matrix of g with respect to this basis is the following diagonal matrix:

$$D = \begin{pmatrix} 1 & 0 & 0 & 0 \\ 0 & 3 & 0 & 0 \\ 0 & 0 & 5 & 0 \\ 0 & 0 & 0 & 22/15 \end{pmatrix}$$

which, by the way, proves that g is non-degenerate. The change-of-basis matrix is

$$P = \begin{pmatrix} 1 & 1 & 0 & -2/3 \\ 0 & 1 & 0 & 1/3 \\ 0 & 0 & 1 & -1/5 \\ 0 & 0 & 0 & 1 \end{pmatrix}$$

Now one could verify by a direct computation that $D = {}^tPGP$.

► **Exercise 5.5.6.** Let U be the vector subspace of $\mathbb{R}^4$ generated by the vectors $u_1 = (1, 2, -3, 1)$ and $u_2 = (0, 2, -1, 2)$. Determine the orthogonal projection of the vector $v = (2, 3, -1, 1)$ onto the subspace U. Also determine a vector w, of minimum norm, such that $v + w \in U$.

Solution. We can begin by determining a basis of the subspace $U^\perp$, orthogonal to U. A generic vector $v = (x_1, x_2, x_3, x_4)$ of $\mathbb{R}^4$ belongs to $U^\perp$ if and only if $v \cdot u_1 = x_1 + 2x_2 - 3x_3 + x_4 = 0$ and $v \cdot u_2 = 2x_2 - x_3 + 2x_4 = 0$. The subspace $U^\perp$ is therefore the set of solutions of the following linear system:

$$U^\perp : \begin{cases} x_1 + 2x_2 - 3x_3 + x_4 = 0 \\ 2x_2 - x_3 + 2x_4 = 0. \end{cases}$$

The solutions are given by

$$\begin{cases} x_1 = 4x_2 + 5x_4 \\ x_3 = 2x_2 + 2x_4, \end{cases}$$

for any $x_2, x_4 \in \mathbb{R}$. The subspace $U^\perp$ therefore has dimension 2 and a basis is given by the vectors $w_1 = (4, 1, 2, 0)$ and $w_2 = (5, 0, 2, 1)$, obtained by setting $x_2 = 1$, $x_4 = 0$ and $x_2 = 0$, $x_4 = 1$, respectively.

Recalling that $\mathbb{R}^4 = U \oplus U^\perp$, we deduce that the vectors u_1, u_2, w_1 and w_2 are a basis of $\mathbb{R}^4$. The vector $v = (2, 3, -1, 1)$ can therefore be written as follows

$$v = \alpha_1 u_1 + \alpha_2 u_2 + \beta_1 w_1 + \beta_2 w_2.$$

This gives us the following system of linear equations

$$\begin{cases} \alpha_1 + 4\beta_1 + 5\beta_2 = 2 \\ 2\alpha_1 + 2\alpha_2 + \beta_1 = 3 \\ -3\alpha_1 - \alpha_2 + 2\beta_1 + 2\beta_2 = -1 \\ \alpha_1 + 2\alpha_2 + \beta_2 = 1 \end{cases}$$

whose solution is

$$\begin{cases} \alpha_1 = \frac{1}{2} \\ \alpha_2 = \frac{1}{2} \\ \beta_1 = 1 \\ \beta_2 = -\frac{1}{2}. \end{cases}$$

Therefore we have

$$v = \tfrac{1}{2}\, u_1 + \tfrac{1}{2}\, u_2 + w_1 - \tfrac{1}{2}\, w_2.$$

Recalling that $U = \langle u_1, u_2 \rangle$ and $U^\perp = \langle w_1, w_2 \rangle$, by setting

$$v' = \tfrac{1}{2}\, u_1 + \tfrac{1}{2}\, u_2 = (1/2, 2, -2, 3/2)$$
$$v'' = w_1 - \tfrac{1}{2}\, w_2 = (3/2, 1, 1, -1/2)$$

we have $v' \in U$, $v'' \in U^\perp$ and $v = v' + v''$. The vector v' is the orthogonal projection of the vector v onto the subspace U (while v'' is the orthogonal projection of v onto $U^\perp$). Now noting that $v - v'' = v' \in U$, from what we have seen above we conclude that the vector w we are looking for (i.e., the vector w of minimum norm such that $v + w \in U$) is given by $w = -v'' = (-3/2, -1, -1, 1/2)$.

There is another method to determine the orthogonal projection of the vector v onto the subspace U. We write $v = v' + v''$, with $v' \in U$ and $v'' \in U^\perp$. Since $v' \in U$ we have $v' = \alpha_1 u_1 + \alpha_2 u_2$, as before. But now we get v'' by writing $v'' = v - v' = (2, 3, -1, 1) - (\alpha_1, 2\alpha_1 + 2\alpha_2, -3\alpha_1 - \alpha_2, \alpha_1 + 2\alpha_2) = (2 - \alpha_1, 3 - 2\alpha_1 - 2\alpha_2, -1 + 3\alpha_1 + \alpha_2, 1 - \alpha_1 - 2\alpha_2)$. Since $v'' \in U^\perp$ we must have $v'' \cdot u_1 = 0$ and $v'' \cdot u_2 = 0$, from which we obtain the following equations:

$$\begin{cases} 12 - 15\alpha_1 - 9\alpha_2 = 0 \\ 9 - 9\alpha_1 - 9\alpha_2 = 0. \end{cases}$$

By solving this system, we find

$$\begin{cases} \alpha_1 = 1/2 \\ \alpha_2 = 1/2 \end{cases}$$

from which we obtain $v' = (1/2, 2, -2, 3/2)$ e $v'' = (3/2, 1, 1, -1/2)$.

We note that this method is much more advantageous than the previous one because we had to solve a system with two equations and two unknowns, while in the previous method we had to solve a system with four equations and four unknowns.

► **Exercise 5.5.7.** In the Euclidean vector space $\mathbb{R}^4$, let U be the subspace with equations $2x_1 + x_2 - 3x_4 = 0$ and $x_1 - 2x_2 + x_3 = 0$. Determine a basis of U and a basis of the subspace $U^\perp$ orthogonal to U. Given the subspace W with equation $x_1 + 3x_2 - x_3 + 2x_4 = 0$, find a basis of $W \cap U^\perp$ and complete this basis to an orthogonal basis of W.

Solution. The subspace U is the set of solutions of the linear system

$$U : \begin{cases} 2x_1 + x_2 - 3x_4 = 0 \\ x_1 - 2x_2 + x_3 = 0 \end{cases}$$

which is equivalent to the following system:

$$U : \begin{cases} x_2 = -2x_1 + 3x_4 \\ x_3 = -5x_1 + 6x_4. \end{cases}$$

This system has infinitely many solutions, depending on two parameters (x_1 and x_4), so the subspace U has dimension 2. A basis is given by the vectors

$$u_1 = (1, -2, -5, 0), \qquad u_2 = (0, 3, 6, 1),$$

obtained by setting $x_1 = 1$, $x_4 = 0$ and $x_1 = 0$, $x_4 = 1$, respectively.

We now look for a basis of the subspace $U^\perp$, orthogonal to U. A vector $w = (x_1, x_2, x_3, x_4)$ belongs to $U^\perp$ if and only if

$$\begin{cases} u_1 \cdot w = x_1 - 2x_2 - 5x_3 = 0 \\ u_2 \cdot w = 3x_2 + 6x_3 + x_4 = 0. \end{cases}$$

From this system we obtain

$$U^\perp : \begin{cases} x_1 = 2x_2 + 5x_3 \\ x_4 = -3x_2 - 6x_3. \end{cases}$$

So these are the equations of the subspace $U^\perp$. This subspace therefore has dimension 2 and a basis is given by the vectors

$$w_1 = (2, 1, 0, -3), \qquad w_2 = (5, 0, 1, -6),$$

obtained by setting $x_2 = 1$, $x_3 = 0$ and $x_2 = 0$, $x_3 = 1$, respectively.

To find a basis of $W \cap U^\perp$ we note that the vectors belonging to $W \cap U^\perp$ must simultaneously satisfy the equations of $U^\perp$ and of W, i.e., they are the solutions of the following system:

$$W \cap U^\perp : \begin{cases} x_1 = 2x_2 + 5x_3 \\ x_4 = -3x_2 - 6x_3 \\ x_1 + 3x_2 - x_3 + 2x_4 = 0. \end{cases}$$

The solutions of this system are given by

$$\begin{cases} x_1 = -11x_3 \\ x_2 = -8x_3 \\ x_4 = 18x_3 \\ x_3 \text{ any value} \end{cases}$$

from which we deduce that $W \cap U^\perp$ has dimension 1 and a basis is given by the vector $v_1 = (-11, -8, 1, 18)$.

The subspace $W : x_1 + 3x_2 - x_3 + 2x_4 = 0$ has dimension 3, so to complete the basis v_1 of $W \cap U^\perp$ to an orthogonal basis of W, we have to determine two vectors $v_2, v_3 \in W$ in such a way that the vectors v_1, v_2, v_3 are linearly independent (and are therefore a basis of W) and are also pairwise perpendicular.

We start with the vector v_2. If we set $v_2 = (x_1, x_2, x_3, x_4)$, we have

$$v_2 \in W \quad \Leftrightarrow \quad x_1 + 3x_2 - x_3 + 2x_4 = 0,$$
$$v_2 \perp v_1 \quad \Leftrightarrow \quad v_1 \cdot v_2 = -11x_1 - 8x_2 + x_3 + 18x_4 = 0.$$

The vector v_2 is therefore one of the solutions of the system

$$\begin{cases} x_1 + 3x_2 - x_3 + 2x_4 = 0 \\ -11x_1 - 8x_2 + x_3 + 18x_4 = 0 \end{cases}$$

which is equivalent to the system

$$\begin{cases} x_2 = -2x_1 + 4x_4 \\ x_3 = -5x_1 + 14x_4. \end{cases}$$

As v_2 we can therefore take the following:

$$v_2 = (1, -2, -5, 0),$$

obtained by setting $x_1 = 1$ and $x_4 = 0$.

Finally we set $v_3 = (x_1, x_2, x_3, x_4)$. This vector must belong to the subspace W and must be orthogonal to the vectors v_1 and v_2:

$$v_3 \in W \quad \Leftrightarrow \quad x_1 + 3x_2 - x_3 + 2x_4 = 0,$$
$$v_3 \perp v_1 \quad \Leftrightarrow \quad v_1 \cdot v_3 = -11x_1 - 8x_2 + x_3 + 18x_4 = 0,$$
$$v_3 \perp v_2 \quad \Leftrightarrow \quad v_2 \cdot v_3 = x_1 - 2x_2 - 5x_3 = 0.$$

The vector v_3 is therefore one of the solutions of the system

$$\begin{cases} x_1 + 3x_2 - x_3 + 2x_4 = 0 \\ -11x_1 - 8x_2 + x_3 + 18x_4 = 0 \\ x_1 - 2x_2 - 5x_3 = 0. \end{cases}$$

Recalling that the system formed by the first two equations was equivalent to

$$\begin{cases} x_2 = -2x_1 + 4x_4 \\ x_3 = -5x_1 + 14x_4 \end{cases}$$

we deduce that v_3 is a solution of the system

$$\begin{cases} x_2 = -2x_1 + 4x_4 \\ x_3 = -5x_1 + 14x_4 \\ x_1 - 2x_2 - 5x_3 = 0. \end{cases}$$

This system has infinitely many solutions (depending on a parameter) and one of them is

$$\begin{cases} x_1 = 13 \\ x_2 = -6 \\ x_3 = 5 \\ x_4 = 5. \end{cases}$$

As third vector of the basis of W we can therefore take the vector $v_3 = (13, -6, 5, 5)$.

▶ **Exercise 5.5.8.** Let $f\colon \mathbb{R}^3 \to \mathbb{R}^3$ be the linear function whose matrix (with respect to the canonical bases) is

$$A = \begin{pmatrix} 2 & -1 & 3 \\ 1 & 2 & 1 \\ 3 & 1 & 4 \end{pmatrix}$$

Let $g\colon \mathbb{R}^3 \to \mathbb{R}^3$ be a linear function satisfying the following equality:

$$g(v) \cdot w = v \cdot f(w), \qquad \forall v, w \in \mathbb{R}^3.$$

(a) Compute $g(2, 3, 1)$.

(b) Write the matrix of g with respect to the canonical bases.

(c) Find a basis of $\mathrm{Ker}(f)$ and a basis of $\mathrm{Im}(f)$.

(d) Find a basis of $\mathrm{Ker}(g)$ and a basis of $\mathrm{Im}(g)$.

(e) Prove that $\mathrm{Im}(g) = \mathrm{Ker}(f)^\perp$ and $\mathrm{Im}(f) = \mathrm{Ker}(g)^\perp$.

Solution. Let $v = (2, 3, 1)$ and $g(v) = (a, b, c)$. If we choose $w = e_1 = (1, 0, 0)$, we have $g(v) \cdot w = (a, b, c) \cdot (1, 0, 0) = a$. On the other hand we have $v \cdot f(w) = (2, 3, 1) \cdot (2, 1, 3) = 10$ and from the equality $g(v) \cdot w = v \cdot f(w)$ we get $a = 10$. The values of b and c can be obtained in a similar way. By setting $w = e_2 = (0, 1, 0)$, we get

$$\begin{aligned} g(v) \cdot w &= (a, b, c) \cdot (0, 1, 0) = b \\ &= v \cdot f(w) = (2, 3, 1) \cdot (-1, 2, 1) = 5. \end{aligned}$$

Finally, setting $w = e_3 = (0, 0, 1)$, we get

$$\begin{aligned} g(v) \cdot w &= (a, b, c) \cdot (0, 0, 1) = c \\ &= v \cdot f(w) = (2, 3, 1) \cdot (3, 1, 4) = 13. \end{aligned}$$

This shows that $g(2, 3, 1) = (10, 5, 13)$.

To find the matrix of g with respect to the canonical bases it would be enough to compute $g(e_1)$, $g(e_2)$ and $g(e_3)$ (these are the three columns of the matrix of g) using the same method just used to compute $g(2, 3, 1)$. Instead, we proceed as follows. Let $v = (x, y, z)$ and set $g(v) = (a, b, c)$. If we denote by B the matrix of g with respect to the canonical bases, we must have

$$g(v) = \begin{pmatrix} a \\ b \\ c \end{pmatrix} = B \begin{pmatrix} x \\ y \\ z \end{pmatrix}$$

Now let $w = (\alpha, \beta, \gamma)$. The equality $g(v) \cdot w = v \cdot f(w)$ can be rewritten as follows:

$$(a, b, c) \begin{pmatrix} \alpha \\ \beta \\ \gamma \end{pmatrix} = v \cdot f(w) = (x, y, z) A \begin{pmatrix} \alpha \\ \beta \\ \gamma \end{pmatrix}$$

Since this equality must hold for every vector w (i.e., for every α, β, γ), it must necessarily be

$$(a, b, c) = (x, y, z) A.$$

By transposing both sides, we get

$$g(v) = \begin{pmatrix} a \\ b \\ c \end{pmatrix} = {}^tA \begin{pmatrix} x \\ y \\ z \end{pmatrix}$$

from which we deduce that $B = {}^tA$. The matrix of g is therefore the transpose of the matrix A of f.

The kernel of f is given by the solutions of the following linear system:

$$\begin{cases} 2x - y + 3z = 0 \\ x + 2y + z = 0 \\ 3x + y + 4z = 0. \end{cases}$$

We obtain

$$\begin{cases} x = -7z/5 \\ y = z/5 \\ z \text{ any value} \end{cases}$$

from which we deduce that the kernel of f has dimension 1 and is generated by the vector $u = (-7, 1, 5)$. The image of f therefore has dimension 2 and a basis is given, for example, by the first two columns of the matrix A, $w_1 = (2, 1, 3)$ and $w_2 = (-1, 2, 1)$.

The kernel and image of g can be determined in an analogous way. The kernel of g is the set of solutions of the following system:

$$\begin{cases} 2x + y + 3z = 0 \\ -x + 2y + z = 0 \\ 3x + y + 4z = 0. \end{cases}$$

We obtain

$$\begin{cases} x = -z \\ y = -z \\ z \text{ any value} \end{cases}$$

from which we deduce that $\mathrm{Ker}(g)$ has dimension 1 and is generated by the vector $u' = (1, 1, -1)$. The image of g therefore has dimension 2 and a basis is given by the first two columns of the matrix ${}^tA$ of g, i.e., by the first two rows of A: $w'_1 = (2, -1, 3)$ and $w'_2 = (1, 2, 1)$.

To prove that $\mathrm{Im}(g) = \mathrm{Ker}(f)^\perp$ we note that they have the same dimension, in fact we have

$$\dim \mathrm{Ker}(f)^\perp = \dim \mathbb{R}^3 - \dim \mathrm{Ker}(f) = 3 - 1 = 2.$$

Thus it suffices to prove that the vectors of a basis of $\mathrm{Im}(g)$ are orthogonal to the vectors of a basis of $\mathrm{Ker}(f)$. In fact we have

$$\begin{cases} w'_1 \cdot u = (2, -1, 3) \cdot (-7, 1, 5) = 0 \\ w'_2 \cdot u = (1, 2, 1) \cdot (-7, 1, 5) = 0, \end{cases}$$

as required. Similarly we have

$$\dim \mathrm{Ker}(g)^\perp = \dim \mathbb{R}^3 - \dim \mathrm{Ker}(g) = 3 - 1 = 2$$

hence $\dim \mathrm{Ker}(g)^\perp = \dim \mathrm{Im}(f)$. In order to prove the equality between these two subspaces, it suffices therefore to prove that the vectors of a basis of $\mathrm{Im}(f)$ are orthogonal to the vectors of a basis of $\mathrm{Ker}(g)$. In fact we have

$$\begin{cases} w_1 \cdot u' = (2, 1, 3) \cdot (1, 1, -1) = 0 \\ w_2 \cdot u' = (-1, 2, 1) \cdot (1, 1, -1) = 0, \end{cases}$$

as required.

► **Exercise 5.5.9.** In the vector space $\mathbb{R}^4$, endowed with the usual scalar product, consider the subspace U generated by the vectors $u_1 = (1, 0, -2, 0)$, $u_2 = (-3, -6, 6, 2)$ and $u_3 = (0, 3, 0, -1)$.

(a) Determine the Cartesian equations and a basis of U and of the subspace $U^\perp$ (orthogonal of U).

(b) We denote by $\pi\colon \mathbb{R}^4 \to \mathbb{R}^4$ the orthogonal projection onto the subspace $U^\perp$. Write the matrix of π with respect to the canonical basis of $\mathbb{R}^4$ and determine the kernel and image of π.

(c) Given the vector $v = (1, 1, 1, 1)$ write v in the form $v = v' + v''$, with $v' \in U$ and $v'' \in U^\perp$.

Solution. To determine the dimension of U we compute the rank of the following matrix:

$$\begin{pmatrix} 1 & 0 & -2 & 0 \\ -3 & -6 & 6 & 2 \\ 0 & 3 & 0 & -1 \end{pmatrix}$$

To the second row we add three times the first, obtaining

$$\begin{pmatrix} 1 & 0 & -2 & 0 \\ 0 & -6 & 0 & 2 \\ 0 & 3 & 0 & -1 \end{pmatrix}$$

Now we divide the second row by 2 and then add the second row to the third row:

$$\begin{pmatrix} 1 & 0 & -2 & 0 \\ 0 & -3 & 0 & 1 \\ 0 & 0 & 0 & 0 \end{pmatrix}$$

This matrix has rank 2, which means that only two of the three vectors u_1, u_2, u_3 are linearly independent. We therefore have $\dim U = 2$ and as the basis of U we can take the vectors $u_1 = (1, 0, -2, 0)$ and $u_3 = (0, 3, 0, -1)$.

A generic vector (x_1, x_2, x_3, x_4) belongs to the subspace U if and only if we have

$$(x_1, x_2, x_3, x_4) = \alpha u_1 + \beta u_3,$$

that is

$$\begin{cases} x_1 = \alpha \\ x_2 = 3\beta \\ x_3 = -2\alpha \\ x_4 = -\beta. \end{cases}$$

By eliminating the two parameters α and β we obtain the following (Cartesian) equations for the subspace U:

$$U : \begin{cases} 2x_1 + x_3 = 0 \\ x_2 + 3x_4 = 0. \end{cases}$$

A vector $w = (x_1, x_2, x_3, x_4)$ belongs to the subspace $U^\perp$ if and only if it is orthogonal to the vectors of a basis of U:

$$\begin{cases} w \cdot u_1 = x_1 - 2x_3 = 0 \\ w \cdot u_3 = 3x_2 - x_4 = 0. \end{cases}$$

So these are the Cartesian equations of $U^\perp$. The solutions of this system are given by

$$U^\perp : \begin{cases} x_1 = 2x_3 \\ x_4 = 3x_2 \\ x_2, x_3 \text{ any value.} \end{cases}$$

We thus conclude that $U^\perp$ has dimension 2 and a basis is given by the vectors $w_1 = (0, 1, 0, 3)$ and $w_2 = (2, 0, 1, 0)$.

Let's now answer the third question (the calculations we will do will be useful for answering the second as well). Since $U \oplus U^\perp = \mathbb{R}^4$, the vectors u_1, u_3, w_1, w_2 are a basis of $\mathbb{R}^4$. So we can write

$$v = \lambda_1 u_1 + \lambda_2 u_3 + \lambda_3 w_1 + \lambda_4 w_2,$$

that is

$$\begin{cases} 1 = \lambda_1 + 2\lambda_4 \\ 1 = 3\lambda_2 + \lambda_3 \\ 1 = -2\lambda_1 + \lambda_4 \\ 1 = -\lambda_2 + 3\lambda_3. \end{cases}$$

The solution of this system is

$$\begin{cases} \lambda_1 = -1/5 \\ \lambda_2 = 1/5 \\ \lambda_3 = 2/5 \\ \lambda_4 = 3/5 \end{cases}$$

hence we have

$$v = -\tfrac{1}{5}\, u_1 + \tfrac{1}{5}\, u_3 + \tfrac{2}{5}\, w_1 + \tfrac{3}{5}\, w_2.$$

If we set

$$v' = -\tfrac{1}{5}\,u_1 + \tfrac{1}{5}\,u_3 \in U, \qquad v'' = \tfrac{2}{5}\,w_1 + \tfrac{3}{5}\,w_2 \in U^\perp$$

we obtain the decomposition of the vector v as the sum $v = v' + v''$, with $v' \in U$ and $v'' \in U^\perp$, as required (the vectors v' and v'' are the orthogonal projections of v onto the subspaces U and $U^\perp$, respectively). We find

$$v' = \left(-\tfrac{1}{5}, \tfrac{3}{5}, \tfrac{2}{5}, -\tfrac{1}{5}\right), \qquad v'' = \left(\tfrac{6}{5}, \tfrac{2}{5}, \tfrac{3}{5}, \tfrac{6}{5}\right).$$

Let us now consider the function $\pi\colon \mathbb{R}^4 \to \mathbb{R}^4$ which associates to each vector $v \in \mathbb{R}^4$ its orthogonal projection onto the subspace $U^\perp$. In other words, given $v \in \mathbb{R}^4$ we need to decompose v as the sum $v = v' + v''$, with $v' \in U$ and $v'' \in U^\perp$, exactly as we just did for the vector $v = (1, 1, 1, 1)$. The function π is defined by setting $\pi(v) = v''$.

From the definition of the function π we can immediately deduce that $\mathrm{Im}(\pi) = U^\perp$ and $\mathrm{Ker}(\pi) = U$ (the only vectors whose orthogonal projection on $U^\perp$ is zero are the vectors orthogonal to $U^\perp$, i.e., the vectors of U). We also recall that the columns of the matrix of π with respect to the canonical basis of $\mathbb{R}^4$ are the images, via π, of the vectors of the canonical basis, i.e., they are the orthogonal projections onto $U^\perp$ of the vectors e_1, e_2, e_3, e_4.

We start with the vector $e_1 = (1, 0, 0, 0)$. Writing

$$e_1 = \lambda_1 u_1 + \lambda_2 u_3 + \lambda_3 w_1 + \lambda_4 w_2$$

we get the system

$$\begin{cases} \lambda_1 + 2\lambda_4 = 1 \\ 3\lambda_2 + \lambda_3 = 0 \\ -2\lambda_1 + \lambda_4 = 0 \\ -\lambda_2 + 3\lambda_3 = 0 \end{cases}$$

whose solution is

$$\begin{cases} \lambda_1 = 1/5 \\ \lambda_2 = 0 \\ \lambda_3 = 0 \\ \lambda_4 = 2/5. \end{cases}$$

Hence we have

$$e_1 = \tfrac{1}{5}\,u_1 + \tfrac{2}{5}\,w_2.$$

From this we deduce that the orthogonal projection of the vector e_1 onto the subspace $U^\perp$ is

$$\pi(e_1) = \tfrac{2}{5}\,w_2 = \left(\tfrac{4}{5}, 0, \tfrac{2}{5}, 0\right).$$

So this is the first column of the matrix of π.

Let us now consider the vector $e_2 = (0, 1, 0, 0)$. Writing

$$e_2 = \lambda_1 u_1 + \lambda_2 u_3 + \lambda_3 w_1 + \lambda_4 w_2$$

we get the system

$$\begin{cases} \lambda_1 + 2\lambda_4 = 0 \\ 3\lambda_2 + \lambda_3 = 1 \\ -2\lambda_1 + \lambda_4 = 0 \\ -\lambda_2 + 3\lambda_3 = 0 \end{cases}$$

whose solution is

$$\begin{cases} \lambda_1 = 0 \\ \lambda_2 = 3/10 \\ \lambda_3 = 1/10 \\ \lambda_4 = 0. \end{cases}$$

Hence we have

$$e_2 = \tfrac{3}{10}\, u_3 + \tfrac{1}{10}\, w_1.$$

From this we deduce that the orthogonal projection of the vector e_2 onto the subspace $U^\perp$ is

$$\pi(e_2) = \tfrac{1}{10}\, w_1 = \left(0, \tfrac{1}{10}, 0, \tfrac{3}{10}\right).$$

So this is the second column of the matrix of π.

Now let's move on to the vector $e_3 = (0, 0, 1, 0)$. Writing

$$e_3 = \lambda_1 u_1 + \lambda_2 u_3 + \lambda_3 w_1 + \lambda_4 w_2$$

we get the system

$$\begin{cases} \lambda_1 + 2\lambda_4 = 0 \\ 3\lambda_2 + \lambda_3 = 0 \\ -2\lambda_1 + \lambda_4 = 1 \\ -\lambda_2 + 3\lambda_3 = 0 \end{cases}$$

whose solution is

$$\begin{cases} \lambda_1 = -2/5 \\ \lambda_2 = 0 \\ \lambda_3 = 0 \\ \lambda_4 = 1/5. \end{cases}$$

Hence we have

$$e_3 = -\tfrac{2}{5}\, u_1 + \tfrac{1}{5}\, w_2.$$

From this we deduce that the orthogonal projection of the vector e_3 onto the subspace $U^\perp$ is

$$\pi(e_3) = \tfrac{1}{5}\, w_2 = \left(\tfrac{2}{5}, 0, \tfrac{1}{5}, 0\right).$$

So this is the third column of the matrix of π.

Finally we consider the vector $e_4 = (0, 0, 0, 1)$. Writing

$$e_4 = \lambda_1 u_1 + \lambda_2 u_3 + \lambda_3 w_1 + \lambda_4 w_2$$

we get the system

$$\begin{cases} \lambda_1 + 2\lambda_4 = 0 \\ 3\lambda_2 + \lambda_3 = 0 \\ -2\lambda_1 + \lambda_4 = 0 \\ -\lambda_2 + 3\lambda_3 = 1 \end{cases}$$

whose solution is

$$\begin{cases} \lambda_1 = 0 \\ \lambda_2 = -1/10 \\ \lambda_3 = 3/10 \\ \lambda_4 = 0. \end{cases}$$

Hence we have

$$e_4 = -\tfrac{1}{10}\, u_3 + \tfrac{3}{10}\, w_1.$$

From this we deduce that the orthogonal projection of the vector e_4 onto the subspace $U^\perp$ is

$$\pi(e_4) = \tfrac{3}{10}\, w_1 = \left(0, \tfrac{3}{10}, 0, \tfrac{9}{10}\right).$$

So this is the fourth column of the matrix of π. Therefore the matrix is

$$A = \begin{pmatrix} \frac{4}{5} & 0 & \frac{2}{5} & 0 \\ 0 & \frac{1}{10} & 0 & \frac{3}{10} \\ \frac{2}{5} & 0 & \frac{1}{5} & 0 \\ 0 & \frac{3}{10} & 0 & \frac{9}{10} \end{pmatrix}.$$

► **Exercise 5.5.10.** In the Euclidean vector space $\mathbb{R}^4$ (with the usual dot product) let V be the subspace generated by the vectors $v_1 = (2, 0, 1, -1)$, $v_2 = (1, -1, 0, 1)$ and $v_3 = (1, -2, -1, 0)$. Write the matrix of the restriction of the dot product to the subspace V with respect to the basis $\{v_1, v_2, v_3\}$. Determine an orthonormal basis of V.

Solution. Recall that the dot product is a symmetric (positive definite) bilinear form. By definition, its matrix $G = (g_{ij})$ with respect to a basis $\{v_1, v_2, \dots, v_n\}$ is obtained by setting $g_{ij} = v_i \cdot v_j$, for $i, j = 1, \dots, n$. In this case we have:

$$\begin{array}{lll} v_1 \cdot v_1 = 6 & v_1 \cdot v_2 = 1 & v_1 \cdot v_3 = 1 \\ v_2 \cdot v_2 = 3 & v_2 \cdot v_3 = 3 & v_3 \cdot v_3 = 6 \end{array}$$

hence the matrix of the restriction of the dot product to the subspace V, with respect to the basis $\{v_1, v_2, v_3\}$, is

$$G = \begin{pmatrix} 6 & 1 & 1 \\ 1 & 3 & 3 \\ 1 & 3 & 6 \end{pmatrix}$$

To determine an orthonormal basis we apply the Gram–Schmidt procedure to the basis $\{v_1, v_2, v_3\}$ of V.

Let $w_1 = v_1 = (2, 0, 1, -1)$; we have $w_1 \cdot w_1 = 6$. Now we set $w_2 = v_2 + \alpha_1 w_1$. Requiring that w_2 is orthogonal to w_1, i.e., that $w_2 \cdot w_1 = 0$, we find

$$\alpha_1 = -\frac{v_2 \cdot w_1}{w_1 \cdot w_1} = -\frac{1}{6}.$$

Hence we have

$$w_2 = v_2 - \tfrac{1}{6}\, v_1 = \left(\tfrac{2}{3}, -1, -\tfrac{1}{6}, \tfrac{7}{6}\right),$$

from which we deduce that

$$w_2 \cdot w_2 = \frac{17}{6}.$$

Finally we set $w_3 = v_3 + \alpha_1 w_1 + \alpha_2 w_2$. By requiring that $w_3 \cdot w_1 = 0$ and $w_3 \cdot w_2 = 0$, we obtain

$$\alpha_1 = -\frac{v_3 \cdot w_1}{w_1 \cdot w_1} = -\tfrac{1}{6}$$

$$\alpha_2 = -\frac{v_3 \cdot w_2}{w_2 \cdot w_2} = -1,$$

hence

$$w_3 = v_3 - \tfrac{1}{6}\, w_1 - w_2 = v_3 - v_2 = (0, -1, -1, -1).$$

Therefore we have

$$w_3 \cdot w_3 = 3.$$

The basis $\{w_1, w_2, w_3\}$ is an orthogonal basis of V and the norms of these vectors are as follows:

$$\|w_1\| = \sqrt{w_1 \cdot w_1} = \sqrt{6},\ \|w_2\| = \sqrt{w_2 \cdot w_2} = \sqrt{\tfrac{17}{6}},\ \|w_3\| = \sqrt{w_3 \cdot w_3} = \sqrt{3}.$$

To obtain an orthonormal basis of V it is now sufficient to divide each vector w_i of the orthogonal basis by its norm. Thus we get the vectors

$$w'_1 = \frac{w_1}{\|w_1\|}, \quad w'_2 = \frac{w_2}{\|w_2\|}, \quad w'_3 = \frac{w_3}{\|w_3\|},$$

which are an orthonormal basis of V.

▶ **Exercise 5.5.11.** In the Euclidean vector space $\mathbb{R}^4$, endowed with the usual scalar product, consider the subspace U with equation $x - 2y + 3z + 2w = 0$.

(a) Find the dimension and a basis of U.

(b) Using the Gram–Schmidt procedure, find an orthonormal basis of U.

(c) Determine the orthogonal projection of the vector $v = (3, 1, 0, -2)$ onto the subspace U.

Solution. From the equation of U we get $x = 2y - 3z - 2w$. Assigning to y, z and w the values $(1, 0, 0)$, $(0, 1, 0)$ and $(0, 0, 1)$ yield the three vectors

$$u_1 = (2, 1, 0, 0), \quad u_2 = (-3, 0, 1, 0), \quad u_3 = (-2, 0, 0, 1)$$

which are a basis of U (hence $\dim U = 3$).

We now apply the Gram–Schmidt procedure to this basis. Let $w_1 = u_1 = (2, 1, 0, 0)$; we have $w_1 \cdot w_1 = 5$. Now we set $w_2 = u_2 + \alpha_1 w_1$. Requiring that w_2 is orthogonal to w_1, i.e., that $w_2 \cdot w_1 = 0$, we find

$$\alpha_1 = -\frac{u_2 \cdot w_1}{w_1 \cdot w_1} = \frac{6}{5}.$$

Hence we have

$$w_2 = u_2 + \tfrac{6}{5}\, w_1 = \left(-\tfrac{3}{5}, \tfrac{6}{5}, 1, 0\right)$$

and $w_2 \cdot w_2 = \frac{14}{5}$. Now we set $w_3 = u_3 + \alpha_1 w_1 + \alpha_2 w_2$. Requiring that w_3 is orthogonal to w_1 and w_2, i.e., that $w_3 \cdot w_1 = w_3 \cdot w_2 = 0$, we find

$$\alpha_1 = -\frac{u_3 \cdot w_1}{w_1 \cdot w_1} = \frac{4}{5}, \qquad \alpha_2 = -\frac{u_3 \cdot w_2}{w_2 \cdot w_2} = -\frac{3}{7}.$$

Hence we get

$$w_3 = u_3 + \tfrac{4}{5}\, w_1 - \tfrac{3}{7}\, w_2 = \left(-\tfrac{1}{7}, \tfrac{2}{7}, -\tfrac{3}{7}, 1\right)$$

and $w_3 \cdot w_3 = \frac{9}{7}$. The vectors w_1, w_2 and w_3 are an orthogonal basis of the subspace U; their norms are given respectively by:

$$\begin{aligned} \|w_1\| &= \sqrt{w_1 \cdot w_1} = \sqrt{5} \\ \|w_2\| &= \sqrt{w_2 \cdot w_2} = \sqrt{14/5} \\ \|w_3\| &= \sqrt{w_3 \cdot w_3} = \sqrt{9/7} = \tfrac{3}{\sqrt{7}}. \end{aligned}$$

To obtain an orthonormal basis of U it is now sufficient to divide the vectors w_1, w_2 and w_3 by their norms:

$$w_1' = \frac{w_1}{\|w_1\|} = \Big(\frac{2}{\sqrt{5}}, \frac{1}{\sqrt{5}}, 0, 0\Big)$$
$$w_2' = \frac{w_2}{\|w_2\|} = \Big(-\frac{3}{5}\sqrt{\frac{5}{14}}, \frac{6}{5}\sqrt{\frac{5}{14}}, \sqrt{\frac{5}{14}}, 0\Big)$$
$$w_3' = \frac{w_3}{\|w_3\|} = \Big(-\frac{\sqrt{7}}{21}, \frac{2\sqrt{7}}{21}, -\frac{\sqrt{7}}{7}, \frac{\sqrt{7}}{3}\Big).$$

To determine the orthogonal projection of the vector $v = (3, 1, 0, -2)$ onto the subspace U we can start by observing that, from the equation of U

$$x - 2y + 3z + 2w = 0,$$

we deduce that the vector $n = (1, -2, 3, 2)$ is a basis of the subspace $U^\perp$. Since $U \oplus U^\perp = \mathbb{R}^4$ and recalling that $u_1 = (2, 1, 0, 0)$, $u_2 = (-3, 0, 1, 0)$, $u_3 = (-2, 0, 0, 1)$ is a basis of U, we deduce that the vectors u_1, u_2, u_3 and n are a basis of $\mathbb{R}^4$. We can therefore write the vector v as a linear combination of these vectors, as follows: $v = \lambda_1 u_1 + \lambda_2 u_2 + \lambda_3 u_3 + \lambda_4 n$.

Thus we obtain the following linear system

$$\begin{cases} 2\lambda_1 - 3\lambda_2 - 2\lambda_3 + \lambda_4 = 3 \\ \lambda_1 - 2\lambda_4 = 1 \\ \lambda_2 + 3\lambda_4 = 0 \\ \lambda_3 + 2\lambda_4 = -2 \end{cases}$$

whose solution is

$$\begin{cases} \lambda_1 = 2/3 \\ \lambda_2 = 1/2 \\ \lambda_3 = -5/3 \\ \lambda_4 = -1/6. \end{cases}$$

Therefore we have

$$v = \tfrac{2}{3}\, u_1 + \tfrac{1}{2}\, u_2 - \tfrac{5}{3}\, u_3 - \tfrac{1}{6}\, n.$$

By setting

$$v' = \tfrac{2}{3}\, u_1 + \tfrac{1}{2}\, u_2 - \tfrac{5}{3}\, u_3 = \Big(\frac{19}{6}, \frac{2}{3}, \frac{1}{2}, -\frac{5}{3}\Big)$$
$$v'' = -\tfrac{1}{6}\, n = \Big(-\frac{1}{6}, \frac{1}{3}, -\frac{1}{2}, -\frac{1}{3}\Big)$$

we obtain the decomposition $v = v' + v''$, with $v' \in U$ and $v'' \in U^{\perp}$; the vectors v' and v'' are therefore the orthogonal projections of v onto the subspaces U and $U^{\perp}$, respectively.

The vector v', the orthogonal projection of v onto U, can also be determined in another way. We have already noted that a basis of the subspace $U^{\perp}$ is given by the vector $n = (1, -2, 3, 2)$. Since the norm of this vector is not 1, we normalize it thus obtaining the unit vector

$$\hat{n} = \frac{n}{\|n\|} = \frac{1}{3\sqrt{2}}(1, -2, 3, 2).$$

Now we can calculate the orthogonal projection of the vector v onto the subspace $U^{\perp}$ generated by $\hat{n}$, i.e., the vector v'', using the following formula:

$$v'' = (v \cdot \hat{n})\, \hat{n} = -\tfrac{1}{6}\, n = \left(-\frac{1}{6}, \frac{1}{3}, -\frac{1}{2}, -\frac{1}{3}\right).$$

Recalling that $v = v' + v''$, we can now compute the orthogonal projection of v onto the subspace U as follows:

$$v' = v - v'' = \left(\frac{19}{6}, \frac{2}{3}, \frac{1}{2}, -\frac{5}{3}\right).$$

▶ **Exercise 5.5.12.** In the Euclidean vector space $\mathbb{R}^4$, endowed with the usual scalar product, let U be the subspace generated by the vector $u = (1, -1, 2, 3)$ and let $W = U^{\perp}$ be the orthogonal subspace of U.

(a) Determine the Cartesian equation and a basis of W.

(b) Using the Gram–Schmidt procedure, transform the basis found in point (a) into an orthogonal basis.

(c) Given the vector $v = (3, -1, 5, 2)$ find two vectors $v_1 \in U$ and $v_2 \in U^{\perp}$ such that $v = v_1 + v_2$.

(d) Let $f\colon \mathbb{R}^4 \to \mathbb{R}^4$ be the function which associates to each vector $v \in \mathbb{R}^4$ its orthogonal projection onto the subspace U. Write the matrix of f with respect to the canonical basis of $\mathbb{R}^4$. Find the kernel of f?

Solution. A generic vector $w = (x_1, x_2, x_3, x_4) \in W$ must be orthogonal to the vector $u = (1, -1, 2, 3)$, which is the generator of the subspace U. The subspace W is therefore the set of solutions of the following (Cartesian) equation:

$$W : u \cdot w = x_1 - x_2 + 2x_3 + 3x_4 = 0.$$

From this equation we obtain $x_2 = x_1 + 2x_3 + 3x_4$ and assigning to x_1, x_3 and x_4 the values $(1,0,0)$, $(0,1,0)$ and $(0,0,1)$ yield the three vectors

$$w_1 = (1,1,0,0), \quad w_2 = (0,2,1,0), \quad w_3 = (0,3,0,1)$$

which are a basis of W (hence $\dim W = 3$).

We now apply the Gram–Schmidt procedure to this basis. Let $w_1' = w_1 = (1,1,0,0)$; we have $w_1' \cdot w_1' = 2$. Now we set $w_2' = w_2 + \alpha_1 w_1'$. Requiring that w_2' is orthogonal to w_1', i.e., that $w_2' \cdot w_1' = 0$, we find

$$\alpha_1 = -\frac{w_2 \cdot w_1'}{w_1' \cdot w_1'} = -1.$$

Hence we get

$$w_2' = w_2 - w_1' = (-1,1,1,0)$$

and $w_2' \cdot w_2' = 3$. Now we set $w_3' = w_3 + \alpha_1 w_1' + \alpha_2 w_2'$. Requiring that w_3' is orthogonal to w_1' and w_2', i.e., that $w_3' \cdot w_1' = w_3' \cdot w_2' = 0$, we find

$$\alpha_1 = -\frac{w_3 \cdot w_1'}{w_1' \cdot w_1'} = -\frac{3}{2}, \qquad \alpha_2 = -\frac{w_3 \cdot w_2'}{w_2' \cdot w_2'} = -1.$$

Hence we get

$$w_3' = w_3 - \tfrac{3}{2}\, w_1' - w_2' = \left(-\tfrac{1}{2}, \tfrac{1}{2}, -1, 1\right).$$

The vectors w_1', w_2' and w_3' are an orthogonal basis of the subspace W.

We now look for two vectors $v_1 \in U$ and $v_2 \in U^\perp$ such that $v = v_1 + v_2$, where $v = (3,-1,5,2)$ (note that this is equivalent to saying that v_1 is the orthogonal projection of the vector v onto the subspace U, while v_2 is the orthogonal projection of v onto the subspace $U^\perp$). We note that it will be sufficient to determine v_1, since then we will have $v_2 = v - v_1$.

Since U is spanned by the vector $u = (1,-1,2,3)$, saying that $v_1 \in U$ is equivalent to saying that $v_1 = \lambda u$, for some scalar λ. Therefore we have $v_2 = v - \lambda u$. We therefore have only one unknown λ and we will therefore need only one equation. This equation is found by observing that requiring v_2 to belong to the subspace $U^\perp$ is equivalent to requiring v_2 to be orthogonal to the vector u, the generator of the subspace U. We must therefore have $v_2 \cdot u = (v - \lambda u) \cdot u = 0$, from which we obtain

$$\lambda = \frac{v \cdot u}{u \cdot u}.$$

We thus find that the vector v_1 (which is the orthogonal projection of v onto the subspace U) is given by the following formula:

$$v_1 = \left(\frac{v \cdot u}{u \cdot u}\right) u.$$

We have

$$v_1 = \tfrac{4}{3}\,u = \left(\tfrac{4}{3}, -\tfrac{4}{3}, \tfrac{8}{3}, 4\right)$$

hence

$$v_2 = v - v_1 = \left(\tfrac{5}{3}, \tfrac{1}{3}, \tfrac{7}{3}, -2\right).$$

To write the matrix of f with respect to the canonical basis of $\mathbb{R}^4$ we recall that the columns of this matrix are the images through f of the vectors of the canonical basis. For each $i = 1, 2, 3, 4$, the vector $f(e_i)$ is the orthogonal projection of e_i onto the subspace U; it can therefore be calculated with a formula similar to the one found previously:

$$f(e_i) = \left(\frac{e_i \cdot u}{u \cdot u}\right) u.$$

We have:

$$\begin{aligned}
f(e_1) &= \tfrac{1}{15}\,u = \left(\tfrac{1}{15}, -\tfrac{1}{15}, \tfrac{2}{15}, \tfrac{1}{5}\right)\\
f(e_2) &= -\tfrac{1}{15}\,u = \left(-\tfrac{1}{15}, \tfrac{1}{15}, -\tfrac{2}{15}, -\tfrac{1}{5}\right)\\
f(e_3) &= \tfrac{2}{15}\,u = \left(\tfrac{2}{15}, -\tfrac{2}{15}, \tfrac{4}{15}, \tfrac{2}{5}\right)\\
f(e_4) &= \tfrac{1}{5}\,u = \left(\tfrac{1}{5}, -\tfrac{1}{5}, \tfrac{2}{5}, \tfrac{3}{5}\right).
\end{aligned}$$

The matrix of f is

$$A = \begin{pmatrix}
\frac{1}{15} & -\frac{1}{15} & \frac{2}{15} & \frac{1}{5}\\
-\frac{1}{15} & \frac{1}{15} & -\frac{2}{15} & -\frac{1}{5}\\
\frac{2}{15} & -\frac{2}{15} & \frac{4}{15} & \frac{2}{5}\\
\frac{1}{5} & -\frac{1}{5} & \frac{2}{5} & \frac{3}{5}
\end{pmatrix}$$

To determine the kernel of f it suffices to note that the only vectors whose orthogonal projection on U is the zero vector are the vectors orthogonal to U, i.e., the vectors belonging to the subspace $U^\perp$. This means that $\operatorname{Ker} f = U^\perp$.

► **Exercise 5.5.13.** In the Euclidean vector space $\mathbb{R}^4$, endowed with the usual scalar product, consider the vectors $u_1 = (2, -1, 0, 3)$ and $u_2 = (1, 3, -1, 2)$. We define $f\colon \mathbb{R}^4 \to \mathbb{R}^2$ by setting $f(v) = (u_1 \cdot v, u_2 \cdot v)$, for any $v \in \mathbb{R}^4$.

(a) Prove that f is linear and determine bases of $\operatorname{Ker} f$ and of $\operatorname{Im} f$.

(b) Write the matrix of f with respect to the canonical bases.

(c) Given the vector $w = (1, 2) \in \mathbb{R}^2$, compute $f^{-1}(w)$.

(d) Find a basis of $(\operatorname{Ker} f)^\perp$.

(e) Prove that for any linear function $g\colon \mathbb{R}^2 \to \mathbb{R}^4$, the composition $g \circ f\colon \mathbb{R}^4 \to \mathbb{R}^4$ always has an eigenvalue equal to 0, with multiplicity ≥ 2.

Solution. The linearity of f is a direct consequence of the bilinearity of the dot product. However, it can be verified directly as follows: given $v_1, v_2 \in \mathbb{R}^4$ and $\lambda_1, \lambda_2 \in \mathbb{R}$, we have

$$\begin{aligned} f(\lambda_1 v_1 + \lambda_2 v_2) &= \big(u_1 \cdot (\lambda_1 v_1 + \lambda_2 v_2), u_2 \cdot (\lambda_1 v_1 + \lambda_2 v_2)\big) \\ &= (\lambda_1 u_1 \cdot v_1 + \lambda_2 u_1 \cdot v_2, \lambda_1 u_2 \cdot v_1 + \lambda_2 u_2 \cdot v_2) \\ &= \lambda_1 (u_1 \cdot v_1, u_2 \cdot v_1) + \lambda_2 (u_1 \cdot v_2, u_2 \cdot v_2) \\ &= \lambda_1 f(v_1) + \lambda_2 f(v_2). \end{aligned}$$

The kernel of f is the set of vectors $v = (x_1, x_2, x_3, x_4) \in \mathbb{R}^4$ such that $f(v) = \mathbf{0}$. We have $f(v) = (2x_1 - x_2 + 3x_4, x_1 + 3x_2 - x_3 + 2x_4)$, hence the equation $f(v) = \mathbf{0}$ is equivalent to the system

$$\begin{cases} 2x_1 - x_2 + 3x_4 = 0 \\ x_1 + 3x_2 - x_3 + 2x_4 = 0, \end{cases}$$

whose solutions are given by

$$\begin{cases} x_2 = 2x_1 + 3x_4 \\ x_3 = 7x_1 + 11x_4. \end{cases}$$

The kernel of f therefore has dimension 2 and is generated by the vectors $v_1 = (1, 2, 7, 0)$ and $v_2 = (0, 3, 11, 1)$, obtained by setting respectively $x_1 = 1$, $x_4 = 0$ and $x_1 = 0$, $x_4 = 1$ in the previous system. Since $\dim(\operatorname{Ker} f) + \dim(\operatorname{Im} f) = 4$, we conclude that the image of f has dimension 2 and therefore $\operatorname{Im} f = \mathbb{R}^2$. As the basis of the image of f one can therefore take any basis of $\mathbb{R}^2$, for example the canonical basis.

From the definition of f it immediately follows that the matrix of f with respect to the canonical bases is

$$A = \begin{pmatrix} 2 & -1 & 0 & 3 \\ 1 & 3 & -1 & 2 \end{pmatrix}$$

whose rows are the vectors u_1 and u_2.

Given the vector $w = (1, 2)$, its inverse image $f^{-1}(w)$ is the set of solutions of the following system:

$$\begin{cases} 2x_1 - x_2 + 3x_4 = 1 \\ x_1 + 3x_2 - x_3 + 2x_4 = 2. \end{cases}$$

We get

$$\begin{cases} x_2 = 2x_1 + 3x_4 - 1 \\ x_3 = 7x_1 + 11x_4 - 5, \end{cases}$$

which means that

$$f^{-1}(w) = (0, -1, -5, 0) + \operatorname{Ker} f.$$

To determine a basis of $(\operatorname{Ker} f)^\perp$ it suffices to observe that, by definition of the function f, we have $u_1 \cdot v = u_2 \cdot v = 0$, for each $v \in \operatorname{Ker} f$. This means that the vectors u_1 and u_2 are orthogonal to $\operatorname{Ker} f$. Now it suffices to note that $\dim(\operatorname{Ker} f)^\perp = 4 - \dim(\operatorname{Ker} f) = 2$ and that the vectors u_1 and u_2 are linearly independent to conclude that a basis of $(\operatorname{Ker} f)^\perp$ is given by the vectors u_1 e u_2.

Finally, given any linear function $g \colon \mathbb{R}^2 \to \mathbb{R}^4$, let us consider the composite function $g \circ f \colon \mathbb{R}^4 \to \mathbb{R}^4$. For each vector $v \in \operatorname{Ker} f$, we have $(g \circ f)(v) = g(f(v)) = g(\mathbf{0}) = \mathbf{0}$, which means that all vectors of the kernel of f are eigenvectors of the composite function $g \circ f$, relative to the eigenvalue 0. Since $\dim(\operatorname{Ker} f) = 2$, there exist at least two linearly independent eigenvectors corresponding to the eigenvalue 0, so the (algebraic) multiplicity of the eigenvalue 0 must be ≥ 2.

► **Exercise 5.5.14.** In the Euclidean vector space $\mathbb{R}^4$, endowed with the usual scalar product, let U be the subspace defined by the equations $x_1 - x_2 + x_4 = 0$, $3x_1 + x_2 - x_3 + 2x_4 = 0$ and $x_1 + 3x_2 - x_3 = 0$.

(a) Find a basis of U and a basis of $U^\perp$.

(b) Given the vector $v_1 = (2, 1, -2, 3)$ find two vectors $u \in U$ and $u' \in U^\perp$ such that $v_1 = u + u'$.

(c) Given the vectors $v_2 = (5, -3, 9, 1)$ and $w = (1, 2, 1, -1)$ determine (if possible) the Cartesian equation of a subspace W, of dimension 3, of $\mathbb{R}^4$, such that w is the orthogonal projection of v_2 onto W.

(d) Determine an orthogonal basis of the subspace W found in point (c).

Solution. The subspace U is the set of solutions of the linear system

$$U : \begin{cases} x_1 - x_2 + x_4 = 0 \\ 3x_1 + x_2 - x_3 + 2x_4 = 0 \\ x_1 + 3x_2 - x_3 = 0. \end{cases}$$

We have

$$U : \begin{cases} x_1 = x_2 - x_4 \\ x_3 = 4x_2 - x_4 \end{cases}$$

from which it follows that U has dimension 2 and a basis is given by the vectors $u_1 = (1, 1, 4, 0)$ and $u_2 = (-1, 0, -1, 1)$, obtained by setting respectively $x_2 = 1$, $x_4 = 0$ and $x_2 = 0$, $x_4 = 1$ in the previous system.

To find a basis of $U^\perp$ we note that a generic vector $u' = (x_1, x_2, x_3, x_4)$ belongs to $U^\perp$ if and only if $u' \cdot u_1 = x_1 + x_2 + 4x_3 = 0$ and $u' \cdot u_2 = -x_1 - x_3 + x_4 = 0$. The subspace $U^\perp$ is therefore given by the solutions of the following linear system:

$$U^\perp : \begin{cases} x_1 + x_2 + 4x_3 = 0 \\ -x_1 - x_3 + x_4 = 0. \end{cases}$$

We obtain

$$U^\perp : \begin{cases} x_2 = -x_1 - 4x_3 \\ x_4 = x_1 + x_3 \end{cases}$$

from which we deduce that $U^\perp$ has dimension 2 and a basis is given by the vectors $u'_1 = (1, -1, 0, 1)$ and $u'_2 = (0, -4, 1, 1)$, obtained by setting respectively $x_1 = 1$, $x_3 = 0$ and $x_1 = 0$, $x_3 = 1$ in the previous system.

Let us now consider the vector $v_1 = (2, 1, -2, 3)$ and write it as a linear combination of the vectors u_1, u_2, u'_1 and u'_2 found previously:

$$\alpha_1 u_1 + \alpha_2 u_2 + \beta_1 u'_1 + \beta_2 u'_2 = v_1.$$

We get the system

$$\begin{cases} \alpha_1 - \alpha_2 + \beta_1 = 2 \\ \alpha_1 - \beta_1 - 4\beta_2 = 1 \\ 4\alpha_1 - \alpha_2 + \beta_2 = -2 \\ \alpha_2 + \beta_1 + \beta_2 = 3 \end{cases}$$

whose solution is

$$\begin{cases} \alpha_1 = 0 \\ \alpha_2 = 1 \\ \beta_1 = 3 \\ \beta_2 = -1 \end{cases}$$

Hence we have

$$v_1 = u_2 + 3u'_1 - u'_2$$

and the vectors $u \in U$ and $u' \in U^\perp$ are $u = u_2 = (-1, 0, -1, 1)$ and $u' = 3u'_1 - u'_2 = (3, 1, -1, 2)$.

Let us now consider the vectors $v_2 = (5, -3, 9, 1)$ and $w = (1, 2, 1, -1)$. Let w' be a vector such that $v_2 = w + w'$: we have $w' = v_2 - w = (4, -5, 8, 2)$ and note that $w' \cdot w = 0$, i.e., the vectors w and w' are orthogonal. This means that the three-dimensional subspace W for which the vector w is the orthogonal projection of v_2 onto W is the subspace orthogonal to the vector w'. This subspace W is therefore the set of vectors (x_1, x_2, x_3, x_4) such that

$$(x_1, x_2, x_3, x_4) \cdot w' = 4x_1 - 5x_2 + 8x_3 + 2x_4 = 0.$$

This is the Cartesian equation of the subspace W.

Finally, we determine an orthogonal basis of W. We decide not to use the Gram–Schmidt procedure. We start by choosing (arbitrarily) a vector $w_1 \in W$; for example $w_1 = (1, 0, 0, -2)$. To choose the second vector w_2 we note that it must satisfy the equation $4x_1 - 5x_2 + 8x_3 + 2x_4 = 0$ of W and must also be orthogonal to the vector w_1, i.e., we must have $w_1 \cdot w_2 = x_1 - 2x_4 = 0$. Therefore, to determine w_2 it is enough to find one of the solutions of the system

$$\begin{cases} 4x_1 - 5x_2 + 8x_3 + 2x_4 = 0 \\ x_1 - 2x_4 = 0. \end{cases}$$

For example, we can choose $w_2 = (2, 2, 0, 1)$.

Since $\dim W = 3$, all we have to do is find a third vector $w_3 \in W$, orthogonal to w_1 and w_2. This vector must be a solution of the following system:

$$\begin{cases} 4x_1 - 5x_2 + 8x_3 + 2x_4 = 0 \\ w_1 \cdot w_3 = x_1 - 2x_4 = 0 \\ w_2 \cdot w_3 = 2x_1 + 2x_2 + x_4 = 0. \end{cases}$$

This system has infinitely many solutions (depending on one parameter) and one of them is given by the vector $w_3 = (32, -40, -45, 16)$. An orthogonal basis of the subspace W is therefore given by the vectors w_1, w_2 and w_3 just found.

► **Exercise 5.5.15.** In the Euclidean vector space $\mathbb{R}^4$, endowed with the usual scalar product, consider the subspace U generated by the vectors $u_1 = (2, 2, 0, 1)$, $u_2 = (0, 1, 1, 0)$, $u_3 = (1, 0, -4, -1)$.

(a) Given the vector $v = (3, -2, 1, 2)$ determine its orthogonal projection onto U.

(b) Find, if possible, a subspace $L \subset \mathbb{R}^4$, $L \neq U^\perp$, such that $U \oplus L = \mathbb{R}^4$.

(c) Determine an orthonormal basis of U.

(d) Let $f\colon \mathbb{R}^4 \to \mathbb{R}^4$ be the linear function which associates to each vector $v \in \mathbb{R}^4$ its orthogonal projection $f(v)$ onto the subspace U. Determine a basis of the kernel of f.

Solution. Let us find a basis of $U^\perp$. If $w = (x_1, x_2, x_3, x_4) \in U^\perp$, we have

$$\begin{cases} w \cdot u_1 = 2x_1 + 2x_2 + x_4 = 0 \\ w \cdot u_2 = x_2 + x_3 = 0 \\ w \cdot u_3 = x_1 - 4x_3 - x_4 = 0 \end{cases}$$

from which we obtain

$$\begin{cases} x_1 = -2x_2 \\ x_2 \text{ any value} \\ x_3 = -x_2 \\ x_4 = 2x_2 \end{cases}$$

This means that $U^\perp$ has dimension 1 and a basis is given by the vector $w = (-2, 1, -1, 2)$.

We now write the vector v as follows: $v = v' + v''$, with $v' \in U$ and $v'' \in U^\perp$. The vector v' is the orthogonal projection of v onto U (while v'' is the orthogonal projection of v onto $U^\perp$).

Since $U^\perp$ is generated by the vector w, we can write $v'' = \lambda w$, hence $v' = v - v'' = v - \lambda w$. Since v' and w are orthogonal, we have $v' \cdot w = 0$, which gives the following equation:

$$v' \cdot w = (v - \lambda w) \cdot w = v \cdot w - \lambda w \cdot w = 0$$

from which we get

$$\lambda = \frac{v \cdot w}{w \cdot w} = -\frac{1}{2}.$$

Hence we have

$$v' = v + \tfrac{1}{2} w = \left(2, -\tfrac{3}{2}, \tfrac{1}{2}, 3\right).$$

Since $\dim U = 3$, to determine a subspace $L \neq U^\perp$, such that $U \oplus L = \mathbb{R}^4$, it suffices to determine a vector ℓ linearly independent of w and such that the vectors $\{u_1, u_2, u_3, \ell\}$ are a basis of $\mathbb{R}^4$. There are infinitely many such vectors ℓ; for example it is easy to verify that the vector $\ell = (1, 0, 0, 0)$ satisfies the required conditions. We can therefore set L equal to the subspace generated by the vector ℓ.

To determine an orthonormal basis of U we can apply the Gram–Schmidt procedure to the basis $\{u_1, u_2, u_3\}$ of U. Let $w_1 = u_1 = (2, 2, 0, 1)$; we have

$w_1 \cdot w_1 = 9$. Now we set $w_2 = u_2 + \alpha_1 w_1$. Requiring that w_2 is orthogonal to w_1, i.e., that $w_2 \cdot w_1 = 0$, we find

$$\alpha_1 = -\frac{u_2 \cdot w_1}{w_1 \cdot w_1} = -\frac{2}{9}.$$

Hence we get

$$w_2 = u_2 - \tfrac{2}{9}\, w_1 = \left(-\tfrac{4}{9}, \tfrac{5}{9}, 1, -\tfrac{2}{9}\right).$$

To simplify the expression of w_2 we decide to multiply everything by 9 and therefore to set

$$w_2 = (-4, 5, 9, -2).$$

Now we have $w_2 \cdot w_2 = 126$. Now we set $w_3 = u_3 + \alpha_1 w_1 + \alpha_2 w_2$. Requiring that w_3 is orthogonal to w_1 and w_2, i.e., that $w_3 \cdot w_1 = w_3 \cdot w_2 = 0$, we find

$$\alpha_1 = -\frac{u_3 \cdot w_1}{w_1 \cdot w_1} = -\frac{1}{9}, \qquad \alpha_2 = -\frac{u_3 \cdot w_2}{w_2 \cdot w_2} = \frac{19}{63}.$$

Hence we get

$$w_3 = u_3 - \tfrac{1}{9}\, w_1 + \tfrac{19}{63}\, w_2 = \left(-\tfrac{3}{7}, \tfrac{9}{7}, -\tfrac{9}{7}, -\tfrac{12}{7}\right).$$

As before, to simplify the expression of w_3 we decide to multiply everything by 7 and divide by 3, thus setting

$$w_3 = (-1, 3, -3, -4).$$

Now we have $w_3 \cdot w_3 = 35$.

The vectors w_1, w_2 and w_3 are an orthogonal basis of the subspace U; their norms are given respectively by:

$$\begin{aligned} \|w_1\| &= \sqrt{w_1 \cdot w_1} = 3 \\ \|w_2\| &= \sqrt{w_2 \cdot w_2} = 3\sqrt{14} \\ \|w_3\| &= \sqrt{w_3 \cdot w_3} = \sqrt{35}. \end{aligned}$$

To obtain an orthonormal basis of U it is now sufficient to divide the vectors w_1, w_2 and w_3 by their norms:

$$\begin{aligned} w_1' &= \frac{w_1}{\|w_1\|} = \left(\frac{2}{3}, \frac{2}{3}, 0, \frac{1}{3}\right) \\ w_2' &= \frac{w_2}{\|w_2\|} = \left(-\frac{4}{3\sqrt{14}}, \frac{5}{3\sqrt{14}}, \frac{9}{3\sqrt{14}}, -\frac{2}{3\sqrt{14}}\right) \\ w_3' &= \frac{w_3}{\|w_3\|} = \left(-\frac{1}{\sqrt{35}}, \frac{3}{\sqrt{35}}, -\frac{3}{\sqrt{35}}, -\frac{4}{\sqrt{35}}\right). \end{aligned}$$

Finally, let us consider the linear function f which associates to each vector $v \in \mathbb{R}^4$ its orthogonal projection $f(v)$ onto the subspace U. The vectors v such that $f(v) = \mathbf{0}$ are the vectors orthogonal to U, therefore we have $\operatorname{Ker}(f) = U^\perp$ and a basis of $\operatorname{Ker}(f)$ is given by the vector $w = (-2, 1, -1, 2)$ (which is the basis of $U^\perp$ found at the beginning).

► **Exercise 5.5.16.** In $\mathbb{R}^4$ endowed with the usual scalar product consider the subspace V given by the equation $3x_1 - x_2 + x_3 - 2x_4 = 0$.

(a) Find a basis of V.

(b) Determine the orthogonal projection of $v = (1, 1, 1, 1)$ onto V.

(c) Given the subspace $\langle(1, 2, -1, 0)\rangle$ of V determine a subspace M of dimension 2 whose vectors are orthogonal to $\langle(1, 2, -1, 0)\rangle$.

(d) Find all vectors of $\mathbb{R}^4$ whose orthogonal projection onto V is the vector $(1, 2, -1, 0)$.

Solution. From the equation of V we get

$$V : x_2 = 3x_1 + x_3 - 2x_4$$

from which we deduce that V has dimension 3 and a basis is given by the vectors $v_1 = (1, 3, 0, 0)$, $v_2 = (0, 1, 1, 0)$, $v_3 = (0, -2, 0, 1)$.

Again from the equation of V it follows that the vector $w = (3, -1, 1, -2)$ is orthogonal to V, indeed it is a basis of $V^\perp$.

We now write the vector v as the sum $v = v' + v''$, with $v' \in V$ and $v'' \in V^\perp$. Since w is a basis of $V^\perp$, we have $v'' = \lambda w$ and therefore $v' = v - v'' = v - \lambda w$. Since $v' \in V$, it must be orthogonal to w, therefore we have $v' \cdot w = 0$. Thus we obtain the equation

$$v' \cdot w = (v - \lambda w) \cdot w = v \cdot w - \lambda w \cdot w = 0$$

from which it follows

$$\lambda = \frac{v \cdot w}{w \cdot w} = \frac{1}{15}.$$

Hence we have

$$\begin{aligned} v'' &= \tfrac{1}{15}\, w = \left(\tfrac{3}{15}, -\tfrac{1}{15}, \tfrac{1}{15}, -\tfrac{2}{15}\right) \\ v' &= v - v'' = \left(\tfrac{12}{15}, \tfrac{16}{15}, \tfrac{14}{15}, \tfrac{17}{15}\right). \end{aligned}$$

The vector v' is the orthogonal projection of $v = (1, 1, 1, 1)$ onto V.

Since M must have dimension 2, a basis must be given by two vectors of V that are orthogonal to the vector $(1, 2, -1, 0)$. If we denote by (x_1, x_2, x_3, x_4)

a generic vector, requiring that this vector belongs to V and is orthogonal to the vector $(1, 2, -1, 0)$, we obtain the system

$$\begin{cases} 3x_1 - x_2 + x_3 - 2x_4 = 0 \\ x_1 + 2x_2 - x_3 = 0 \end{cases}$$

from which we obtain

$$\begin{cases} x_2 = -4x_1 + 2x_4 \\ x_3 = -7x_1 + 4x_4 \end{cases}$$

A basis of M is therefore given by the vectors $m_1 = (1, -4, -7, 0)$ and $m_2 = (0, 2, 4, 1)$.

Finally, to find all the vectors whose orthogonal projection onto V is the vector $(1, 2, -1, 0)$, it suffices to remember that $w = (3, -1, 1, -2)$ is a basis of $V^\perp$. The vectors are therefore of the form

$$u_t = (1, 2, -1, 0) + t(3, -1, 1, -2) = (1 + 3t, 2 - t, -1 + t, -2t),$$

for every value of the parameter t.

▶ **Exercise 5.5.17.** In the Euclidean vector space $\mathbb{R}^4$, endowed with the usual scalar product, consider the vectors $u = (3, -2, 1, -2)$ and $w = (1, -1, 0, 1)$.

(a) Determine the equation of a subspace W, of dimension 3, such that the vector w is the orthogonal projection of u onto W. Also determine an orthogonal basis $\{w_1, w_2, w_3\}$ of W, such that $w_1 = w$.

(b) Let $f \colon \mathbb{R}^4 \to \mathbb{R}^4$ be defined by $f(v) = (v \cdot u)w + (v \cdot w)u$. Find a basis of the kernel and a basis of the image of f.

(c) Let U_1 and U_2 be the subspaces generated, respectively, by $u_1 = (0, 0, 1, 0)$ and $u_2 = (0, 1, 0, 1)$. Let us denote by p_1 and p_2 the orthogonal projections onto U_1 and U_2. Find the set F of all vectors $v \in \mathbb{R}^4$ such that $\|p_1(v)\| = \|p_2(v)\|$ and determine if F is a vector subspace of $\mathbb{R}^4$.

Solution. If w is to be the orthogonal projection of u onto a subspace W, we must have $u = w + w'$, with $w \in W$ and $w' \in W^\perp$. Since we know u and w, we can compute $w' = u - w = (2, -1, 1, -3)$. Obviously, for a subspace W with the required properties to exist, the vectors w and w' must be orthogonal. This is true, since we have $w \cdot w' = 0$.

Since W must have dimension 3, we have $\dim W^\perp = 1$, hence the vector $w' = (2, -1, 1, -3)$ is a basis of $W^\perp$. But then the components of w' are precisely the coefficients appearing in the Cartesian equation of W. Hence the equation of W is

$$W : 2x_1 - x_2 + x_3 - 3x_4 = 0.$$

Now let us look for an orthogonal basis of W whose first vector is $w_1 = w = (1, -1, 0, 1)$. If $w_2 = (x_1, x_2, x_3, x_4)$ is a generic vector, we need to require that $w_2 \in W$ and that w_1 and w_2 are orthogonal, i.e., that $w_1 \cdot w_2 = 0$. This gives us the system

$$\begin{cases} 2x_1 - x_2 + x_3 - 3x_4 = 0 \\ x_1 - x_2 + x_4 = 0 \end{cases}$$

which is equivalent to

$$\begin{cases} x_2 = x_1 + x_4 \\ x_3 = -x_1 + 4x_4 \end{cases}$$

A solution of this system (in this case the solution obtained by setting $x_1 = 1$ and $x_4 = 0$) gives the vector $w_2 = (1, 1, -1, 0)$.

The vector w_3 is found in a similar way. Let $w_3 = (x_1, x_2, x_3, x_4)$ be a generic vector. We need to require that $w_3 \in W$ and that w_3 is orthogonal to w_1 and w_2, i.e., $w_1 \cdot w_3 = 0$ and $w_2 \cdot w_3 = 0$. We obtain the system

$$\begin{cases} 2x_1 - x_2 + x_3 - 3x_4 = 0 \\ x_1 - x_2 + x_4 = 0 \\ x_1 + x_2 - x_3 = 0 \end{cases}$$

which is equivalent to

$$\begin{cases} x_1 = x_4 \\ x_2 = 2x_4 \\ x_3 = 3x_4 \end{cases}$$

A solution of this system gives the vector $w_3 = (1, 2, 3, 1)$.

From the definition of the function f

$$f(v) = (v \cdot u)w + (v \cdot w)u$$

we can see that the image of each vector v is a linear combination of the vectors w and u. Since w and u are linearly independent, they are a basis of the image of f (which therefore has dimension 2).

To determine the kernel of f we note that $f(v) = \mathbf{0}$ if and only if $v \cdot u = 0$ and $v \cdot w = 0$ (because w and u are linearly independent), so $\mathrm{Ker}(f)$ is given by the solutions of the following system

$$\begin{cases} v \cdot u = 3x_1 - 2x_2 + x_3 - 2x_4 = 0 \\ v \cdot w = x_1 - x_2 + x_4 = 0 \end{cases}$$

We get

$$\begin{cases} x_2 = x_1 + x_4 \\ x_3 = -x_1 + 4x_4 \end{cases}$$

from which we deduce that a basis of $\mathrm{Ker}(f)$ is given by $(1, 1, -1, 0)$ and $(0, 1, 4, 1)$.

If U is a subspace spanned by a vector $u \neq \mathbf{0}$, the orthogonal projection of a generic vector v onto U is given by the following formula:

$$p_U(v) = \left(\frac{v \cdot u}{u \cdot u}\right) u$$

Denoting with p_1 and p_2 the orthogonal projections onto U_1 and U_2 respectively, and setting $v = (x_1, x_2, x_3, x_4)$, we then have

$$p_1(v) = \left(\frac{v \cdot u_1}{u_1 \cdot u_1}\right) u_1 = x_3 u_1 = (0, 0, x_3, 0)$$

$$p_2(v) = \left(\frac{v \cdot u_2}{u_2 \cdot u_2}\right) u_2 = \frac{x_2 + x_4}{2} u_2 = \left(0, \frac{x_2 + x_4}{2}, 0, \frac{x_2 + x_4}{2}\right)$$

Therefore we have

$$\|p_1(v)\| = \sqrt{x_3^2}, \qquad \|p_2(v)\| = \sqrt{\frac{(x_2 + x_4)^2}{2}},$$

and the equality $\|p_1(v)\| = \|p_2(v)\|$ is equivalent to the following equation which describes the set F

$$F : x_3^2 = \frac{1}{2}(x_2 + x_4)^2$$

which is equivalent to

$$F : x_3 = \pm\frac{\sqrt{2}}{2}(x_2 + x_4)$$

Since the equation of F is not linear, we conclude that F is not a vector subspace of $\mathbb{R}^4$. In fact we note that the vectors $v_1 = (0, 2, \sqrt{2}, 0)$ and $v_2 = (0, 2, -\sqrt{2}, 0)$ belong to F, but their sum is the vector $v_1 + v_2 = (0, 4, 0, 0)$ which does not satisfy the equation of F.

► **Exercise 5.5.18.** In the Euclidean vector space $\mathbb{R}^3$ we consider the subspaces V_1, generated by the vector $v_1 = (1, 2, -1)$ and V_2, generated by the vector $v_2 = (1, 1, -1)$.

(a) Find a basis of $(V_1 + V_2)^\perp$ and a basis of $V_1^\perp \cap V_2^\perp$.

(b) Determine the orthogonal projection of the vector $w = (1, -1, 4)$ onto the subspace $V_1 + V_2$.

(c) Let p_{V_1} and p_{V_2} be the orthogonal projections onto V_1 and V_2, respectively. Find all vectors $v \in \mathbb{R}^3$ such that $p_{V_1}(v) = p_{V_2}(v)$.

(d) Determine the set S of all vectors of $\mathbb{R}^3$ whose orthogonal projection onto V_1 is the vector $(2, 4, -2)$.

Solution. The vectors v_1 and v_2 are linearly independent, hence a basis of $V_1 + V_2$ is given by $\{v_1, v_2\}$. The vectors of $(V_1 + V_2)^\perp$ are therefore the vectors that are orthogonal to v_1 and v_2. In $\mathbb{R}^3$ a vector orthogonal to v_1 and v_2 is given by their cross product

$$v_3 = v_1 \times v_2 = (-1, 0, -1).$$

We conclude that $v_3 = (-1, 0, -1)$ is a basis of $(V_1 + V_2)^\perp$.

Similarly, the vectors of $V_1^\perp \cap V_2^\perp$ must belong to $V_1^\perp$ and $V_2^\perp$, so they must be orthogonal to v_1 and v_2. But that means that

$$V_1^\perp \cap V_2^\perp = (V_1 + V_2)^\perp.$$

Let us now consider the vector $w = (1, -1, 4)$ and write $w = w' + w''$, with $w' \in V_1 + V_2$ and $w'' \in (V_1 + V_2)^\perp$. Since $(V_1 + V_2)^\perp$ is spanned by v_3, we have $w'' = \lambda v_3$, hence $w' = w - w'' = w - \lambda v_3$. Since w' is orthogonal to v_3, we have

$$w' \cdot v_3 = (w - \lambda v_3) \cdot v_3 = w \cdot v_3 - \lambda v_3 \cdot v_3 = 0,$$

from which we obtain

$$\lambda = \frac{w \cdot v_3}{v_3 \cdot v_3} = -\frac{5}{2}.$$

Therefore we have

$$\begin{aligned} w'' &= -\tfrac{5}{2}\, v_3 = \left(\tfrac{5}{2}, 0, \tfrac{5}{2}\right) \\ w' &= w - w'' = \left(-\tfrac{3}{2}, -1, \tfrac{3}{2}\right) \end{aligned}$$

and w' is the orthogonal projection of w onto the subspace $V_1 + V_2$.

The vector $p_{V_1}(v)$ is the orthogonal projection of v onto the subspace V_1, so it is a multiple of v_1. Similarly, the vector $p_{V_2}(v)$ is the orthogonal projection of v onto the subspace V_2, so it is a multiple of v_2. Since v_1 and v_2 are linearly independent, we can have $p_{V_1}(v) = p_{V_2}(v)$ if and only if $p_{V_1}(v) = p_{V_2}(v) = \mathbf{0}$, which is equivalent to saying that v must be orthogonal to v_1 and v_2. But we have already seen that the vectors orthogonal to v_1 and v_2 are the multiples of v_3. We therefore conclude that the vectors $v \in \mathbb{R}^3$ such that $p_{V_1}(v) = p_{V_2}(v)$ are the vectors of the type $v = \lambda v_3$, for each $\lambda \in \mathbb{R}$.

The set S of all vectors of $\mathbb{R}^3$ whose orthogonal projection onto V_1 is the vector $(2, 4, -2)$ is

$$S = (2, 4, -2) + V_1^{\perp}.$$

The subspace $V_1^{\perp}$ is described by the equation

$$V_1^{\perp} : x_1 + 2x_2 - x_3 = 0$$

hence a basis is given by the vectors $(-2, 1, 0)$ and $(1, 0, 1)$. Therefore we have

$$S = \{(2, 4, -2) + \alpha(-2, 1, 0) + \beta(1, 0, 1) \mid \forall \alpha, \beta \in \mathbb{R}\}.$$

▶ **Exercise 5.5.19.** In the Euclidean vector space $\mathbb{R}^4$, endowed with the usual scalar product, let U be the subspace of equations $2x_1 - x_3 + 3x_4 = 0$ and $x_1 + x_2 - 2x_4 = 0$.

(a) Find a basis of U and a basis of $U^{\perp}$.

(b) Given the vector $v = (4, 2, 4, 2)$, find a vector w of minimum norm such that $v + w \in U$.

(c) Does there exist a subspace $L \subset \mathbb{R}^4$ such that $L \oplus U = \mathbb{R}^4$ and $L \oplus U^{\perp} = \mathbb{R}^4$? If the answer is yes, determine a basis of L.

(d) Let $f \colon \mathbb{R}^4 \to \mathbb{R}^4$ be the linear function which associates, to a vector of $\mathbb{R}^4$, its orthogonal projection onto U. Find the eigenvalues and eigenspaces of f and determine whether f is diagonalizable.

Solution. A basis of U is found by solving the system

$$U : \begin{cases} 2x_1 - x_3 + 3x_4 = 0 \\ x_1 + x_2 - 2x_4 = 0 \end{cases}$$

This system is equivalent to

$$\begin{cases} x_3 = 2x_1 + 3x_4 \\ x_2 = -x_1 + 2x_4 \end{cases}$$

from which we deduce that a basis of U is given by $u_1 = (1, -1, 2, 0)$ and $u_2 = (0, 2, 3, 1)$.

A vector $w = (x_1, x_2, x_3, x_4)$ belongs to $U^\perp$ if and only if it is orthogonal to the vectors of a basis of U: $w \cdot u_1 = 0$ and $w \cdot u_2 = 0$. Thus we find the following equations of the subspace $U^\perp$:

$$U^\perp : \begin{cases} x_1 - x_2 + 2x_3 = 0 \\ 2x_2 + 3x_3 + x_4 = 0 \end{cases}$$

This system is equivalent to

$$\begin{cases} x_1 = x_2 - 2x_3 \\ x_4 = -2x_2 - 3x_3 \end{cases}$$

from which we deduce that a basis of $U^\perp$ is given by $w_1 = (1, 1, 0, -2)$ and $w_2 = (-2, 0, 1, -3)$.

Obviously, to find a basis of $U^\perp$, it would have sufficed to note that it is formed by the two vectors whose components are the coefficients of the unknowns in the two equations which define the subspace U.

Given the vector $v = (4, 2, 4, 2)$, requiring w to be the vector of *minimum norm* such that $v + w \in U$ is equivalent to requiring that $w \in U^\perp$ (see figure below).

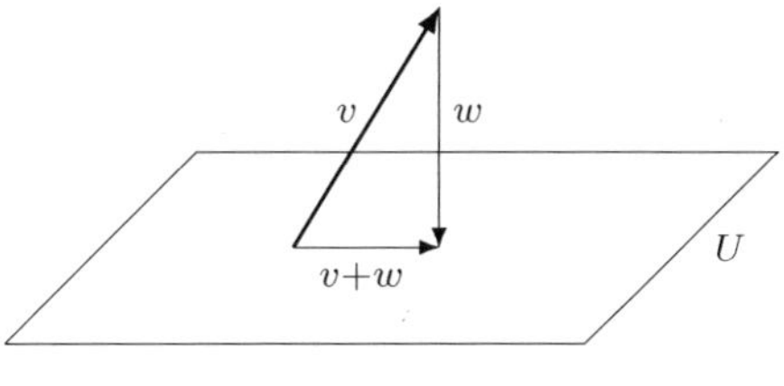

In this case w is the opposite of the orthogonal projection of v onto $U^\perp$, while the vector $v + w$ is the orthogonal projection of v onto U.

Since $w \in U^\perp$, we can write

$$w = \alpha_1 w_1 + \alpha_2 w_2 = (\alpha_1 - 2\alpha_2, \alpha_1, \alpha_2, -2\alpha_1 - 3\alpha_2),$$

from which we obtain

$$v + w = (4 + \alpha_1 - 2\alpha_2, 2 + \alpha_1, 4 + \alpha_2, 2 - 2\alpha_1 - 3\alpha_2).$$

The two equations necessary to determine the value of the unknowns α_1 and α_2 can be obtained by noting that the vector $v + w$ must be orthogonal to the two vectors of the basis of $U^\perp$ found previously:

$$\begin{cases} (v+w)\cdot w_1 = 0 \\ (v+w)\cdot w_2 = 0 \end{cases}$$

We get the system

$$\begin{cases} 6\alpha_1 + 4\alpha_2 + 2 = 0 \\ 4\alpha_1 + 14\alpha_2 - 10 = 0 \end{cases}$$

whose solution is $\alpha_1 = -1$, $\alpha_2 = 1$. Hence we have $w = -w_1 + w_2 = (-3, -1, 1, -1)$.

If L must be such that $L \oplus U = \mathbb{R}^4$ and $L \oplus U^\perp = \mathbb{R}^4$, since $\dim U = \dim U^\perp = 2$, we must have $\dim L = 2$. A basis of L must therefore be given by two linearly independent vectors ℓ_1 and ℓ_2 such that $\{\ell_1, \ell_2, u_1, u_2\}$ and $\{\ell_1, \ell_2, w_1, w_2\}$ are bases of $\mathbb{R}^4$. It is easy to check that a possible choice is given by the vectors $\ell_1 = (1, 0, 0, 0)$, $\ell_2 = (0, 1, 0, 0)$ (but there are infinitely many others). Therefore a subspace L with the required properties exists, but it is certainly not unique.

The function f is defined by setting, for each $v \in \mathbb{R}^4$, $f(v) = u$, where $u \in U$ is the orthogonal projection of the vector v onto the subspace U.

To determine the eigenvalues and eigenvectors of f it suffices to note that, if $v \in U$ we have $f(v) = v$, while if $v \in U^\perp$ we have $f(v) = \mathbf{0}$. From this we deduce that $\lambda = 1$ and $\lambda = 0$ are eigenvalues of f; the eigenspace corresponding to the eigenvalue 1 is the subspace U, while the eigenspace corresponding to the eigenvalue 0 is $U^\perp$.

Since $U \oplus U^\perp = \mathbb{R}^4$, the vectors u_1, u_2, w_1, w_2 are a basis of $\mathbb{R}^4$ formed by eigenvectors of f. This means that f is diagonalizable. The matrix of f with respect to the basis $\{u_1, u_2, w_1, w_2\}$ is

$$\begin{pmatrix} 1 & 0 & 0 & 0 \\ 0 & 1 & 0 & 0 \\ 0 & 0 & 0 & 0 \\ 0 & 0 & 0 & 0 \end{pmatrix}.$$

▶ **Exercise 5.5.20.** Let V be the vector subspace of $\mathbb{R}^5$ with equation $x_1 + x_2 + x_3 + x_4 + x_5 = 0$, endowed with the usual scalar product. Let $U \subset V$ be the subspace generated by the vectors $u_1 = (2, -1, 0, -1, 0)$, $u_2 = (1, 3, 0, 0, -4)$ and let $U^\perp \subset V$ be the orthogonal of U in V.

(a) Find an orthogonal basis of U.

(b) Find a basis of $U^\perp$.

(c) Given the vector $v = (4, 3, -4, 0, -3) \in V$ determine the orthogonal projections of v onto U and $U^\perp$.

(d) Determine whether there exists a nonzero vector subspace $L \subset \mathbb{R}^5$ which is orthogonal to both U and $U^\perp$. If such a subspace exists, determine a basis for it and whether it is unique.

Solution. The subspace U has dimension 2, so to find an orthogonal basis of U it is enough to find two vectors of U that are orthogonal to each other. As the first vector we can take u_1. Now, to find a vector of U that is orthogonal to u_1, we set $u_2' = u_2 + \alpha u_1$ and require that $u_1 \cdot u_2' = 0$. We get $u_1 \cdot u_2' = u_1 \cdot u_2 + \alpha u_1 \cdot u_1 = 0$, hence $\alpha = -\frac{u_1 \cdot u_2}{u_1 \cdot u_1} = \frac{1}{6}$. So the vector we are looking for is $u_2' = u_2 + \frac{1}{6}u_1 = (4/3, 17/6, 0, -1/6, -4)$. An orthogonal basis of U is therefore given by the vectors u_1 and u_2'.

Since $U^\perp \subset V$, the vectors of $U^\perp$ must be vectors of V orthogonal to the vectors u_1 and u_2 of the basis of U. If we denote by $w = (x_1, x_2, x_3, x_4, x_5)$ the generic vector of $U^\perp$, we must therefore have

$$\begin{cases} x_1 + x_2 + x_3 + x_4 + x_5 = 0 \\ w \cdot u_1 = 2x_1 - x_2 - x_4 = 0 \\ w \cdot u_2 = x_1 + 3x_2 - 4x_5 = 0 \end{cases}$$

This system is equivalent to

$$\begin{cases} x_1 = 4x_5 - 3x_2 \\ x_3 = 9x_2 - 13x_5 \\ x_4 = 8x_5 - 7x_2 \end{cases}$$

Since there are two free parameters, we have $\dim U^\perp = 2$ and a basis of $U^\perp$ is given by $w_1 = (-3, 1, 9, -7, 0)$ and $w_2 = (4, 0, -13, 8, 1)$.

To determine the orthogonal projections of the vector v onto U and $U^\perp$, we write $v = v' + v''$, with $v' \in U$ and $v'' \in U^\perp$. Then we have $v' = \lambda_1 u_1 + \lambda_2 u_2$ and $v'' = v - v' = v - \lambda_1 u_1 - \lambda_2 u_2$.

Since we have introduced the two unknowns λ_1 and λ_2, we need two equations. Note that requiring $v'' \in U^\perp$ is equivalent to requiring it to be orthogonal to the vectors u_1 and u_2 (which generate U). Hence we must have

$$\begin{cases} v'' \cdot u_1 = v \cdot u_1 - \lambda_1 u_1 \cdot u_1 - \lambda_2 u_2 \cdot u_1 = 0 \\ v'' \cdot u_2 = v \cdot u_2 - \lambda_1 u_1 \cdot u_2 - \lambda_2 u_2 \cdot u_2 = 0 \end{cases}$$

This gives the system

$$\begin{cases} 5 - 6\lambda_1 + \lambda_2 = 0 \\ 25 + \lambda_1 - 26\lambda_2 = 0 \end{cases}$$

whose solution is

$$\begin{cases} \lambda_1 = 1 \\ \lambda_2 = 1 \end{cases}$$

Therefore we have $v' = u_1 + u_2 = (3, 2, 0, -1, -4)$ and $v'' = v - v' = (1, 1, -4, 1, 1)$.

Since $U^\perp$ is the orthogonal of U in V, we have $U \oplus U^\perp = V$ (i.e., the vectors u_1, u_2, w_1, w_2 are a basis of the subspace V). If $L \subset \mathbb{R}^5$ must be orthogonal to both U and $U^\perp$, the vectors of L must be orthogonal to the vectors u_1, u_2, w_1, w_2, and therefore also to all vectors of V. This means that L is the orthogonal of V in $\mathbb{R}^5$. This proves that L exists and is unique, and $\dim L = 1$, since $\dim V = 4$.

To find a basis of L it suffices to note that the coefficients of the unknowns in the equation $x_1 + x_2 + x_3 + x_4 + x_5 = 0$ of V are the components of a vector $\ell = (1, 1, 1, 1, 1)$ which is orthogonal to all vectors of V. This vector ℓ is therefore a basis of the subspace L.

6

Affine Geometry

6.1 Lines and planes

► **Exercise 6.1.1.** In the affine space $\mathbb{A}^3(\mathbb{R})$ determine the line r passing through the point $P = (p_1, p_2, p_3)$ and parallel to the vector $v = (v_1, v_2, v_3)$.

Solution. The line r is the set of points X such that the vector $X - P$ is a multiple of the vector v. So we have

$$r = \{X \in \mathbb{A}^3 \,|\, X - P = \lambda v, \lambda \in \mathbb{R}\}.$$

If we set $X = (x_1, x_2, x_3)$, we obtain the equations

$$\begin{cases} x_1 - p_1 = \lambda v_1 \\ x_2 - p_2 = \lambda v_2 \\ x_3 - p_3 = \lambda v_3 \end{cases}$$

These are the parametric equations of the line r.

If we eliminate the parameter λ from these equations, and if we assume that $v_1 \neq 0$, $v_2 \neq 0$ and $v_3 \neq 0$, we obtain the following two equations

$$\frac{x_1 - p_1}{v_1} = \frac{x_2 - p_2}{v_2} = \frac{x_3 - p_3}{v_3}$$

which are the Cartesian equations of the line r.

► **Exercise 6.1.2.** In the affine space $\mathbb{A}^3(\mathbb{R})$ determine the line r passing through the points $P = (p_1, p_2, p_3)$ and $Q = (q_1, q_2, q_3)$.

Solution. The line r is the line passing through the point P parallel to the vector $v = Q - P = (q_1 - p_1, q_2 - p_2, q_3 - p_3)$. Now we can apply what we saw in the previous exercise. The parametric equations of r are therefore

$$\begin{cases} x_1 - p_1 = \lambda(q_1 - p_1) \\ x_2 - p_2 = \lambda(q_2 - p_2) \\ x_3 - p_3 = \lambda(q_3 - p_3) \end{cases}$$

If we set $X = (x_1, x_2, x_3)$, these equations can be written as follows:

$$X = \lambda Q + (1 - \lambda)P, \quad \lambda \in \mathbb{R}.$$

► **Exercise 6.1.3.** In the affine space $\mathbb{A}^3(\mathbb{R})$ determine the plane π passing through the point $P = (p_1, p_2, p_3)$ and parallel to the vectors $v = (v_1, v_2, v_3)$ and $w = (w_1, w_2, w_3)$.

Solution. The plane π is the set of points X such that the vector $X - P$ belongs to the vector subspace generated by v and w. So we have

$$\pi = \{X \in \mathbb{A}^3 \mid X - P = \lambda v + \mu w, \lambda, \mu \in \mathbb{R}\}.$$

If we set $X = (x_1, x_2, x_3)$, we obtain the equations

$$\begin{cases} x_1 - p_1 = \lambda v_1 + \mu w_1 \\ x_2 - p_2 = \lambda v_2 + \mu w_2 \\ x_3 - p_3 = \lambda v_3 + \mu w_3 \end{cases}$$

These are the parametric equations of the plane π.

Another description of the plane π can be obtained as follows. Let u be a vector perpendicular to v and w (for example, take $u = v \times w$). The plane π is the set of points X such that the vector $X - P$ is perpendicular to u. So we have

$$\pi = \{X \in \mathbb{A}^3 \mid (X - P) \cdot u = 0\}.$$

If we set $u = (u_1, u_2, u_3)$, we obtain the equation

$$u_1(x_1 - p_1) + u_2(x_2 - p_2) + u_3(x_3 - p_3) = 0.$$

This is the (Cartesian) equation of the plane π.

► **Exercise 6.1.4.** In the affine space $\mathbb{A}^3(\mathbb{R})$ determine the plane π passing through the three (non-aligned) points $P = (p_1, p_2, p_3)$, $Q = (q_1, q_2, q_3)$ and $R = (r_1, r_2, r_3)$.

Solution. The plane π is the plane passing through the point P and parallel to the vectors $v = Q - P = (q_1 - p_1, q_2 - p_2, q_3 - p_3)$ and $w = R - P = (r_1 - p_1, r_2 - p_2, r_3 - p_3)$. Now we can apply what we saw in the previous exercise. The parametric equations of π are

$$\begin{cases} x_1 - p_1 = \lambda(q_1 - p_1) + \mu(r_1 - p_1) \\ x_2 - p_2 = \lambda(q_2 - p_2) + \mu(r_2 - p_2) \\ x_3 - p_3 = \lambda(q_3 - p_3) + \mu(r_3 - p_3) \end{cases}$$

If we set $X = (x_1, x_2, x_3)$, these equations can be written as follows:

$$X = \lambda Q + \mu R + (1 - \lambda - \mu)P, \quad \lambda, \mu \in \mathbb{R}.$$

► **Exercise 6.1.5.** In the affine space $\mathbb{A}^3(\mathbb{R})$ let r be the line of equations

$$\begin{cases} 2x - 3y + 1 = 0 \\ 2x - 2y + z - 1 = 0. \end{cases}$$

Determine the line s passing through the origin and parallel to r, and the plane of the pencil with axis r passing through the origin.

Solution. To determine the line s it is sufficient to know a vector parallel to r. To determine this vector it is enough to know any two points of the line r; their difference will provide the vector we are looking for.

It is easy to check that the points $P = (0, \frac{1}{3}, \frac{5}{3})$ and $Q = (-\frac{1}{2}, 0, 2)$ belong to r, so as a vector parallel to r we can take $Q - P = (-\frac{1}{2}, -\frac{1}{3}, \frac{1}{3})$. The parametric equations of s are

$$\begin{cases} x = -\frac{1}{2}\lambda \\ y = -\frac{1}{3}\lambda \\ z = \frac{1}{3}\lambda \end{cases}$$

By eliminating the parameter λ we obtain the Cartesian equations of s, which are (for example)

$$\begin{cases} 2x - 3y = 0 \\ y + z = 0. \end{cases}$$

We now turn to the determination of the plane. The equation of the pencil of planes with axis r is

$$\lambda(2x - 3y + 1) + \mu(2x - 2y + z - 1) = 0.$$

By requesting that it pass through the origin, we get $\lambda - \mu = 0$, hence we can choose, for instance, $\lambda = \mu = 1$. This gives the equation of the plane, which is therefore $4x - 5y + z = 0$.

► **Exercise 6.1.6.** Let r_1 and r_2 be the lines passing through $P_1 = (0, 1, 2)$ and $P_2 = (2, -1, 0)$ and parallel to the vectors $v_1 = (2, 1, 1)$ and $v_2 = (-2, 3, 0)$, respectively. Let $\mathscr{F}_1$ and $\mathscr{F}_2$ be the pencils of planes with axis r_1 and r_2, respectively. Find the planes π_1 of $\mathscr{F}_1$ and π_2 of $\mathscr{F}_2$ passing through the point $P = (1, 1, 1)$, and the intersections of the line $r = \pi_1 \cap \pi_2$ with the coordinates planes. Finally, determine the triangle given by the intersections of the lines r, r_1 and r_2 with the plane of equation $x = 0$.

Solution. The Cartesian equations of r_1 are

$$\frac{x-0}{2} = \frac{y-1}{1} = \frac{z-2}{1}$$

from which we get

$$r_1 : \begin{cases} x - 2y + 2 = 0 \\ y - z + 1 = 0. \end{cases}$$

The Cartesian equations of r_2 are

$$\begin{cases} x = -2t + 2 \\ y = 3t - 1 \\ z = 0 \end{cases}$$

from which we get

$$r_2 : \begin{cases} 3x + 2y - 4 = 0 \\ z = 0. \end{cases}$$

The equations of the pencils $\mathscr{F}_1$ and $\mathscr{F}_2$ are

$$\mathscr{F}_1 : \lambda(x - 2y + 2) + \mu(y - z + 1) = 0$$

and

$$\mathscr{F}_2 : \lambda'(3x + 2y - 4) + \mu' z = 0.$$

Imposing the passage through the point $P = (1, 1, 1)$ we get $\lambda = -\mu$ and $\lambda' = -\mu'$. Setting, for example, $\lambda = \lambda' = 1$ we find the equations of the planes π_1 and π_2:

$$\pi_1 : x - 3y + z + 1 = 0, \qquad \pi_2 : 3x + 2y - z - 4 = 0.$$

Now we have

$$r = \pi_1 \cap \pi_2 : \begin{cases} x - 3y + z + 1 = 0 \\ 3x + 2y - z - 4 = 0. \end{cases}$$

The rest of the exercise is now very easy.

► **Exercise 6.1.7.** Let r be the line passing through the points $P = (1, 0, -2)$ and $Q = (0, -1, 3)$, and let s be the line passing through the point $R = (-1, 1, 0)$ and parallel to the vector $v = (1, -1, -\frac{1}{2})$.

Determine the plane π_1, passing through the points P, Q and R, the plane π_2 containing the line s and passing through the midpoint of the segment PQ, and finally a vector u parallel to the line $\pi_1 \cap \pi_2$.

Solution. A vector parallel to the line r is $Q - P = (-1, -1, 5)$; so the equations of r are

$$\frac{x-1}{-1} = \frac{y-0}{-1} = \frac{z+2}{5}$$

from which we obtain

$$r : \begin{cases} -x + y + 1 = 0 \\ 5y + z + 2 = 0. \end{cases}$$

The equations of s are

$$\frac{x+1}{1} = \frac{y-1}{-1} = \frac{z-0}{-\frac{1}{2}}$$

from which we obtain

$$s : \begin{cases} x + y = 0 \\ y - 2z - 1 = 0. \end{cases}$$

The plane π_1 is the plane of the pencil with axis r that contains the point R. The equation of the pencil of planes with axis r is

$$\lambda(-x + y + 1) + \mu(5y + z + 2) = 0.$$

Requiring that R belongs to this plane we get $\lambda = -\frac{7}{3}\,\mu$. Setting $\mu = 3$ we find the equation of the plane π_1:

$$\pi_1 : 7x + 8y + 3z - 1 = 0.$$

The midpoint M of the segment PQ is

$$M = \frac{P+Q}{2} = (1/2, -1/2, 1/2)$$

Now the plane π_2 can be determined with the same procedure used previously to determine π_1 (we consider the pencil of planes with axis s and we require that M belongs to this plane). We obtain

$$\pi_2 : x + y = 0.$$

Hence the equations of the line $\pi_1 \cap \pi_2$ are

$$\pi_1 \cap \pi_2 : \begin{cases} 7x + 8y + 3z - 1 = 0 \\ x + y = 0, \end{cases}$$

and to find a vector parallel to this line it is enough to find two points, for example the points $A = (0, 0, \frac{1}{3})$ and $B = (-1, 1, 0)$. Hence the vector is $u = B - A = (-1, 1, \frac{1}{3})$.

► **Exercise 6.1.8.** Find the equation of the plane containing the lines r and s with equations

$$r : \begin{cases} y + 1 = 0 \\ 2x + 2z - 3 = 0 \end{cases} \qquad s : \begin{cases} 2x - 3y + 2z - 1 = 0 \\ x - 2y + z = 0. \end{cases}$$

Solution. The pencil of planes with axis r has the following equation:

$$\lambda(2x + 2z - 3) + \mu(y + 1) = 0.$$

To find the plane containing r and s it suffices to determine the plane of this pencil which passes through a point of s; obviously it will then be necessary to verify that the whole line s is contained in this plane.

A point on the line s is $P = (2, 1, 0)$; by requiring that P belongs to the previous pencil of planes we obtain $\lambda = -2\mu$, hence, setting $\mu = 1$, we find the equation of the plane π containing r and P:

$$\pi : -4x + y - 4z + 7 = 0.$$

To verify that the whole line s is contained in π it is sufficient to verify that π contains another point of s (different from P); in fact if a plane contains two points of a line then it contains the whole line.

It is easy to verify that $Q = (0, 1, 2)$ is a point of s and that $Q \in \pi$. This proves that π is the plane containing the two lines r and s.

6.2 Distances and angles

► **Exercise 6.2.1.** Determine the distance of the point $P = (2, 0, -1)$ from the line r whose parametric equations are

$$\begin{cases} x = t + 1 \\ y = 2 \\ z = t - 1. \end{cases}$$

Solution. We recall the formulas that allow us to determine the distance of a point from a line in the Euclidean affine space $\mathbb{A}^3$.

Let r be the line passing through a point Q and parallel to the vector v, and let P be another point. The distance $d = d(P, r)$ of P from the line r is given by:

$$d^2 = (P - Q) \cdot (P - Q) - \frac{((P - Q) \cdot v)^2}{v \cdot v}.$$

Equivalently, we have the following formula:

$$d = \frac{\|(P - Q) \times v\|}{\|v\|},$$

where $\|w\| = \sqrt{w \cdot w}$ is the norm of the vector w.

To determine the distance of P from r it is therefore sufficient to know a point Q of r and a vector v parallel to r. In this case it is easy to verify that one can take $Q = (1, 2, -1)$ and $v = (1, 0, 1)$ and therefore, applying for example the first formula, one finds $d(P, r)^2 = \frac{9}{2}$, from which we obtain $d(P, r) = \frac{3}{2}\sqrt{2}$.

► **Exercise 6.2.2.** Determine the lines passing through $Q = (0, 1, -1)$, whose distance from the point $P = (1, 1, 0)$ is 1 and which are contained in the plane π with equation $y + z = 0$.

Solution. Let $X = (x, y, z)$ be a generic point and consider the line r, passing through Q and X. A vector parallel to this line is $v = X - Q = (x, y - 1, z + 1)$. Using the formula for the distance of a point from a line, we have:

$$d(P, r)^2 = \|P - Q\|^2 - \frac{((P - Q) \cdot v)^2}{\|v\|^2}.$$

By requiring that $d(P, r) = 1$, this equation becomes

$$1 = 2 - \frac{(x + z + 1)^2}{x^2 + (y - 1)^2 + (z + 1)^2}$$

from which we obtain

$$y^2 - 2xz - 2x - 2y + 1 = 0.$$

This equation describes the locus of points X such that the line passing through Q and X has a distance equal to 1 from the point P; it is therefore a cone whose vertex is Q and whose axis is the line through Q and P. Since we are looking for the lines contained in the plane π, they are found by intersecting the cone with the plane π. Note that the plane π contains the vertex Q of the cone, so the intersection of π with the cone consists of two lines (at least if the plane is not tangent to the cone). The intersection between the plane π and the cone is given by the following system:

$$\begin{cases} y^2 - 2xz - 2x - 2y + 1 = 0 \\ y + z = 0. \end{cases}$$

Substituting $y = -z$ in the first equation, we get

$$(z + 1)(z + 1 - 2x) = 0,$$

from which we obtain the two systems

$$\begin{cases} z + 1 = 0 \\ y + z = 0 \end{cases} \qquad \begin{cases} z + 1 - 2x = 0 \\ y + z = 0. \end{cases}$$

So these are the equations of the two lines obtained by intersecting π with the cone, which are the lines we are looking for.

► **Exercise 6.2.3.** Determine the equation of the locus of points in the Euclidean affine space $\mathbb{A}^3(\mathbb{R})$ that belong to the lines passing through the point $P = (2, 1, 1)$ which form an angle equal to $\frac{\pi}{6}$ with the plane π with equation $2x + y - z = 0$.

Solution. From the equation of π we deduce that the vector $n = (2, 1, -1)$ is orthogonal to the plane π. A line makes an angle of $\frac{\pi}{6}$ with π if and only if it makes an angle of $\frac{\pi}{3}$ with the vector n, so the problem reduces to find all the lines passing through P which form an angle equal to $\frac{\pi}{3}$ with the vector n. These lines therefore form a cone with vertex P, with axis the line through P parallel to n, and with half-opening angle $\frac{\pi}{3}$. The equation of this cone can easily be found as follows: let $X = (x, y, z)$, this point belongs to the cone if and only if the vector $X - P$ makes an angle of $\frac{\pi}{3}$ with the vector n; one must therefore have

$$|(X - P) \cdot n| = \|X - P\| \, \|n\| \, \cos \frac{\pi}{3}.$$

We obtain

$$|2(x-2) + (y-1) - (z-1)| = \frac{\sqrt{6}}{2}\sqrt{(x-2)^2 + (y-1)^2 + (z-1)^2}$$

from which, squaring and simplifying, we finally get

$$5x^2 - y^2 - z^2 + 8xy - 8xz - 4yz - 26x - 10y + 22z + 14 = 0.$$

This is the equation of the locus we are looking for.

► **Exercise 6.2.4.** Determine the equation of the locus of points in the affine Euclidean space $\mathbb{A}^3(\mathbb{R})$ belonging to the lines passing through the point $P = (3, 3, 3)$ which intersect the curve C with equation

$$\begin{cases} x^2 + 2y^2 + 3z^2 = 4 \\ x + z = 0. \end{cases}$$

Solution. This locus is the set of all lines joining P with a point Q of the curve C, as the point Q varies in the curve C. It is therefore a *cone* with vertex P projecting the curve C.

Let $S = (x_S, y_S, z_S)$ and consider the line through P and S, whose parametric equations are

$$\begin{cases} x = tx_S + 3(1-t) \\ y = ty_S + 3(1-t) \\ z = tz_S + 3(1-t). \end{cases}$$

Requiring that this line intersects the curve C, we obtain the following system

$$\begin{cases} (tx_S + 3(1-t))^2 + 2(ty_S + 3(1-t))^2 + 3(tz_S + 3(1-t))^2 = 4 \\ (tx_S + 3(1-t)) + (tz_S + 3(1-t)) = 0. \end{cases}$$

Computing t from the second equation

$$t = \frac{6}{6 - x_S - z_S}$$

and substituting in the first, we get

$$\left(\frac{3x_S - 3z_S}{6 - x_S - z_S}\right)^2 + 2\left(\frac{-3x_S + 6y_S - 3z_S}{6 - x_S - z_S}\right)^2 + 3\left(\frac{-3x_S + 3z_S}{6 - x_S - z_S}\right)^2 = 4$$

from which, neglecting the index S, we finally obtain

$$25x^2 + 36y^2 + 25z^2 - 36xy - 22xz - 36yz + 24x + 24z - 72 = 0.$$

This is the equation of the locus we are looking for.

► **Exercise 6.2.5.** Determine the lines of the plane π with equation $x - y + 2z = 0$ parallel to the vector $v = (1, 3, 1)$ and whose distance from the point $P = (1, 0, 0)$ is $\sqrt{6}$.

Solution. The lines parallel to the vector v and whose distance from the point P is $\sqrt{6}$ form a cylinder with axis the line passing through P and parallel to v and base radius $\sqrt{6}$. The lines we are looking for are therefore obtained by intersecting this cylinder with the plane π.

Let r be the line passing through P and parallel to v; it has parametric equations

$$r : \begin{cases} x = t + 1 \\ y = 3t \\ z = t. \end{cases}$$

Let $X = (x, y, z)$ be a generic point of the cylinder: it must be $d(X, r) = \sqrt{6}$. Recalling the formula that provides the distance of a point from a line, we have:

$$\|X - P\|^2 - \frac{((X - P) \cdot v)^2}{\|v\|^2} = 6$$

from which we obtain

$$(x - 1)^2 + y^2 + z^2 - \frac{(x - 1 + 3y + z)^2}{11} = 6.$$

Simplifying, we get:

$$5x^2 + y^2 + 5z^2 - 3xy - xz - 3yz - 10x + 3y + z - 28 = 0.$$

This is the equation of the cylinder with axis r and base radius $\sqrt{6}$. By intersecting this cylinder with the plane π we get the system

$$\begin{cases} 5x^2 + y^2 + 5z^2 - 3xy - xz - 3yz - 10x + 3y + z - 28 = 0 \\ x - y + 2z = 0, \end{cases}$$

from which, computing y from the second equation and replacing in the first, we obtain

$$\begin{cases} 3(x-z)^2 - 7(x-z) - 28 = 0 \\ y = x + 2z. \end{cases}$$

The two solutions of the first equation are $x - z = \frac{7-\sqrt{385}}{6}$ and $x - z = \frac{7+\sqrt{385}}{6}$; therefore we obtain the two systems

$$\begin{cases} x - z = \dfrac{7-\sqrt{385}}{6} \\ y = x + 2z. \end{cases} \qquad \begin{cases} x - z = \dfrac{7+\sqrt{385}}{6} \\ y = x + 2z. \end{cases}$$

that are the equations of the two lines we are looking for.

► **Exercise 6.2.6.** In the three-dimensional Euclidean affine space, determine the equation of the locus of the points belonging to the lines whose distance from the point $P = (1, 1, 2)$ is 1 and which form an angle equal to $\frac{\pi}{3}$ with the planes $\pi_1 : x + y = 0$ and $\pi_2 : x - z = 0$.

Solution. From the equations of the planes it can be deduced that the vectors perpendicular to the planes π_1 and π_2 are respectively $n_1 = (1, 1, 0)$ and $n_2 = (1, 0, -1)$. Note that a line forms an angle equal to $\frac{\pi}{3}$ with the planes π_1 and π_2 if and only if it forms an angle equal to $\frac{\pi}{6}$ with the vectors n_1 and n_2. Let $v = (x, y, z)$ be a generic vector: imposing v to form an angle equal to $\frac{\pi}{6}$ with the vectors n_1 and n_2 we get the following system:

$$\begin{cases} v \cdot n_1 = \|v\| \, \|n_1\| \cos \frac{\pi}{6} \\ v \cdot n_2 = \|v\| \, \|n_2\| \cos \frac{\pi}{6} \end{cases}$$

whose solution (counted with multiplicity 2) is

$$\begin{cases} x = -2z \\ y = -z. \end{cases}$$

Setting, for example, $z = -1$, we obtain the vector $v = (2, 1, -1)$. The lines we are looking for are therefore all the lines parallel to the vector v and whose distance from the point P is 1: they form a cylinder whose axis is the line passing through P and parallel to v and whose base radius 1. The equation of this cylinder is found using the formula for the distance of a point from a line (see the previous exercise). After some calculations, we get

$$2x^2 + 5y^2 + 5z^2 - 4xy + 4xz + 2yz - 8x - 10y - 26z + 29 = 0.$$

► **Exercise 6.2.7.** In the Euclidean affine space $\mathbb{A}^3(\mathbb{R})$ let π be the plane containing the line r with equations $\frac{1}{2}(x-1) = y+1 = \frac{1}{3}z$ and the point $P=(0,1,1)$.

Find the lines contained in the plane π, passing through P and whose distance from the point $Q=(1,1,-1)$ is $\sqrt{3}$. Also determine the orthogonal projection Q_0 of Q onto π and the lines of π passing through Q_0 and orthogonal to the lines found previously.

Solution. The equations of the line r are

$$\begin{cases} x-2y-3=0 \\ 3y-z+3=0. \end{cases}$$

The pencil of planes with axis r has the following equation:

$$\lambda(x-2y-3)+\mu(3y-z+3)=0.$$

Requiring that P belongs to this plane we get $\lambda=\mu$; hence we can set $\lambda=\mu=1$, so we obtain the equation of π:

$$\pi : x+y-z=0.$$

Now let $X=(x,y,z)$ and let r_X be the line passing through P and X. Recalling the formula for the distance of a point from a line, we have

$$d(Q,r_X)^2 = \|Q-P\|^2 - \frac{((Q-P)\cdot(X-P))^2}{\|X-P\|^2}$$

from which we obtain

$$3 = 5 - \frac{(x-2z+2)^2}{x^2+(y-1)^2+(z-1)^2}.$$

Simplifying, we get

$$x^2+2y^2-2z^2+4xz-4x-4y+4z=0.$$

This is the equation of a cone with vertex P whose axis is the line through P and Q. The lines we are looking for are found intersecting this cone with the plane π:

$$\begin{cases} x^2+2y^2-2z^2+4xz-4x-4y+4z=0 \\ x+y-z=0. \end{cases}$$

Computing $y = z - x$ from the second equation and replacing in the first, we get $3x^2 = 0$, hence $x = 0$. Note that there is only one solution (with multiplicity 2); therefore there is only one line r_1 contained in the plane π, passing through P and whose distance from the point Q is $\sqrt{3}$ (this means that the plane π is tangent to the cone). The equations of this line are

$$r_1 : \begin{cases} x = 0 \\ y - z = 0. \end{cases}$$

To determine Q_0 it suffices to find the line through Q perpendicular to π and intersect it with π. From the equation of π it is easy to see that a vector perpendicular to π is $w = (1, 1, -1)$, therefore the equations of the line are

$$\begin{cases} x = t + 1 \\ y = t + 1 \\ z = -t - 1. \end{cases}$$

By intersecting this line with the plane π we find the point $Q_0 = (0, 0, 0)$.

Finally, to determine the line s it is sufficient to find a vector v parallel to s, i.e., perpendicular to w and also to the line r_1. It is easy to see that the points $A = (0, 1, 1)$ and $B = (0, 0, 0)$ belong to r_1, so a vector parallel to r_1 is $u = A - B = (0, 1, 1)$. Consequently, a vector v perpendicular to u and w is $v = u \times w = (-2, 1, -1)$. The equations of the line s are

$$\begin{cases} x = -2t \\ y = t \\ z = -t. \end{cases}$$

Finally, note that the line s can also be determined as the intersection of the plane π with the plane passing through Q_0 and perpendicular to r_1, whose equation is $y + z = 0$. The Cartesian equations of s are therefore

$$\begin{cases} x + y - z = 0 \\ y + z = 0. \end{cases}$$

► **Exercise 6.2.8.** In the Euclidean affine space $\mathbb{A}^3(\mathbb{R})$ determine the equations of the lines passing through $P = (1, 0, 0)$, intersecting the line r with equations

$$r : \begin{cases} y + z - 1 = 0 \\ x = 0 \end{cases}$$

and such that they form an angle $\frac{\pi}{6}$ with r.

Solution. The points $A = (0, 0, 1)$ and $B = (0, 1, 0)$ belong to the line r, whose parametric equations are given by $X = A + t(B - A)$, that is

$$\begin{cases} x = 0 \\ y = t \\ z = 1 - t. \end{cases}$$

The lines passing through P and intersecting the line r are the lines passing through P and the generic point $X_t = (0, t, 1 - t)$ of r. Now it suffices to require that the angle formed by the vectors $X_t - P$ and $B - A$ be equal to $\frac{\pi}{6}$. Therefore we have

$$|(X_t - P) \cdot (B - A)| = \|X_t - P\| \, \|B - A\| \cos \tfrac{\pi}{6}$$

which gives

$$|2t - 1| = \sqrt{1 + t^2 + (1 - t)^2}\,\sqrt{2}\,\frac{\sqrt{3}}{2}.$$

By squaring and simplifying we get

$$t^2 - t - 2 = 0,$$

whose solutions are $t = -1$ and $t = 2$.

For these values of t we obtain two points of r, $X_{-1} = (0, -1, 2)$ and $X_2 = (0, 2, -1)$. The lines we are looking for are therefore the lines s and s' passing through P and X_{-1} and through P and X_2, respectively. Their Cartesian equations are therefore

$$s : \begin{cases} x - y - 1 = 0 \\ 2y + z = 0, \end{cases} \qquad s' : \begin{cases} x - z - 1 = 0 \\ y + 2z = 0. \end{cases}$$

► **Exercise 6.2.9.** In the Euclidean affine plane $\mathbb{A}^2(\mathbb{R})$ let r_1 and r_2 be the lines with equation $x - 2y + 1 = 0$ and $3x - y + 2 = 0$, respectively. Find the equations of the lines s_1 and s_2 bisectors of the angles formed by r_1 and r_2.

Solution. Recall that the bisecting lines of the angles formed by r_1 and r_2 have the property that their points are equidistant from r_1 and r_2.

So let $X = (x, y)$ be a generic point of the plane. We now impose the condition that the distance of X from r_1 is equal to the distance of X from r_2: $d(X, r_1) = d(X, r_2)$. We obtain the following equation:

$$\frac{|x - 2y + 1|}{\sqrt{5}} = \frac{|3x - y + 2|}{\sqrt{10}}$$

By eliminating the denominators and the absolute values, we obtain the two equations:

$$\sqrt{2}\,(x - 2y + 1) = \pm(3x - y + 2)$$

from which it follows that the equations of the two lines are

$$s_1 : (3 - \sqrt{2})x + (2\sqrt{2} - 1)y + 2 - \sqrt{2} = 0$$

and

$$s_2 : (3 + \sqrt{2})x - (2\sqrt{2} + 1)y + 2 + \sqrt{2} = 0.$$

6.3 Recap exercises

► **Exercise 6.3.1.** In the three-dimensional Euclidean space we consider the points $A = (0, -1, 1)$, $B = (-1, 0, 2)$ and $C = (1, -1, -4)$.

(a) Find the equation of the plane π passing through A, B and C.

(b) Determine the line r passing through A and through the midpoint M of the segment BC.

(c) Determine the line s passing through A, contained in the plane π, and orthogonal to the line r.

Solution. To find the equation of the plane π we compute the vectors $v_1 = B - A = (-1, 1, 1)$ and $v_2 = C - A = (1, 0, -5)$; π will therefore be the plane passing through A and parallel to the vectors v_1 and v_2. Its parametric equation is

$$\pi : X = A + \lambda_1 v_1 + \lambda_2 v_2,$$

that is

$$\pi : \begin{cases} x = -\lambda_1 + \lambda_2 \\ y = -1 + \lambda_1 \\ z = 1 + \lambda_1 - 5\lambda_2 \end{cases}$$

These are the parametric equations of the plane π. To find the Cartesian equation we just eliminate the parameters λ_1 and λ_2. We find $\lambda_1 = y + 1$,

$\lambda_2 = x + y + 1$ and substituting these expressions in the third equation we obtain the Cartesian equation of π:

$$\pi : 5x + 4y + z + 3 = 0.$$

From this equation we deduce that a vector orthogonal to the plane π is

$$n_\pi = (5, 4, 1).$$

Let us now determine the midpoint M of the segment BC:

$$M = \frac{B + C}{2} = (0, -1/2, -1).$$

Since the line r passes through the points A and M, one of its vectors is

$$v_r = M - A = (0, 1/2, -2).$$

The parametric equations of r are therefore $X = A + tv_r$, i.e.,

$$r : \begin{cases} x = 0 \\ y = -1 + \frac{1}{2}t \\ z = 1 - 2t \end{cases}$$

If we want its Cartesian equations it is enough to compute $t = 2y + 2$ from the second equation and replace this expression in the third. We obtain

$$r : \begin{cases} x = 0 \\ 4y + z + 3 = 0. \end{cases}$$

To find the line s we only need to know its vector v_s. This vector must be orthogonal to v_r (since the lines r and s must be perpendicular) and "contained" in the plane π (i.e., parallel to π). Requiring that the vector v_s is parallel to the plane π is equivalent to requiring that this vector is orthogonal to n_π (since n_π is a vector orthogonal to the plane π). We therefore conclude that v_s must be a vector orthogonal to both the vectors n_π and v_r. We can therefore take $v_s = n_\pi \times v_r$ (the cross product of n_π and v_r). We get

$$v_s = (-17/2, 10, 5/2).$$

The line s is therefore given by the parametric equation $X = A + tv_s$, i.e.,

$$s : \begin{cases} x = -\frac{17}{2}t \\ y = -1 + 10t \\ z = 1 + \frac{5}{2}t. \end{cases}$$

► **Exercise 6.3.2.** In the three-dimensional Euclidean space we consider the lines r and s with equations

$$r : \begin{cases} x + 3y + 1 = 0 \\ x + 3y + z - 2 = 0 \end{cases} \qquad s : \begin{cases} x - 2z - 7 = 0 \\ y - z - 4 = 0 \end{cases}$$

Verify that the lines r and s are skew. Find the equations of the line ℓ incident to the lines r and s and orthogonal to both. Determine the distance between r and s. Finally, find the points $R \in r$ and $S \in s$ of minimum distance.

Solution. Let's check if the two lines are intersecting:

$$r \cap s : \begin{cases} x + 3y + 1 = 0 \\ x + 3y + z - 2 = 0 \\ x - 2z - 7 = 0 \\ y - z - 4 = 0 \end{cases}$$

It is easy to verify that this system has no solution, hence $r \cap s = \varnothing$ and therefore the lines r and s are not intersecting. To check if they are parallel we look for a vector of each line. Using the equations of r one can easily find two points $A, B \in r$; for example $A = (-1, 0, 3)$ and $B = (2, -1, 3)$. Hence a vector of r is $v_r = B - A = (3, -1, 0)$.

Two points $C, D \in s$ can be determined in a completely analogous way, for example $C = (1, 1, -3)$ and $D = (-1, 0, -4)$; one can then take the vector $v_s = C - D = (2, 1, 1)$ as the vector of the line s. Now we note that the vectors v_r and v_s are not multiples of each other, so the lines r and s are not parallel. This proves that r and s are skew.

To find the equation of the line ℓ intersecting the lines r and s and orthogonal to both, we proceed as follows: we try to find a point $P \in r$ and a point $Q \in s$ such that the vector $w = P - Q$ is orthogonal to the vectors of r and s; in this way the line passing through the points P and Q will be the line ℓ that we are looking for. In doing so we will also answer the next two questions; in fact the distance between the lines r and s coincides with the distance between the points P and Q, which are the points of minimum distance of r and s.

Since the line r passes through the point $A = (-1, 0, 3)$ and is parallel to the vector $v_r = (3, -1, 0)$, a generic point $P \in r$ can be written in the form $P = A + \lambda v_r = (-1 + 3\lambda, -\lambda, 3)$, for $\lambda \in \mathbb{R}$. Similarly, since the line s passes through the point $D = (-1, 0, -4)$ and is parallel to the vector $v_s = (2, 1, 1)$, a generic point $Q \in s$ can be written in the form $Q = D +$

$\mu v_s = (-1 + 2\mu, \mu, -4 + \mu)$, for $\mu \in \mathbb{R}$. Hence the vector $w = P - Q$ is $w = (3\lambda - 2\mu, -\lambda - \mu, 7 - \mu)$. Requiring that w is orthogonal to the vectors v_r and v_s we obtain the following system:

$$\begin{cases} w \cdot v_r = 10\lambda - 5\mu = 0 \\ w \cdot v_s = 5\lambda - 6\mu + 7 = 0 \end{cases}$$

whose solution is

$$\begin{cases} \lambda = 1 \\ \mu = 2. \end{cases}$$

The points P and Q are therefore

$$P = (2, -1, 3) \qquad \text{and} \qquad Q = (3, 2, -2).$$

The line ℓ is the line passing through P and Q, whose parametric equation is $X = P + t(Q - P)$, which is equivalent to the following system:

$$\ell : \begin{cases} x = 2 + t \\ y = -1 + 3t \\ z = 3 - 5t \end{cases}$$

If we want to find the Cartesian equations of ℓ we can get $t = x - 2$ from the first equation and substitute it in the other two, obtaining the following system:

$$\ell : \begin{cases} 3x - y - 7 = 0 \\ 5x + z - 13 = 0. \end{cases}$$

Finally, note that the points $R \in r$ and $S \in s$ of minimum distance are $R = P = (2, -1, 3)$ and $S = Q = (3, 2, -2)$, and the distance of r from s is given by: $\operatorname{dist}(r, s) = \operatorname{dist}(P, Q) = \|w\| = \sqrt{35}$.

► **Exercise 6.3.3.** In the three-dimensional Euclidean space we consider the line

$$r : \begin{cases} x - 2y - 3 = 0 \\ 2x + y + z + 1 = 0 \end{cases}$$

and the point $P = (1, 3, -2)$.

(a) Find the equation of the plane π containing the line r and passing through the point P.

(b) Find the equation of the line s passing through the point P, perpendicular to the line r and contained in the plane π.

(c) Finally, determine the point R of intersection of the lines r and s and the distance of the point P from the line r.

Solution. To find the equation of the plane π containing the line r and passing through the point P we consider the pencil of planes with axis r, given by

$$\lambda(x - 2y - 3) + \mu(2x + y + z + 1) = 0$$

and we require that P belongs to the plane:

$$-8\lambda + 4\mu = 0.$$

We obtain $\mu = 2\lambda$ hence we can set $\lambda = 1$ and $\mu = 2$. Substituting these values into the equation of the pencil of planes, we obtain the equation of the plane π:

$$\pi : 5x + 2z - 1 = 0.$$

A vector perpendicular to this plane is $n_\pi = (5, 0, 2)$.

We now look for a vector v_r of the line r. To do this we determine two (arbitrary) points of r, for example the points $A = (1, -1, -2)$ and $B = (3, 0, -7)$, and compute their difference:

$$v_r = B - A = (2, 1, -5).$$

Since the line s must be orthogonal to r and contained in the plane π, one of its vectors v_s must be orthogonal to the vector v_r and also to the vector n_π, which is orthogonal to the plane π. We can therefore take v_s equal to the cross product of the vectors v_r and n_π:

$$v_s = v_r \times n_\pi = (2, -29, -5).$$

The line s is therefore the line passing through the point P and parallel to the vector v_s, hence its parametric equations are given by $X = P + tv_s$, i.e.,

$$s : \begin{cases} x = 1 + 2t \\ y = 3 - 29t \\ z = -2 - 5t. \end{cases}$$

The point $R = r \cap s$ can be found as follows:

$$R = r \cap s : \begin{cases} x - 2y - 3 = 0 \\ 2x + y + z + 1 = 0 \\ x = 1 + 2t \\ y = 3 - 29t \\ z = -2 - 5t. \end{cases}$$

By solving this system we find $R = \left(\frac{19}{15}, -\frac{13}{15}, -\frac{8}{3}\right)$. The distance of P from the line r is the distance of P from the point R, i.e., the norm of the vector $R - P = \left(\frac{4}{15}, -\frac{58}{15}, -\frac{2}{3}\right)$. Therefore we have

$$\operatorname{dist}(P, r) = \operatorname{dist}(P, R) = \|R - P\| = \tfrac{2}{15}\sqrt{870}.$$

▶ **Exercise 6.3.4.** In the three-dimensional Euclidean space we consider the lines r and s with equations

$$r : \begin{cases} x = 2 - t \\ y = -1 + 2t \\ z = t \end{cases} \qquad s : \begin{cases} x - y + 4z - 1 = 0 \\ 2x + y - 2 = 0 \end{cases}$$

Verify that the lines r and s are skew and determine their distance. Find the line ℓ passing through the point $P = (2, 0, 1)$ and intersecting the lines r and s. Determine the coordinates of the intersection points $R = r \cap \ell$ and $S = s \cap \ell$.

Solution. Let's check if the lines r and s are intersecting:

$$r \cap s : \begin{cases} x = 2 - t \\ y = -1 + 2t \\ z = t \\ x - y + 4z - 1 = 0 \\ 2x + y - 2 = 0 \end{cases}$$

It is easy to verify that this system has no solutions, so $r \cap s = \varnothing$. The two lines are therefore parallel or skew.

We now look for vectors of r and s. From the parametric equations of the line r it is easy to deduce that one of its vectors is $v_r = (-1, 2, 1)$. To determine a vector of the line s we first look for two points of this line. From the equations of s we deduce that the points $A = (1,0,0)$ and $B = (2,-2,-3/4)$ belong to this line, therefore a vector of s is $v_s = B - A = (1,-2,-3/4)$. The vectors v_r and v_s are not proportional, so the lines r and s are not parallel. We have thus proved that r and s are skew.

To determine the distance of r from s we proceed as follows. We choose a point on the line r, for example the point $C = (2,-1,0)$, and a point on the line s, for example $A = (1,0,0)$. Now we consider the vector $w = C - A = (1,-1,0)$. Recalling that v_r and v_s are vectors of the lines r and s respectively, we have:

$$\operatorname{dist}(r,s) = \frac{|w \cdot (v_r \times v_s)|}{\|v_r \times v_s\|}.$$

By computing the cross product of v_r and v_s we obtain

$$v_r \times v_s = (1/2, 1/4, 0),$$

hence $\|v_r \times v_s\| = \sqrt{5}/4$. Then we have

$$w \cdot (v_r \times v_s) = \det \begin{pmatrix} 1 & -1 & 0 \\ -1 & 2 & 1 \\ 1 & -2 & -\frac{3}{4} \end{pmatrix} = \frac{1}{4}.$$

From the previous formula we get

$$\operatorname{dist}(r,s) = \frac{\sqrt{5}}{5}.$$

To find the line ℓ passing through the point $P = (2,0,1)$ and intersecting the lines r and s let's consider a generic point of r, given by $R_t = (2-t, -1+2t, t)$, and denote by ℓ_t the line passing through the points P and R_t:

$$\ell_t : \begin{cases} x = 2 + \lambda(-t) \\ y = \lambda(-1+2t) \\ z = 1 + \lambda(t-1) \end{cases}$$

This is actually a "family" of infinite lines ℓ_t, parametrized by $t \in \mathbb{R}$. All these lines pass through the point P and intersect the line r (at the point R_t). All that is needed is to find out which of these lines also intersects the line s, i.e., for which t we have $\ell_t \cap s \neq \varnothing$. So let's look for the intersection between the lines ℓ_t and s:

$$\ell_t \cap s : \begin{cases} x = 2 - \lambda t \\ y = (-1 + 2t)\lambda \\ z = 1 + (t-1)\lambda \\ x - y + 4z - 1 = 0 \\ 2x + y - 2 = 0 \end{cases}$$

This system has a unique solution, given by

$$\begin{cases} t = 1/2 \\ \lambda = 2 \\ x = 1 \\ y = 0 \\ z = 0. \end{cases}$$

This means that the value of t for which the line ℓ_t intersects the lines r and s is $t = 1/2$ and in this case the intersection point S between s and ℓ_t has coordinates $S = (1, 0, 0)$. The line ℓ therefore has the following parametric equations:

$$\ell = \ell_{1/2} : \begin{cases} x = 2 - \frac{1}{2}\lambda \\ y = 0 \\ z = 1 - \frac{1}{2}\lambda \end{cases}$$

and the point $R = r \cap \ell$ is $R = R_{1/2} = (3/2, 0, 1/2)$.

Finally we note that the line ℓ could also be determined in the following way. Let us consider the plane π containing the line s and the point P. To find its equation we consider the pencil of planes with axis s

$$\lambda(x - y + 4z - 1) + \mu(2x + y - 2) = 0$$

and require that P belongs to the plane

$$5\lambda + 2\mu = 0.$$

Setting $\lambda = -2$ and $\mu = 5$ and substituting these values in the equation of the pencil of planes, we find the plane we are looking for

$$\pi : 8x + 7y - 8z - 8 = 0.$$

Since the line ℓ must pass through the point P and must intersect the line s, it must necessarily be contained in the plane π. Let us now find the intersection between the plane π and the line r:

$$\pi \cap r : \begin{cases} x = 2 - t \\ y = -1 + 2t \\ z = t \\ 8x + 7y - 8z - 8 = 0. \end{cases}$$

The solution of this system is

$$\begin{cases} t = 1/2 \\ x = 3/2 \\ y = 0 \\ z = 1/2. \end{cases}$$

Therefore the plane π and the line r intersect at the point of coordinates $(3/2, 0, 1/2)$. From what we have seen previously, it can be deduced that the line ℓ must necessarily intersect the line r at the point $R = (3/2, 0, 1/2)$; it is therefore the line passing through P and parallel to the vector $R - P = (-1/2, 0, -1/2)$. Thus the equations of the line ℓ written previously are found.

► **Exercise 6.3.5.** In the three-dimensional Euclidean space we consider the points $A = (0, -1, 1)$, $B = (-1, 0, 2)$ and $C = (1, -1, -4)$.

(a) Find the equation of the plane π passing through A, B and C.

(b) Determine the line r passing through A and through the midpoint M of the segment BC.

(c) Determine the line s passing through A, contained in the plane π and orthogonal to the line r.

(d) Finally, given the point $P = (1, 1, 1)$, determine the distances of P from the line r and from the plane π.

Solution. The plane π has an equation of the form $ax + by + cz + d = 0$. Imposing that the plane passes through the points A, B and C, the following system is obtained

$$\begin{cases} -b + c + d = 0 \\ -a + 2c + d = 0 \\ a - b - 4c + d = 0 \end{cases}$$

whose solutions are

$$\begin{cases} a = 5c \\ b = 4c \\ d = 3c \\ c \text{ any value.} \end{cases}$$

By setting $c = 1$ the following Cartesian equation of the plane π is obtained:

$$\pi : 5x + 4y + z + 3 = 0.$$

Let's compute the coordinates of the midpoint M of the segment BC:

$$M = \frac{B + C}{2} = \left(0, -\tfrac{1}{2}, -1\right).$$

The vector of the line r is

$$v_r = M - A = \left(0, \tfrac{1}{2}, -2\right)$$

hence the parametric equations of r are $X = A + tv_r$, i.e.,

$$r : \begin{cases} x = 0 \\ y = -1 + \tfrac{1}{2}t \\ z = 1 - 2t. \end{cases}$$

If we want the Cartesian equations of r we can derive t from the second equation, $t = 2y + 2$, and substitute this value into the third, thus obtaining

$$r : \begin{cases} x = 0 \\ 4y + z + 3 = 0. \end{cases}$$

Let v_s be a vector of the line s. Since s must be contained in the plane π, the vector v_s must be orthogonal to the vector $n_\pi = (5, 4, 1)$, orthogonal to the plane π. Moreover v_s must also be orthogonal to the vector v_r, since the line s must be orthogonal to the line r. So for v_s we need a vector that is orthogonal to both n_π and v_r; for example we can take their cross product:

$$v_s = n_\pi \times v_r = \left(-\tfrac{17}{2}, 10, \tfrac{5}{2}\right).$$

The parametric equations of the line s are therefore $X = A + tv_s$, i.e.,

$$s : \begin{cases} x = -\tfrac{17}{2}t \\ y = -1 + 10t \\ z = 1 + \tfrac{5}{2}t. \end{cases}$$

To compute the distance of the point P from the line r we consider the vector $u = P - A = (1,2,0)$. We can now use the following formula:

$$\operatorname{dist}(P,r) = \frac{\|u \times v_r\|}{\|v_r\|}.$$

We get $u \times v_r = (-4,2,1/2)$, $\|u \times v_r\| = 9/2$ e $\|v_r\| = \frac{1}{2}\sqrt{17}$. Therefore we have

$$\operatorname{dist}(P,r) = \frac{9\sqrt{17}}{17}.$$

Finally, for the distance of P from the plane π, we have

$$\operatorname{dist}(P,\pi) = \frac{13}{\sqrt{42}} = \frac{13\sqrt{42}}{42}.$$

► **Exercise 6.3.6.** In the three-dimensional Euclidean space we consider the points $A = (-1,2,-3)$, $B = (2,-1,3)$, $C = (-1,0,0)$.

(a) Find the Cartesian equation of the plane π containing the triangle ABC and calculate the area of this triangle.

(b) Determine the Cartesian equations of the line s, perpendicular to the plane π and passing through the point $P = (0,-3,0)$.

(c) Compute the distance between the line s and the line r passing through A and B.

(d) Find the orthogonal projection of the line OP onto the plane π (where O is the point $(0,0,0)$).

(e) Compute the angle between π and the plane XY.

Solution. Let $v = B - A = (3,-3,6)$ and $w = C - A = (0,-2,3)$. The parametric equations of the plane π are given by $X = A + \lambda v + \mu w$, i.e.,

$$\pi : \begin{cases} x = -1 + 3\lambda \\ y = 2 - 3\lambda - 2\mu \\ z = -3 + 6\lambda + 3\mu. \end{cases}$$

By eliminating the two parameters λ and μ we find the following Cartesian equation for the plane π:

$$\pi : x - 3y - 2z + 1 = 0.$$

The area of the triangle of vertices A, B and C can be calculated using the following formula:

$$\text{Area}(ABC) = \frac{1}{2}\,\|v \times w\|.$$

Computing the cross product of v and w we find $v \times w = (3, -9, -6)$, whose norm is $\|v \times w\| = 3\sqrt{14}$. Hence we have

$$\text{Area}(ABC) = \frac{3}{2}\,\sqrt{14}.$$

A vector orthogonal to the plane π is $n_\pi = (1, -3, -2)$. The line s, perpendicular to the plane π and passing through the point $P = (0, -3, 0)$ therefore has parametric equations given by $X = P + tn_\pi$, i.e.,

$$s : \begin{cases} x = t \\ y = -3 - 3t \\ z = -2t. \end{cases}$$

By eliminating the parameter t the following Cartesian equations are found:

$$s : \begin{cases} 3x + y + 3 = 0 \\ 2x + z = 0. \end{cases}$$

To compute the distance between the lines r and s we proceed as follows. We choose a point on the line r, for example $A = (-1, 2, -3)$, a point on the line s, for example $P = (0, -3, 0)$, and compute the vector $w = A - P = (-1, 5, -3)$. Let us now consider a vector of the line r, for example $v_r = (1, -1, 2)$ (it is the vector $B - A$ divided by 3), and a vector of the line s, $v_s = n_\pi = (1, -3, -2)$. The distance between the lines r and s is given by the following formula:

$$\text{dist}(r, s) = \frac{|(v_r \times v_s) \cdot w|}{\|v_r \times v_s\|}.$$

We have

$$|(v_r \times v_s) \cdot w| = \left| \det \begin{pmatrix} 1 & -1 & 2 \\ 1 & -3 & -2 \\ -1 & 5 & -3 \end{pmatrix} \right| = 18,$$

$v_r \times v_s = (8, 4, -2)$, hence $\|v_r \times v_s\| = 2\sqrt{21}$. Thus we have

$$\text{dist}(r, s) = \frac{3}{7}\,\sqrt{21}.$$

To find the orthogonal projection of the line OP onto the plane π it is sufficient to find the orthogonal projection onto π of two points of this line (for example the points O and P); the line we are looking for is the line passing through the points thus found. Otherwise we can proceed as follows. Since the point O has coordinates $O = (0, 0, 0)$, the parametric equations of the line OP are

$$\begin{cases} x = 0 \\ y = -3t \\ z = 0. \end{cases}$$

Intersecting with the plane π we find the point $T = (0, 1/3, 0)$. Since $T \in \pi$, its orthogonal projection onto the plane π is the point T itself.

To determine the orthogonal projection of the point P onto the plane π we consider the line passing through P and orthogonal to π; this is the line s determined earlier. Now we look for the point $P' = s \cap \pi$:

$$s \cap \pi : \begin{cases} 3x + y + 3 = 0 \\ 2x + z = 0 \\ x - 3y - 2z + 1 = 0. \end{cases}$$

By solving this system, we get $P' = (-5/7, -6/7, 10/7)$. The vector $P' - T$ is

$$P' - T = \left(-\frac{5}{7}, -\frac{25}{21}, \frac{10}{7}\right)$$

hence the line passing through the points T and P' has the following parametric equations

$$\begin{cases} x = -\frac{5}{7}\,\lambda \\ y = \frac{1}{3} - \frac{25}{21}\,\lambda \\ z = \frac{10}{7}\,\lambda. \end{cases}$$

This line is the orthogonal projection of the line OP onto the plane π.

The plane π, with equation $x - 3y - 2z + 1 = 0$, has a normal vector given by $n_\pi = (1, -3, -2)$. Let σ be the plane XY; it has equation $z = 0$. A vector perpendicular to this plane is therefore the vector $n_\sigma = (0, 0, 1)$. We now note that the angle α formed by the two planes π and σ coincides with the angle formed by the two vectors n_π and n_σ. Hence we have

$$\cos\alpha = \frac{n_\pi \cdot n_\sigma}{\|n_\pi\|\,\|n_\sigma\|} = -\frac{1}{7}\sqrt{14},$$

from which we obtain $\alpha = \arccos(-\sqrt{14}/7)$.

► **Exercise 6.3.7.** In the three-dimensional Euclidean affine space consider the plane π with equation $3x - y + z + 2 = 0$ and the points $A = (0, 0, -2)$ and $B = (0, 2, 0)$.

(a) Find the Cartesian equations of the line r passing through the point A and orthogonal to the plane π.

(b) Find the Cartesian equations of the line s passing through the point B, contained in the plane π and orthogonal to the line passing through A and B.

(c) Given the point $P = (3, 2, -1)$, determine the distances of P from the line r and from the plane π. Also determine the orthogonal projection P' of P onto the plane π.

(d) Find all points $C \in \pi$ such that the triangle ABC is equilateral.

Solution. A vector orthogonal to the plane π is $n_\pi = (3, -1, 1)$. This vector is the vector v_r of the line r, which therefore has the following parametric equations:

$$r : \begin{cases} x = 3t \\ y = -t \\ z = -2 + t. \end{cases}$$

From the second equation we have $t = -y$ and substituting in the other two we obtain the following Cartesian equations for the line r:

$$r : \begin{cases} x + 3y = 0 \\ y + z + 2 = 0. \end{cases}$$

We have $B - A = (0, 2, 2) = 2(0, 1, 1)$, hence we can take $w = (0, 1, 1)$ as the vector of the line passing through A and B. Since the line s must be orthogonal to the line through A and B, and must be contained in the plane π, one of its vectors v_s must be orthogonal to the vector w and to the vector n_π, so we can take $v_s = w \times n_\pi = (2, 3, -3)$. The line s is therefore the line passing through the point B having v_s as vector, and therefore its parametric equations are

$$s : \begin{cases} x = 2t \\ y = 2 + 3t \\ z = -3t. \end{cases}$$

From the first equation we obtain $t = x/2$ and substituting in the other two we obtain the following Cartesian equations for the line s:

$$s : \begin{cases} y = 2 + \frac{3}{2}\, x \\ z = -\frac{3}{2}\, x. \end{cases}$$

To compute the distance of the point P from the line r, we set $u = P - A = (3, 2, 1)$ and compute the cross product of u and $v_r = n_\pi$:

$$u \times v_r = (3, 2, 1) \times (3, -1, 1) = (3, 0, -9).$$

The distance of P from r is given by the following formula:

$$\text{dist}(P, r) = \frac{\|u \times v_r\|}{\|v_r\|} = \frac{3\sqrt{10}}{\sqrt{11}}.$$

For the computation of the distance of the point P from the plane π, we have:

$$\text{dist}(P, \pi) = \frac{8}{\sqrt{11}}.$$

To determine the orthogonal projection of the point P onto the plane π, we consider the line passing through P and orthogonal to π, whose parametric equations are

$$\begin{cases} x = 3 + 3t \\ y = 2 - t \\ z = -1 + t. \end{cases}$$

By intersecting this line with the plane π we obtain the system

$$\begin{cases} x = 3 + 3t \\ y = 2 - t \\ z = -1 + t \\ 3x - y + z + 2 = 0, \end{cases}$$

whose solution is

$$\begin{cases} t = -8/11 \\ x = 9/11 \\ y = 30/11 \\ z = -19/11. \end{cases}$$

The point P', the orthogonal projection of P onto the plane π, therefore has the following coordinates:

$$P' = \left(\frac{9}{11}, \frac{30}{11}, -\frac{19}{11}\right).$$

Let us now consider a generic point $C = (a, b, c)$. The request that C belongs to the plane π gives the equation $3a - b + c + 2 = 0$. Now let's compute the squares of the distances between the three points A, B and C:

$$\begin{aligned}
\operatorname{dist}(A, B)^2 &= 8 \\
\operatorname{dist}(A, C)^2 &= a^2 + b^2 + c^2 + 4c + 4 \\
\operatorname{dist}(B, C)^2 &= a^2 + b^2 + c^2 - 4b + 4.
\end{aligned}$$

The equality $\operatorname{dist}(A, C)^2 = \operatorname{dist}(B, C)^2$ gives another equation, $4c = -4b$, while the equality $\operatorname{dist}(A, C)^2 = \operatorname{dist}(A, B)^2$ gives a third equation, $a^2 + b^2 + c^2 + 4c + 4 = 8$. With these three equations we obtain the following system

$$\begin{cases} 3a - b + c + 2 = 0 \\ 4c = -4b \\ a^2 + b^2 + c^2 + 4c + 4 = 8, \end{cases}$$

whose (two) solutions are given by

$$\begin{cases} a = 2\sqrt{\frac{3}{11}} \\ b = 1 + 3\sqrt{\frac{3}{11}} \\ c = -1 - 3\sqrt{\frac{3}{11}} \end{cases} \quad \text{and} \quad \begin{cases} a = -2\sqrt{\frac{3}{11}} \\ b = 1 - 3\sqrt{\frac{3}{11}} \\ c = -1 + 3\sqrt{\frac{3}{11}} \end{cases}$$

We therefore conclude that there exist two points $C \in \pi$ such that the triangle of vertices A, B and C is equilateral; they are

$$C_1 = \left(2\sqrt{\tfrac{3}{11}}, 1+3\sqrt{\tfrac{3}{11}}, -1-3\sqrt{\tfrac{3}{11}}\right) \quad C_2 = \left(-2\sqrt{\tfrac{3}{11}}, 1-3\sqrt{\tfrac{3}{11}}, -1+3\sqrt{\tfrac{3}{11}}\right).$$

► **Exercise 6.3.8.** In the three-dimensional Euclidean affine space we consider the lines

$$r: \begin{cases} x - 2y = 1 \\ 2x - y + z = 3 \end{cases} \qquad s: \begin{cases} y - z = 2 \\ 2x - y - z = 1 \end{cases}$$

(a) Determine if the lines r and s are intersecting, parallel or skew.

(b) Find the Cartesian equation of the plane π containing the line s and parallel to the line r.

(c) Determine the parametric equations of the line r', the orthogonal projection of r onto the plane π.

(d) Given the vector $v = (1, 4, -2)$ determine the line t parallel to the vector v and intersecting the lines r and s.

(e) Let $R = r \cap t$ and $S = s \cap t$. Compute the area of the triangle PRS, where $P = (2, 2, 1)$.

Solution. To find out if the lines r and s are intersecting, we solve the following system:

$$r \cap s : \begin{cases} x - 2y = 1 \\ 2x - y + z = 3 \\ y - z = 2 \\ 2x - y - z = 1 \end{cases}$$

It is easy to find that this system has no solutions, so $r \cap s = \varnothing$: the lines r and s do not intersect.

From the equations of the line r we determine two points of r, for example the points $R_1 = (1, 0, 1)$ and $R_2 = (3, 1, -2)$. A vector of the line r is $v_r = R_2 - R_1 = (2, 1, -3)$. Similarly, we can find two points of s, for example the points $S_1 = (3/2, 2, 0)$ and $S_2 = (1/2, 1, -1)$. A vector of the line s is $v_s = S_1 - S_2 = (1, 1, 1)$. Since the vectors v_r and v_s are not multiples of each other, the lines r and s are not parallel. It follows that the two lines are skew.

The plane π containing the line s and parallel to the line r is the plane passing through a point of s, for example $S_1 = (3/2, 2, 0)$, and parallel to the vectors v_r and v_s. Its parametric equations are then given by

$$\pi : \begin{cases} x = 3/2 + 2\lambda + \mu \\ y = 2 + \lambda + \mu \\ z = -3\lambda + \mu \end{cases}$$

By eliminating the parameters λ and μ from the previous equations, the following Cartesian equation of the plane π is obtained:

$$4x - 5y + z + 4 = 0.$$

To determine the line r', the orthogonal projection of r onto the plane π, let's consider the point $R_1 = (1,0,1)$ of r and calculate its orthogonal projection R_1' onto π. To do this we consider the line ℓ passing through R_1 and perpendicular to the plane π, whose parametric equations are

$$\ell : \begin{cases} x = 1 + 4\lambda \\ y = -5\lambda \\ z = 1 + \lambda \end{cases}$$

and we determine its intersection with π

$$R_1' = \ell \cap \pi : \begin{cases} x = 1 + 4\lambda \\ y = -5\lambda \\ z = 1 + \lambda \\ 4x - 5y + z + 4 = 0. \end{cases}$$

By solving the system we find $R_1' = (1/7, 15/14, 11/14)$. Now it suffices to note that the line r', the orthogonal projection of r onto the plane π, is the line passing through the point R_1' and parallel to the vector $v_r = (2,1,-3)$. Its parametric equations are

$$r' : \begin{cases} x = 1/7 + 2\lambda \\ y = 15/14 + \lambda \\ z = 11/14 - 3\lambda \end{cases}$$

To determine the line t parallel to the vector $v = (1,4,-2)$ and intersecting the lines r and s we can proceed as follows. We denote by R_λ the generic point of r, given by $R_\lambda = R_1 + \lambda v_r = (1 + 2\lambda, \lambda, 1 - 3\lambda)$, and consider the line t_λ passing through R_λ and parallel to the vector v, whose parametric equations are given by

$$t_\lambda : \begin{cases} x = 1 + 2\lambda + \mu \\ y = \lambda + 4\mu \\ z = 1 - 3\lambda - 2\mu. \end{cases}$$

We note that, for each fixed λ, the previous parametric equations describe the points of the line t_λ, while the same equations considered as both parameters λ and μ vary, describe the points of a *plane*, which is the plane containing the line r and parallel to the vector v.

By computing the intersection between the line t_λ and the line s we obtain the system

$$t_\lambda \cap s : \begin{cases} x = 1 + 2\lambda + \mu \\ y = \lambda + 4\mu \\ z = 1 - 3\lambda - 2\mu \\ y - z = 2 \\ 2x - y - z = 1 \end{cases}$$

from which we obtain

$$\begin{cases} \lambda = 0 \\ \mu = 1/2 \\ x = 3/2 \\ y = 2 \\ z = 0 \end{cases}$$

This means that the line t is the line that corresponds to the value $\lambda = 0$ and therefore its parametric equations are

$$t = t_0 : \begin{cases} x = 1 + \mu \\ y = 4\mu \\ z = 1 - 2\mu. \end{cases}$$

The points of intersection between the line t and the lines r and s are, respectively,

$$R = t \cap r = (1, 0, 1), \qquad S = t \cap s = (3/2, 2, 0).$$

To compute the area of the triangle PRS we determine the vectors $\overrightarrow{PR} = R - P = (-1, -2, 0)$ e $\overrightarrow{PS} = S - P = (-1/2, 0, -1)$. The cross product of these two vectors is $\overrightarrow{PR} \times \overrightarrow{PS} = (2, -1, -1)$, whose norm is $\|\overrightarrow{PR} \times \overrightarrow{PS}\| = \sqrt{6}$. Therefore we have

$$\text{Area}(PRS) = \frac{1}{2} \|\overrightarrow{PR} \times \overrightarrow{PS}\| = \frac{\sqrt{6}}{2}.$$

► **Exercise 6.3.9.** In the three-dimensional Euclidean affine space consider the points $A = (2, 3, 1)$, $B = (1, -2, -1)$ and the vector $n = (2, 1, 1)$.

(a) Find the Cartesian equation of the plane π passing through the midpoint M of the segment AB and orthogonal to the vector n.

(b) Determine the angle α formed by the line r, passing through the points A and B, and the plane π.

(c) Given the point $C = (2, -3, 4)$, determine the orthogonal projections C', on the line r, and C'', on the plane π.

(d) Find the equations of the line s passing through the point M, contained in the plane π and orthogonal to the line r.

Solution. The coordinates of the midpoint M of the segment AB are

$$M = \frac{A+B}{2} = \left(\frac{3}{2}, \frac{1}{2}, 0\right).$$

The equation of the generic plane orthogonal to the vector $n = (2, 1, 1)$ is

$$2x + y + z + d = 0.$$

Requiring that M belongs to the plane we find $d = -7/2$, hence the equation of the plane π is

$$\pi : 2x + y + z - 7/2 = 0.$$

To determine the angle α formed by the line r and the plane π we need a vector v_r of r and a vector n orthogonal to π. We have $v_r = A - B = (1, 5, 2)$, while $n = (2, 1, 1)$ is assigned.

If we denote the angle formed by the vectors v_r and n by β, we have

$$\cos\beta = \frac{n \cdot v_r}{\|n\| \, \|v_r\|} = \frac{9}{6\sqrt{5}} = \frac{3\sqrt{5}}{10}.$$

If α is the angle formed by the line r and the plane π, we have $\alpha + \beta = 90°$, therefore

$$\sin\alpha = \cos\beta = \frac{3\sqrt{5}}{10}$$

hence

$$\alpha = \arcsin\frac{3\sqrt{5}}{10}.$$

To determine the orthogonal projection of the point C onto the line r we proceed as follows. Having previously determined a vector of the line r, $v_r = (1, 5, 2)$, we can write the parametric equations of r:

$$r : \begin{cases} x = 2 + t \\ y = 3 + 5t \\ z = 1 + 2t \end{cases}$$

hence the generic point of r is $X = (2+t, 3+5t, 1+2t)$. We now compute the vector $u = X - C = (t, 5t+6, 2t-3)$ and require that u be orthogonal to the line r:

$$u \cdot v_r = 30t + 24 = 0,$$

from which we get $t = -4/5$. By substituting the value of t just found in the coordinates of the point X we obtain the coordinates of the point C' which is the orthogonal projection of C onto r:

$$C' = \left(\frac{6}{5}, -1, -\frac{3}{5}\right).$$

Let us now determine the point C'', the orthogonal projection of C onto the plane π. To do this we write the parametric equations of the line r' passing through the point C and orthogonal to π:

$$r' : \begin{cases} x = 2 + 2t \\ y = -3 + t \\ z = 4 + t \end{cases}$$

The point C'' is the intersection between r' and π and its coordinates are therefore obtained by solving the system

$$\begin{cases} x = 2 + 2t \\ y = -3 + t \\ z = 4 + t \\ 2x + y + z - 7/2 = 0 \end{cases}$$

from which we get

$$C'' = \left(\frac{3}{2}, -\frac{13}{4}, \frac{15}{4}\right).$$

Finally, to find the equations of the line s we need to determine its vector v_s. This vector must be orthogonal to the vectors n and v_r, so we can take as v_s the cross product of n and v_r

$$v_s = n \times v_r = (-3, -3, 9).$$

The parametric equations of s are therefore

$$s : \begin{cases} x = 3/2 - 3t \\ y = 1/2 - 3t \\ z = 9t \end{cases}$$

► **Exercise 6.3.10.** In the three-dimensional Euclidean affine space we consider the points $A = (2, 1, -1)$, $B = (4, -2, 0)$ and the line r with equations $x - 2y - 5 = 0$ and $2y - z = 0$.

(a) Find the Cartesian equation of the plane π passing through the points A and B and intersecting the line r at a point C equidistant from A and B.

(b) Find the Cartesian equations of the line s passing through the point C, contained in the plane π and orthogonal to the line r.

(c) Determine the orthogonal projection of the point A onto the line r.

Solution. The equations of the line r can be rewritten as follows

$$r : \begin{cases} x = 2y + 5 \\ z = 2y \end{cases}$$

from which we deduce that the generic point X of r has coordinates $X = (2y + 5, y, 2y)$. By requiring that the distance of X from A is equal to the distance of X from B, we obtain the equation

$$(2y + 3)^2 + (y - 1)^2 + (2y + 1)^2 = (2y + 1)^2 + (y + 2)^2 + (2y)^2$$

whose unique solution is $y = -1$. We therefore deduce that the only point C of r that is equidistant from A and B is $C = (3, -1, -2)$. The plane π is therefore the plane passing through the points A, B and C. To find its Cartesian equation we can start by determining the vectors $C - A = (1, -2, -1)$ and $C - B = (-1, 1, -2)$. Their cross product

$$n = (C - A) \times (C - B) = (5, 3, -1)$$

is a vector orthogonal to the plane π and therefore the equation of π must have the following form

$$\pi : 5x + 3y - z + d = 0,$$

for some d. Now imposing that the plane passes through one of the three points (for example, through A), we get $d = -14$. Hence the equation of π is

$$\pi : 5x + 3y - z - 14 = 0.$$

Since the line s must be contained in the plane π, its vector v_s must be orthogonal to the vector n (which is orthogonal to the plane π). Moreover

v_s must also be orthogonal to the vector of the line r, which can easily be seen to be $v_r = (2, 1, 2)$. Hence we can take v_s equal to the cross product of n and v_r

$$v_s = n \times v_r = (7, -12, -1).$$

Since s must pass through C, its parametric equations are

$$s : \begin{cases} x = 3 + 7t \\ y = -1 - 12t \\ z = -2 - t \end{cases}$$

From these, by eliminating the parameter t, we obtain the equations

$$s : \begin{cases} x + 7z + 11 = 0 \\ y - 12z - 23 = 0. \end{cases}$$

To determine the orthogonal projection of the point A onto the line r, we can proceed as follows. Given that the vector of r is $v_r = (2, 1, 2)$, the equation

$$2x + y + 2z + d = 0$$

represents the generic plane orthogonal to the line r (it is the equation of the pencil of planes parallel to each other and all orthogonal to r). By requiring that $A = (2, 1, -1)$ belongs to the plane we find $d = -3$, hence

$$\sigma : 2x + y + 2z - 3 = 0$$

is the plane perpendicular to r passing through A. The orthogonal projection A' of the point A onto the line r is the point of intersection between the line r and the plane σ

$$A' = r \cap \sigma : \begin{cases} x - 2y - 5 = 0 \\ 2y - z = 0 \\ 2x + y + 2z - 3 = 0. \end{cases}$$

By solving this system we find

$$A' = \left(\frac{31}{9}, -\frac{7}{9}, -\frac{14}{9}\right).$$

▶ **Exercise 6.3.11.** In the three-dimensional Euclidean affine space we consider $A = (3, -1, 1)$, $B = (2, 1, 3)$ and the line r with equations $x - 3y = 2$ and $x + y - 2z = 6$.

(a) Find the Cartesian equations of the line s passing through A and B.

(b) Determine if the lines r and s are intersecting, parallel or skew.

(c) Determine the distance of the line r from the line s.

(d) Find the Cartesian equation of the plane π passing through the points A and B and intersecting the line r at a point C such that the triangle ABC is right-angled, with the right angle in A.

(e) Given the point $P = (1, -3, 5)$ determine its orthogonal projection onto the plane π.

Solution. A vector of the line s is $v_s = B - A = (-1, 2, 2)$ and therefore the parametric equations of s are

$$s : \begin{cases} x = 3 - t \\ y = -1 + 2t \\ z = 1 + 2t \end{cases}$$

By eliminating the parameter t we obtain the equations of the line s

$$s : \begin{cases} 2x + y = 5 \\ 2x + z = 7 \end{cases}$$

Let us now check if the lines r and s are intersecting:

$$r \cap s : \begin{cases} x - 3y = 2 \\ x + y - 2z = 6 \\ 2x + y = 5 \\ 2x + z = 7 \end{cases}$$

Solving this system we find that it has no solutions, so $r \cap s = \varnothing$. This means that r and s do not intersect.

From the equations of r we can determine two points of r, for example $A' = (2, 0, -2)$ and $B' = (5, 1, 0)$. This allows us to determine a vector of the line r, $v_r = B' - A' = (3, 1, 2)$. Since the vectors v_r and v_s are not parallel (they are not multiples of each other), the lines r and s are not parallel. This proves that r and s are skew.

To determine the distance between the two lines r and s, let us consider the vector $u = A' - A = (-1, 1, -3)$. The distance is given by the following formula:

$$\operatorname{dist}(r, s) = \frac{|u \cdot (v_r \times v_s)|}{\|v_r \times v_s\|}$$

We have $v_r \times v_s = (-2, -8, 7)$, hence $\|v_r \times v_s\| = \sqrt{117}$ and $u \cdot (v_r \times v_s) = -27$. Therefore we obtain

$$\operatorname{dist}(r, s) = \frac{27}{\sqrt{117}} = \frac{9\sqrt{13}}{13}.$$

To find the plane π we first determine the coordinates of the point C. The parametric equations of r are

$$r : \begin{cases} x = 2 + 3t \\ y = t \\ z = -2 + 2t \end{cases}$$

hence the coordinates of a generic point X of the line r are $X = (2+3t, t, -2+2t)$. Now we can compute the vector $X - A = (3t - 1, t + 1, 2t - 3)$. This vector must be orthogonal to $B - A = v_s = (-1, 2, 2)$, so we must have

$$(X - A) \cdot v_s = 3t - 3 = 0,$$

from which we obtain $t = 1$. By substituting the value of t into the coordinates of X, the point C is obtained: $C = (5, 1, 0)$.

To find the Cartesian equation of the plane π passing through the points A, B and C, we can write the equation of the pencil of planes with axis s (each plane passing through the points A and B must contain the line s)

$$\alpha(2x + y - 5) + \beta(2x + z - 7) = 0.$$

By requiring that C belongs to the plane we obtain $\beta = -2\alpha$, hence we can set $\alpha = -1$ and $\beta = 2$. We find the following equation:

$$\pi : 2x - y + 2z = 9.$$

To determine the orthogonal projection of the point $P = (1, -3, 5)$ onto the plane π, we consider the vector $n = (2, -1, 2)$ orthogonal to π and write the parametric equations of the line ℓ passing through P and parallel to n (that is, orthogonal to π):

$$\ell : \begin{cases} x = 1 + 2t \\ y = -3 - t \\ z = 5 + 2t \end{cases}$$

The orthogonal projection of P onto π is the point R of intersection between the line ℓ and the plane π:

$$R = \ell \cap \pi : \begin{cases} x = 1 + 2t \\ y = -3 - t \\ z = 5 + 2t \\ 2x - y + 2z = 9 \end{cases}$$

By solving this system we find

$$R = \left(-\frac{1}{3}, -\frac{7}{3}, \frac{11}{3}\right).$$

▶ **Exercise 6.3.12.** In the three-dimensional Euclidean affine space let π be the plane with equation $x-2y+2z = 1$ and let $A = (3, 0, -1)$ and $B = (1, 2, 2)$ be two points of π. Let $\mathscr{C}$ be the circle contained in the plane π, passing through B and whose center is A.

(a) Find the parametric equations of the line s, contained in the plane π, passing through B and tangent to $\mathscr{C}$.

(b) Find the point $B' \in \mathscr{C}$ diametrically opposite to the point B.

(c) Determine the two points P_1 and P_2 whose distance from the plane π is $d = 6$ and whose orthogonal projection onto π is the point A.

(d) Determine the distances of the points P_1 and P_2 from the line s found in point (a).

Solution. A vector orthogonal to the plane π is $n = (1, -2, 2)$. Let us now consider the vector $\overrightarrow{AB} = B - A = (-2, 2, 3)$ (this vector determines the radius of the circle $\mathscr{C}$). The line s, being contained in the plane π, must be orthogonal to the vector n; moreover, being tangent to the circle $\mathscr{C}$ at the point B, it must be orthogonal to the vector $\overrightarrow{AB}$. Therefore, as a vector of s we can take

$$v_s = \overrightarrow{AB} \times n = (10, 7, 2).$$

Hence the parametric equations of s are

$$s : \begin{cases} x = 1 + 10\,t \\ y = 2 + 7\,t \\ z = 2 + 2\,t \end{cases}$$

The point B' diametrically opposite to the point B is given by

$$B' = A - \overrightarrow{AB} = (5, -2, -4).$$

The points P_1 and P_2 are on the line orthogonal to the plane π passing through A, on opposite sides of the plane. Since the norm of the normal vector n is $\|n\| = 3$, it follows that the two points are

$$P_1 = A + 2n = (5, -4, 3), \qquad P_2 = A - 2n = (1, 4, -5).$$

Finally, we note that the distances of the points P_1 and P_2 from the line s are equal and coincide with the distance of one of them from the point B. Hence we have

$$\text{dist}(P_1, s) = \text{dist}(P_2, s) = \text{dist}(P_1, B) = \|B - P_1\| = \sqrt{53}.$$

► **Exercise 6.3.13.** In the three-dimensional Euclidean affine space we consider the point $P = (1, 1, 5)$ and the line

$$r : \begin{cases} 2x - y + 2 = 0 \\ y + z - 3 = 0 \end{cases}$$

(a) Determine the point P', symmetrical of P with respect to the line r, and the point H where the line r intersects the line PP'.

(b) Determine two points A and A' on the line r such that the quadrilateral $APA'P'$ is a square.

(c) Find the Cartesian equation of the plane π such that the orthogonal projection of P on this plane coincides with the point H.

(d) Find the parametric equations of the line s contained in the plane π, passing through H and orthogonal to the line r.

Solution. A generic point $R \in r$ has coordinates $R = (t, 2t+2, 1-2t)$, from which we can also deduce that a vector of the line r is $v_r = (1, 2, -2)$. The vector $R-P$ has coordinates $R-P = (t-1, 2t+1, -4-2t)$ and the point H (the orthogonal projection of P onto r) is the point $R \in r$ for which the vector $R-P$ is orthogonal to v_r. Hence we have

$$0 = (R-P) \cdot v_r = 9t+9,$$

from which we get $t=-1$. It follows that $H = (-1, 0, 3)$.

Now it is easy to find the point P', symmetrical of P with respect to the line r, simply by noting that H must be the midpoint of the segment PP', i.e., $H = \frac{1}{2}(P+P')$. From this equality we get $P' = (-3, -1, 1)$.

To determine the points A and A' we observe that H must be the center (the point of intersection of the diagonals) of the square $APA'P'$, therefore the distance of H from P must be equal to the distance of H from A and from A'.

The vector $P-H = (2, 1, 2)$ has norm 3, consequently the points A and A' are obtained as follows: $A = H + v$ and $A' = H - v$, where v is a vector with norm 3 parallel to the vector v_r of the line r. But the norm of $v_r = (1, 2, -2)$ is exactly 3, therefore $A = H + v_r = (0, 2, 1)$ and $A' = H - v_r = (-2, -2, 5)$.

Let $ax+by+cz+d=0$ be the equation of the plane π. Requiring that the point H is the orthogonal projection of P onto the plane π implies that the vector $P-H = (2, 1, 2)$ must be orthogonal to π, from which we easily deduce that $a=2$, $b=1$ and $c=2$. Hence the equation of π is

$$\pi : 2x+y+2z+d=0,$$

for some d. To find the value of d, note that π must pass through H. Therefore we have $-2+0+6+d=0$, from which it follows that $d=-4$. Hence the equation of π is

$$\pi : 2x+y+2z-4=0.$$

Let v_s be a vector of the line s. This vector must be orthogonal to the vector $n = (2, 1, 2)$, which is perpendicular to the plane π (since s must be contained in the plane π). Since v_s must also be orthogonal to the vector v_r, we can take $v_s = n \times v_r = (-6, 6, 3)$ or, more simply, $v_s = (-2, 2, 1)$. The line s is the line passing through H and parallel to v_s. Its parametric equations are therefore

$$s : \begin{cases} x = -1-2t \\ y = 2t \\ z = 3+t. \end{cases}$$

► **Exercise 6.3.14.** In the three-dimensional Euclidean affine space, we consider the points $A = (0, 1, -2)$, $B = (3, 2, 2)$ and the line r with equations $x + 3y - 1 = 0$ and $z + 1 = 0$.

(a) Find the point A', the orthogonal projection of A onto the line r.

(b) Find the Cartesian equation of the plane π such that the orthogonal projection of A onto π is the point B.

(c) Among the lines passing through A and coplanar to r, determine the one that has the minimum distance from the point B, and compute this distance.

Solution. From the equations of r it is easy to see that the points $R_1 = (1, 0, -1)$ and $R_2 = (-2, 1, -1)$ belong to this line, therefore the vector $v_r = R_2 - R_1 = (-3, 1, 0)$ is a vector of r. A generic point of r can therefore be written in the form $X = R_1 + tv_r = (1 - 3t, t, -1)$.

We denote by w the vector $X - A = (1 - 3t, t - 1, 1)$; the orthogonality condition between w and v_r is $w \cdot v_r = 10t - 4 = 0$, from which we obtain $t = 2/5$. By substituting this value in the coordinates of the point X we obtain the coordinates of the point A', which is the orthogonal projection of A onto the line r:

$$A' = \left(-\tfrac{1}{5}, \tfrac{2}{5}, -1\right).$$

If the point B is to be the orthogonal projection of A onto π, the vector $n = B - A = (3, 1, 4)$ must be orthogonal to the plane π. Consequently, this plane has an equation of the form $3x + y + 4z + d = 0$. To determine the unknown coefficient d it suffices to note that the plane π must pass through the point B. Thus we find $d = -19$ and the equation of π is

$$\pi : 3x + y + 4z - 19 = 0.$$

The lines passing through A and coplanar to r are lines lying on the plane σ passing through A and containing the line r. To find the equation of this plane we consider the pencil of planes with axis r, whose equation is

$$\lambda(x + 3y - 1) + \mu(z + 1) = 0$$

and we require that A belongs to the plane. We find $2\lambda - \mu = 0$, hence we can set $\lambda = 1$ and $\mu = 2$. By substituting these values into the equation of the pencil of planes, we obtain the equation of the plane σ:

$$\sigma : x + 3y + 2z + 1 = 0.$$

The line s passing through A, coplanar with r, which has the minimum distance from B is the line passing through A and the point B', which is the orthogonal projection of B onto σ. We then proceed to determine B'.

The vector orthogonal to the plane σ is $(1, 3, 2)$ and, consequently, the line orthogonal to this plane and passing through B has parametric equations

$$\begin{cases} x = 3 + t \\ y = 2 + 3t \\ z = 2 + 2t. \end{cases}$$

By intersecting this line with the plane σ we find the point $B' = (2, -1, 0)$. We can now compute the vector of the line s passing through the points A and B': $v_s = B' - A = (2, -2, 2)$ (we can also set $v_s = (1, -1, 1)$). The line s is therefore $A + \langle(1, -1, 1)\rangle$, whose parametric equations are

$$s : \begin{cases} x = t \\ y = 1 - t \\ z = -2 + t. \end{cases}$$

Finally, the distance between the point B and the line s coincides with the distance between the points B and B':

$$\text{dist}(B, s) = \text{dist}(B, B') = \|B - B'\| = \sqrt{14}.$$

► **Exercise 6.3.15.** In the three-dimensional Euclidean affine space we consider the line r with equations $x + 2y = 0$ and $y + z - 1 = 0$ and the plane $\pi : x + y - z = 2$.

(a) Find the parametric equations of the line s, the orthogonal projection of r onto the plane π, and determine the distance between r and s.

(b) Determine if there exist two lines ℓ_1 and ℓ_2, contained in the plane π, parallel to s and such that $\text{dist}(r, \ell_1) = \text{dist}(r, \ell_2) = \text{dist}(\ell_1, \ell_2)$. If so, find the distance between the two lines and the equations of these lines.

Solution. We start by determining if the line r is intersecting or parallel to the plane π:

$$r \cap \pi : \begin{cases} x + 2y = 0 \\ y + z - 1 = 0 \\ x + y - z = 2 \end{cases}$$

Trying to solve this system, it turns out that it has no solutions, so $r \cap \pi = \varnothing$; the line r is therefore parallel to the plane π.

From the equations of r it is easy to see that the points $R_1 = (0, 0, 1)$ and $R_2 = (-2, 1, 0)$ belong to this line, therefore the vector $v_r = R_1 - R_2 = (2, -1, 1)$ is a vector of r. Since r is parallel to the plane π, the line s, which is the orthogonal projection of r onto π, is parallel to r and therefore as a vector of s we can take $v_s = v_r$.

To find a point of s it suffices to determine the orthogonal projection S_1 of the point $R_1 \in r$ onto the plane π.

A vector orthogonal to the plane π is $n_\pi = (1, 1, -1)$, therefore the line passing through R_1 and perpendicular to π has the following parametric equations

$$\begin{cases} x = t \\ y = t \\ z = 1 - t \end{cases}$$

By intersecting this line with the plane π we find $t = 1$ and therefore the point S_1, which is the orthogonal projection of R_1 onto π, has coordinates $S_1 = (1, 1, 0)$. The parametric equations of the line s are therefore $s : X = S_1 + \alpha v_s$, i.e.,

$$s : \begin{cases} x = 1 + 2\alpha \\ y = 1 - \alpha \\ z = \alpha \end{cases}$$

Since the lines r and s are parallel, we have

$$\operatorname{dist}(r, s) = \operatorname{dist}(R_1, S_1) = \|S_1 - R_1\| = \sqrt{3}.$$

As a consequence of the fact that r and s are parallel, it follows that there certainly exist two lines ℓ_1 and ℓ_2, contained in the plane π, parallel to s (and therefore also to r) and such that $\operatorname{dist}(r, \ell_1) = \operatorname{dist}(r, \ell_2) = \operatorname{dist}(\ell_1, \ell_2)$. To find them we proceed as follows. Let us consider the plane σ passing through R_1 and orthogonal to the line r. This plane intersects the two lines ℓ_1 and ℓ_2 at the points $L_1 \in \ell_1$ and $L_2 \in \ell_2$ such that $\operatorname{dist}(r, \ell_1) = \operatorname{dist}(R_1, L_1)$, $\operatorname{dist}(r, \ell_2) = \operatorname{dist}(R_1, L_2)$ and $\operatorname{dist}(\ell_1, \ell_2) = \operatorname{dist}(L_1, L_2)$. Therefore the triangle contained in the plane σ, with vertices R_1, L_1 and L_2 is equilateral and the segment R_1S_1 is the height relative to the base L_1L_2. If we denote by $\ell = \sigma \cap \pi$ the line passing through the points L_1 and L_2, this line is orthogonal to both v_r and n_π, therefore as the vector v_ℓ of this line we can take $v_\ell = v_r \times n_\pi = (0, 3, 3)$ or, simply, $v_\ell = (0, 1, 1)$. The points L_1 and L_2 are on the line ℓ, in a symmetrical position with respect to the point S_1; so we can write

$$L_1 = S_1 + tv_\ell = (1, 1+t, t), \quad L_2 = S_1 - tv_\ell = (1, 1-t, -t),$$

for some $t \in \mathbb{R}$.

Now we have $\operatorname{dist}(L_1, L_2)^2 = 8t^2$, $\operatorname{dist}(R_1, L_1)^2 = 2t^2 + 3$, hence the equality $\operatorname{dist}(R_1, L_1) = \operatorname{dist}(L_1, L_2)$ gives the equation $2t^2 + 3 = 8t^2$, from which we get $t = \pm\sqrt{2}/2$. Therefore, the points L_1 and L_2 have the following coordinates:

$$L_1 = \left(1, 1 + \tfrac{\sqrt{2}}{2}, \tfrac{\sqrt{2}}{2}\right), \quad L_2 = \left(1, 1 - \tfrac{\sqrt{2}}{2}, -\tfrac{\sqrt{2}}{2}\right).$$

The distance between the parallel lines ℓ_1 and ℓ_2 is

$$\operatorname{dist}(\ell_1, \ell_2) = \operatorname{dist}(L_1, L_2) = 2$$

and the equations of ℓ_1 and ℓ_2 are

$$\ell_1 : X = L_1 + \alpha v_r, \quad \ell_2 : X = L_2 + \beta v_r,$$

from which we obtain

$$\ell_1 : \begin{cases} x = 1 + 2\alpha \\ y = 1 + \frac{\sqrt{2}}{2} - \alpha \\ z = \frac{\sqrt{2}}{2} + \alpha \end{cases} \qquad \ell_2 : \begin{cases} x = 1 + 2\beta \\ y = 1 - \frac{\sqrt{2}}{2} - \beta \\ z = -\frac{\sqrt{2}}{2} + \beta. \end{cases}$$

▶ **Exercise 6.3.16.** In the affine space $\mathbb{A}^3_{\mathbb{R}}$, let $\mathscr{S}$ be the sphere with center $C = (2, -1, 1)$, passing through the point $A = (3, 1, 0)$.

(a) Find the Cartesian equation of the plane tangent to $\mathscr{S}$ at the point A.

(b) Let $\mathscr{C}$ be the circle obtained by intersecting the sphere $\mathscr{S}$ with the plane $\pi : 3x + y + 2z - 2 = 0$. Find the center and the radius of $\mathscr{C}$.

(c) After verifying that the point $P = (0, -2, 2)$ belongs to the circle $\mathscr{C}$, write the parametric equations of the line r tangent to $\mathscr{C}$ at P.

(d) Determine whether there are lines through $Q = (1, 0, 2)$ that do not intersect the sphere $\mathscr{S}$.

Solution. The radius of the sphere $\mathscr{S}$ is $r_{\mathscr{S}} = \text{dist}(C, A) = \|A - C\| = \|(1, 2, -1)\| = \sqrt{6}$.

The tangent plane to $\mathscr{S}$ at the point A is the plane passing through A, orthogonal to the vector $A - C = (1, 2, -1)$. Its Cartesian equation is therefore of the form $x + 2y - z + d = 0$. By requiring that A belongs to the plane, we get $d = -5$, hence the equation of the tangent plane is $x + 2y - z - 5 = 0$.

The center C' of the circle $\mathscr{C}$ is the orthogonal projection of the point C onto the plane π. The vector orthogonal to the plane π is $n_\pi = (3, 1, 2)$. Let us consider the line passing through C and parallel to the vector n_π, whose parametric equations are

$$\begin{cases} x = 2 + 3t \\ y = -1 + t \\ z = 1 + 2t \end{cases}$$

Intersecting with the plane π, we get $t = -5/14$, from which we obtain $C' = (13/14, -19/14, 2/7)$.

The radius of the circle $\mathscr{C}$ can be found simply using the Pythagorean Theorem. In fact, we have: $r_{\mathscr{S}}^2 = r_{\mathscr{C}}^2 + \text{dist}(C, C')^2$. The distance between the points C and C' is $\text{dist}(C, C') = \|C - C'\| = \frac{5}{14}\,\|(3, 1, 2)\| = 5\sqrt{14}/14$. Therefore the radius of $\mathscr{C}$ is

$$r_{\mathscr{C}} = \sqrt{6 - \frac{25}{14}} = \sqrt{\frac{59}{14}}.$$

To verify that the point P belongs to $\mathscr{C} = \mathscr{S} \cap \pi$, it is sufficient to verify that $P \in \mathscr{S}$ and $P \in \pi$. We have $\text{dist}(C, P) = \|P - C\| = \|(-2, -1, 1)\| = \sqrt{6}$, hence $P \in \mathscr{S}$. To verify that $P \in \pi$ it is sufficient to check that the coordinates of P satisfy the equation of the plane π.

Since the circle is contained in the plane π, the line r tangent to $\mathscr{C}$ at the point P must also be contained in this plane. Therefore, if v_r is a vector of r, this vector must be orthogonal to $n_\pi = (3, 1, 2)$. Furthermore, the vector v_r must also be orthogonal to the vector $P - C'$, which is proportional to the vector $(-13, -9, 24)$. Hence as v_r we can take the cross product of the vectors $(-13, -9, 24)$ and $(3, 1, 2)$:

$$v_r = (-13, -9, 24) \times (3, 1, 2) = (-42, 98, 14) = 14\,(-3, 7, 1).$$

We can therefore take the vector $(-3, 7, 1)$ as the vector of the line r. The parametric equations of r are

$$r : \begin{cases} x = -3t \\ y = -2 + 7t \\ z = 2 + t \end{cases}$$

Finally we note that $\mathrm{dist}(C, Q) = \|Q - C\| = \|(-1, 1, 1)\| = \sqrt{3} < r_{\mathscr{S}}$, so the point Q is inside the sphere. This means that every line passing through Q must necessarily intersect the sphere $\mathscr{S}$ in two points.

▶ **Exercise 6.3.17.** In the affine space $\mathbb{A}^3_{\mathbb{R}}$, let ℓ be the line passing through the point $(1, 1, -2)$ and parallel to the vector $v_\ell = (1, 0, -1)$ and let $\mathscr{S}$ be the sphere with center $C = (1, -2, 1)$ and radius 3.

(a) Determine the distance of C from the line ℓ, and the point C' symmetrical of C with respect to ℓ.

(b) Find the parametric equations of two skew lines r and s passing respectively through the points $R_1 = (4, 2, -1)$ and $S_1 = (2, 3, 1)$ and intersecting the line ℓ at the points R_2 and S_2, so that R_2 and S_2 are the minimum distance points of the lines r and s.

(c) Find the parametric equations of the lines parallel to $(2, -1, 2)$, tangent to the sphere $\mathscr{S}$ and intersecting the line ℓ.

Solution. We start by determining the point H, the orthogonal projection of the point C onto the line ℓ. The parametric equations of the line ℓ are

$$\ell : \begin{cases} x = 1 + t \\ y = 1 \\ z = -2 - t \end{cases}$$

therefore the coordinates of a generic point $X \in \ell$ are $X = (1 + t, 1, -2 - t)$. The vector $X - C$ is $X - C = (t, 3, -3 - t)$ and, imposing the condition of orthogonality between $X - C$ and the vector of ℓ, we get $(X - C) \cdot v_\ell = 2t + 3 = 0$, from which we obtain $t = -3/2$. Substituting this value into the parametric equations of ℓ, we obtain the coordinates of the point H:

$$H = (-1/2, 1, -1/2).$$

The distance of C from the line ℓ is the distance of C from the point H, i.e., the norm of the vector $H - C$. We get

$$\operatorname{dist}(C, \ell) = \|H - C\| = \|(-3/2, 3, -3/2)\| = \frac{3}{2}\sqrt{6}.$$

To determine the point C', symmetrical of C with respect to ℓ, it suffices to note that H must be the midpoint of the segment CC', i.e., we must have $H = \frac{1}{2}(C + C')$. It follows that the coordinates of C' are $(-2, 4, -2)$.

The requirement that $R_2, S_2 \in \ell$ be the points of minimum distance of the skew lines r and s means that r and s are both orthogonal to ℓ. This, in turn, means that R_2 is the orthogonal projection of R_1 onto ℓ and, similarly, S_2 is the orthogonal projection of S_1 onto ℓ. The points R_2 and S_2 can therefore be found with the same method used to determine the point H. We find $R_2 = (2, 1, -3)$ and $S_2 = (0, 1, -1)$. The vectors of the lines r and s are $v_r = R_1 - R_2 = (2, 1, 2)$ and $v_s = S_1 - S_2 = (2, 2, 2) = 2(1, 1, 1)$ and, consequently, the parametric equations of the two lines are

$$r : \begin{cases} x = 4 + 2t \\ y = 2 + t \\ z = -1 + 2t \end{cases} \qquad s : \begin{cases} x = 2 + t \\ y = 3 + t \\ z = 1 + t \end{cases}$$

To find the equations of the lines parallel to the vector $(2, -1, 2)$, tangent to the sphere $\mathscr{S}$ and intersecting the line ℓ, we proceed as follows. We recall that a generic point $X \in \ell$ is $X = (1 + t, 1, -2 - t)$. We can therefore write the parametric equations of the line γ_t passing through X and parallel to the vector $(2, -1, 2)$:

$$\gamma_t : \begin{cases} x = 1 + t + 2u \\ y = 1 - u \\ z = -2 - t + 2u \end{cases}$$

We note that for each fixed value of t, these equations represent the line γ_t passing through the point of ℓ with coordinates $(1+t, 1, -2-t)$ and parallel to the vector $(2, -1, 2)$, while if we also consider t variable, the same equations describe the plane containing the line ℓ and parallel to the vector $(2, -1, 2)$.

Of all the lines γ_t, those tangent to the sphere $\mathscr{S}$ are those whose distance from the center C of the sphere is equal to the radius of $\mathscr{S}$. We set $v = (2, -1, 2)$, $w = C - X = (-t, -3, 3 + t)$ and compute the cross product $w \times v = (t - 3, 4t + 6, t + 6)$. The square of the norm of this vector is

$\|w \times v\|^2 = 9(2t^2 + 6t + 9)$. The distance of the point C from the line γ_t is given by the following formula:

$$\text{dist}(C, \gamma_t) = \frac{\|w \times v\|}{\|v\|}$$

By requiring that this distance be equal to 3, squaring and simplifying, we obtain the equation $2t^2 + 6t = 0$, whose solutions are $t = 0$ and $t = -3$. These are the values of t which determine the two tangent lines, whose parametric equations are:

$$\gamma_0 : \begin{cases} x = 1 + 2u \\ y = 1 - u \\ z = -2 + 2u \end{cases} \qquad \gamma_{-3} : \begin{cases} x = -2 + 2u \\ y = 1 - u \\ z = 1 + 2u. \end{cases}$$

▶ **Exercise 6.3.18.** In the affine space $\mathbb{A}^3_{\mathbb{R}}$ we consider the plane $\pi_\alpha : \alpha x + 2y - 2\alpha z + 4 = 0$ and the line

$$s : \begin{cases} 3x - y + 5 = 0 \\ 2x + z + 1 = 0 \end{cases}$$

(a) Determine if there exist a line contained in all planes π_α (for each $\alpha \in \mathbb{R}$). If such a line exists, write its equations.

(b) Determine for what value of α the line s is parallel to the plane π_α and for what value of α the line s is perpendicular to π_α.

(c) Now set $\alpha = 1$. Determine the point P' symmetrical of $P = (2, 3, -3)$ with respect to the plane π_1.

Solution. The equation of the plane π_α can be rewritten in the form

$$\alpha(x - 2z) + (2y + 4) = 0,$$

from which we deduce that the line r with Cartesian equations

$$r : \begin{cases} x - 2z = 0 \\ 2y + 4 = 0 \end{cases}$$

is contained in the plane π_α, for all $\alpha \in \mathbb{R}$.

The points $S_1 = (0, 5, -1)$ and $S_2 = (1, 8, -3)$ belong to the line s and therefore a vector of s is $v_s = S_2 - S_1 = (1, 3, -2)$. A vector perpendicular to the plane π_α is $n_\alpha = (\alpha, 2, -2\alpha)$. The line s is parallel to the plane π_α if and only if $v_s \cdot n_\alpha = 0$, from which we get $\alpha = -\frac{6}{5}$. The line s is perpendicular to π_α if and only if $n_\alpha = \lambda v_s$, from which we obtain $\alpha = \lambda = \frac{2}{3}$.

Now we set $\alpha = 1$. The equation of the plane π_α becomes

$$\pi = \pi_1 : x + 2y - 2z + 4 = 0$$

and a vector orthogonal to this plane is $n = (1, 2, -2)$. The parametric equations of the line ℓ passing through the point P and perpendicular to the plane π are

$$\ell : \begin{cases} x = 2 + t \\ y = 3 + 2t \\ z = -3 - 2t \end{cases}$$

The point of intersection between the line ℓ and the plane π is $H = (0, -1, 1)$ (H is the orthogonal projection of P onto π). Noting that H is the midpoint of the segment PP', i.e., $H = \frac{P+P'}{2}$, we deduce that $P' = 2H - P = (-2, -5, 5)$.

► **Exercise 6.3.19.** In the affine space $\mathbb{A}^3_{\mathbb{R}}$ we consider the lines

$$r_{\alpha,\beta} : \begin{cases} x + \alpha\, y + \alpha\, z = 2 \\ \beta\, x + 2y - 2\beta\, z = -2 \end{cases}$$

(a) Verify that the lines $r_{\alpha,\beta}$ ($\forall \alpha, \beta \in \mathbb{R}$) all pass through the same point P (and find the coordinates of P).

(b) Now we set $\alpha = 1$. Determine the value of β so that the line $r_{1,\beta}$ is parallel to the plane $\pi : x + y - z = 1$.

Solution. If all lines $r_{\alpha,\beta}$ pass through a point P then we can determine that point by intersecting two of these lines. Let us therefore consider the line $r_{0,0}$ (i.e., let's set $\alpha = \beta = 0$) and the line $r_{1,0}$:

$$r_{0,0} : \begin{cases} x = 2 \\ 2y = -2 \end{cases} \qquad r_{1,0} : \begin{cases} x + y + z = 2 \\ 2y = -2 \end{cases}$$

By intersecting these two lines we find the point $P = (2, -1, 1)$. Now we have to verify that this point also belongs to all lines $r_{\alpha,\beta}$ (to do this, just substitute the coordinates of P in the equations of $r_{\alpha,\beta}$).

By setting $\alpha = 1$ we obtain the lines

$$r_{1,\beta} : \begin{cases} x + y + z = 2 \\ \beta\, x + 2y - 2\beta\, z = -2 \end{cases}$$

To determine a vector $v_{1,\beta}$ of the line $r_{1,\beta}$ we proceed as follows: let us consider the vectors $n_1 = (1, 1, 1)$ (which is orthogonal to the plane with equation $x + y + z = 2$) and $n_2 = (\beta, 2, -2\beta)$ (which is orthogonal to the plane with equation $\beta\, x + 2y - 2\beta\, z = -2$). The vector $v_{1,\beta}$ is the cross product of n_1 and n_2:

$$v_{1,\beta} = n_1 \times n_2 = (-2\beta - 2, 3\beta, 2 - \beta).$$

The line $r_{1,\beta}$ is parallel to the plane $\pi : x+y-z = 1$ if and only if the vectors $v_{1,\beta}$ and $n_\pi = (1, 1, -1)$ are orthogonal, i.e., if and only if $v_{1,\beta} \cdot n_\pi = 0$, from which we obtain $\beta = 2$.

▶ **Exercise 6.3.20.** In the affine space $\mathbb{A}^3_{\mathbb{R}}$ we consider the lines

$$r_\alpha : \begin{cases} x - 2\alpha z - 2 + 2\alpha^2 = 0 \\ y - \alpha z + \alpha^2 = 0 \end{cases}$$

(a) Determine if the lines r_α all pass through the same point. If so, find this point.

(b) Prove that all lines r_α are contained in the same plane and write the Cartesian equation of this plane.

(c) Let $A = (4, 6, 3)$. Find the point H, the orthogonal projection of A onto the plane found at point (b) and, among all the lines r_α, determine those which have the minimum distance from the point A.

Solution. If all lines r_α pass through a point P then we can determine that point by intersecting two of these lines. Let us therefore consider the line r_0 (i.e., let's set $\alpha = 0$) and the line r_1:

$$r_0 : \begin{cases} x - 2 = 0 \\ y = 0 \end{cases} \qquad r_1 : \begin{cases} x - 2z = 0 \\ y - z + 1 = 0 \end{cases}$$

By intersecting these two lines we find the point $P = (2, 0, 1)$. To check if this point belongs to all lines r_α we substitute the coordinates of P in the equations of r_α. Therefore we find the equation $\alpha^2 - \alpha = 0$, whose solutions are $\alpha = 0$ and $\alpha = 1$. This means that the point P belongs only to the lines r_0 and r_1, hence there is no point that belongs to all lines r_α.

Since the lines r_0 and r_1 intersect, they are contained in a plane π. It follows that if all lines r_α are contained in the same plane, this must necessarily be the plane π. To find the equation of this plane we consider the vectors v_0 and v_1 (the vectors of the lines r_0 and r_1) and their cross product $n = v_0 \times v_1$. We have:

$$v_0 = (0,0,1), \qquad v_1 = (2,1,1), \qquad n = v_0 \times v_1 = (-1,2,0).$$

The vector n is orthogonal to the plane π and therefore the equation of π has the form $-x+2y+d=0$. To find the value of d we require that $P = (2,0,1)$ belongs to the plane. We obtain $d = 2$, hence the plane that contains the lines r_0 and r_1 is $\pi : -x+2y+2=0$.

Now we need to verify that all lines r_α are contained in this plane. To do this we consider the parametric equations of the lines r_α

$$r_\alpha : \begin{cases} x = 2\alpha t + 2 - 2\alpha^2 \\ y = \alpha t - \alpha^2 \\ z = t \end{cases}$$

and substitute into the equation of π: we obtain the identity $0 = 0$, which confirms that, for each $\alpha \in \mathbb{R}$, the line r_α is contained in the plane π.

To determine the orthogonal projection of the point A onto the plane π we consider the line ℓ passing through A and orthogonal to this plane:

$$\ell : \begin{cases} x = 4 - t \\ y = 6 + 2t \\ z = 3 \end{cases}$$

The point H, the orthogonal projection of A onto π, is the point of intersection of the line ℓ and the plane π: $H = (6,2,3)$. The lines r_α which have the minimum distance from the point A are those which pass through the point H just found. Substituting the coordinates of H in the equations of r_α we obtain the equation $\alpha^2 - 3\alpha + 2 = 0$, whose solutions are $\alpha = 1$ and $\alpha = 2$. This means that there are two lines r_α which have minimum distance from A and these are r_1 and r_2.